TRANSPLANT PRODUCTION SYSTEMS

TRANSPLANT PRODUCTION SYSTEMS

Proceedings of the International Symposium on Transplant Production Systems, Yokohama, Japan, 21–26 July 1992

Edited by

K. KURATA

Department of Agricultural Engineering, University of Tokyo, Japan

and

T. KOZAI

Faculty of Horticulture, Chiba University, Japan

SPRINGER-SCIENCE+BUSINESS MEDIA, B.V.

Library of Congress Cataloging-in-Publication Data

International Symposium on Transplant Production Systems (1992 :
Yokohama-shi, Japan)
Transplant production systems : proceedings of the International
Symposium on Transplant Production Systems, Yokohama, Japan, 21-26
July 1992 / edited by K. Kurata and T. Kozai.
p. cm.
Includes index.
ISBN 0-7923-1797-1 (acid-free paper)
1. Transplanting (Plant culture)--Congresses. 2. Plant
propagation--In vitro--Congresses. I. Kurata, K. (Kenji), 1947-
. II. Kozai, T. (Toyoki), 1943- .
SB121.I57 1992
631.5'36--dc20 92-15554

ISBN 978-94-010-5237-5 ISBN 978-94-011-2785-1 (eBook)
DOI 10.1007/ 978-94-011-2785-1

Table of Contents

Preface

The International Symposium on Transplant Production Systems will be held in Yokohama, Japan on July 21-26, 1992. The Symposium will cover biological, engineering, and socioeconomic aspects of transplant production systems. For the Symposium, 'transplant' will be defined as any plant propagated by any means and requiring positional and/or environmental transfer to promote further growth and development. Transplants could include seedlings, grafted seedlings, plantlets, bulblets, artificial (synthetic) seeds, and microtubers.

About half of the Symposium time will be devoted to the special sessions, in which distinguished researchers in this field and persons from leading transplant industries will give invited lectures, in addition to lectures by members of the Symposium Organizing Committee. This volume contains most of the proceedings of the special sessions.

Transplant production systems are now attracting much attention, presumably because of three primary reasons: specialization of labor in plant production, global issues, and advancements in biotechnology. The first reason is related to transplants production for agriculture. In some developed countries, most of the transplant production is performed by the nursery industry. Growers buy transplants from nurseries and raise the transplants to the end products. In Japan, the production of many horticultural plants is also following this move towards specialization of transplant production. Labor shortage is the main driving force behind this trend. Growers are suffering from a shortage of young successors, employees, and part-time helpers. In particular, young labor is streaming out into other industries, and farming relies more and more on aged labor. Under these situations, the farm labor tasks need to be simplified and less strenuous. Buying transplants is one of the ways to meet these needs.

Another factor which promotes separation of the grower from the transplant production process is the consumers' demand for high quality end products, which requires starting with high quality transplants such as disease-free micropropagated plantlets or grafted seedlings. Production of these high quality transplants requires costly facilities and skilled labor and is often beyond an individual grower's abilities. In addition, specialization of the transplant production process enables intensive utilization of skills, techniques, and labor particular to this process, bringing about production of standardized, high quality transplants. Growers, on the other hand, will be able to make a more precise production schedule according to the consumers' demands, provided that a stable supply of standardized high quality transplants from the nursery industry is available. For these reasons, many opportunities for growth remain in the nursery industry. However, the nursery industry itself is suffering from a labor shortage and high labor costs as well. Some nurseries produce transplants in foreign developing counties where labor is much cheaper. However, labor cost is also increasing in these developing countries. These situations demand transplant production systems which are more simple, rational, automated, efficient, and systematic.

The second reason why transplant production systems are now highlighted is related to the global environmental problems, the predicted population explosion, and the people's desire for more amenities. Deforestation and desertification of the earth's surface are progressing at remarkable rates, and the increase in the carbon dioxide concentration in the atmosphere is accelerating. FAO/UNEP estimated the deforestation rate on the earth to be 11,300,000 ha/year (1982). UNEP estimated that 5,000,000 to 7,000,000 ha of agricultural land is lost due to desertification on the earth every year. Texas A & M University estimated that the desertification rate is 20,000,000 ha/year, three times that of the UNEP estimation. The present atmospheric carbon dioxide concentration is 353 ppm (1990) and is increasing by 1.8 ppm every year. Afforestation and reforestation are required for maintaining a 'green earth', maintaining biological diversity, and also for augmentation of a sink for carbon dioxide. Planting of vegetation will play an important role in preventing the expansion of deserts and in converting deserts to agricultural land or forests. The population of the earth is presently 5,112,000,000 (1988) and is predicted to rise to 6,250,000,000 in 2000. More agricultural products will be required for the increasing population. Thus, many of the measures against these environmental and population problems involve creating greenery. In addition to the above-mentioned problems, people's desire for more flowers and greenery in streets, parks, gardens, on building rooftops, in offices, and in open spaces in buildings increase the need for creating greenery. Thus, the demand for greenery is ever increasing and consequently, the effective production of transplants is essential to meet the demand. In particular, for afforestation and reforestation, mass production of standardized high quality transplants is required.

The third reason why transplant production systems are now attracting attention is related to the progress in biotechnology. Progress in biotechnology has produced a variety of new plants. We can anticipate that more plants endowed with useful characteristics such as higher productivity, more tolerance against environmental stress, and more resistance to diseases will be created through gene manipulation in the future. Some people say that the 21st century will be the century of biotechnology. However, the creation of a new plant alone does not necessarily lead to progress in agriculture or forestry. First, in comparison with existing plants, the new plants must show some improvements such as more yield, higher quality product, more tolerance against stresses, less labor requirement, safer product, or more stability in yield. Conventional technology is used in the cultivation of new plants. Furthermore, new achievements in biotechnology will be put into practice using conventional technology such as transplant production technology. Therefore, development of effective transplant production systems is of great importance in introducing new plants into practical production and for afforestation.

The International Symposium on Transplant Production Systems will be held to promote research on transplant production systems and transplant production industries. For this objective, the promotion of cooperation among researchers from various fields is particularly important. There has been little information exchange between vegetable researchers, flower researchers, and forest tree researchers. Likewise, only a small amount of information has been exchanged among various types of researchers such as seedlings researchers, micropropagated transplants researchers, and grafted seedlings researchers. Plant physiologists and engineers also have had little exchange of information. Researchers in all of these areas, particularly the last two mentioned, should become more involved in cooperative studies to promote transplant production research. This is why researchers from various fields related to transplant production will be invited to the Symposium and thus, this volume also contains contributions from various fields. This volume contains a total of 17 contributions on topics such as plug seedling production, micropropagation, robotization, measurements, model development, environmental control, and practical analysis of transplant production industries. Readers of this volume will find useful overviews for each area, practical considerations on transplant production systems, and various ideas for developing transplant production systems.

We wish to express our sincere appreciation to all the contributors to this volume. They have all completed articles for this volume in a very limited amount of time.

Tokyo, March 1992

Kenji Kurata
Toyoki Kozai

Contributors

J. AITKEN-CHRISTIE: Forest Research Institute, Private Bag 3020, Rotorua, New Zealand

J. T. ARNOTT: Forestry Canada, 506 W. Burnside Road, Victoria, B.C. V8Z 1M5, Canada

PH. BOXUS: Centre de Recherches Agronomiques, Station des cultures fruitières et maraîchères, B-5030 Gembloux, Belgium

F. R. BROWN: Mechanical Engineering Division, Silsoe Research Institute, Wrest Park, Silsoe, Bedford, MK45 4HS, U.K.

I. Y. E. CHU: Twyford International Inc., 15245 Telegraph Road, Santa Paula, California 93060, U.S.A.

M. CONNETT: Department of Biological Sciences, Waikato University, Private Bag, Hamilton, New Zealand

K. FUJIWARA: Faculty of Horticulture, Chiba University, Matsudo, Chiba 271, Japan

Y. HASHIMOTO: Department of Biomechanical Systems, Ehime University, Tarumi, Matsuyama, 790, Japan

M. HAYASHI: Department of Biological Science and Technology, Tokai University, Nishino, Numazu, Shizuoka 410-03, Japan

R. D. HEINS: Department of Horticulture, Michigan State University, East Lansing, Michigan 48824-1325, U.S.A.

T. HIROSAWA: Nursery Technology Inc., 3377 Kitsuregawa, Tochigi 329-14, Japan

D. P. HOLDGATE: PhytoNova B.V., P.O. Box 98, 2230 AB Rijnsburg, The Netherlands

T. ITO: Faculty of Horticulture, Chiba University, 648, Matsudo, Chiba, Japan

G. A. GIACOMELLI: Bioresource Engineering Department, Rutgers University-Cook Collge, New Brunswick, NJ 08903-0231, U.S.A.

T. KOZAI: Faculty of Horticulture, Chiba University, Matsudo, Chiba 271, Japan

K. KURATA: Department of Agricultural Engineering, University of Tokyo, Yayoi 1-1-1, Bunkyo-ku, Tokyo 113, Japan

N. LANGE: Department of Horticulture, Michigan State University, East Lansing, Michigan 48824-1325, U.S.A.

R. H. LAWSON: United States Department of Agriculture, Agricultural Research Service, Florist and Nursery Crops Laboratory, 10300 Baltimore Avenue, B-004, Room 208, Beltsville, Maryland 20705-2350, U.S.A.

M. I. MENZIES: Forest Research Institute, Private Bag 3020, Rotorua, New Zealand

H. NONAMI: Department of Biomechanical Systems, Ehime University, Tarumi, Matsuyama, 790, Japan

P. E. READ: Department of Horticulture, University of Nebraska-Lincoln, Lincoln, NE 68583-0724, U.S.A.

M. S. ROH: United States Department of Agriculture, Agricultural Research Service, Florist and Nursery Crops Laboratory, 10300 Baltimore Avenue, B-004, Room 208, Beltsville, Maryland 20705-2350, U.S.A.

W. SIMONTON: The University of Georgia, Georgia Experiment Station, 1109 Experiment Street, Griffin, GA 30223-1797, U.S.A.

K. C. TING: Bioresource Engineering Department, Rutgers University-Cook Collge, New Brunswick, NJ 08903-0231, U.S.A.

T. F. WALLACE, JR.: Department of Horticulture, Michigan State University, East Lansing, Michigan 48824-1325, U.S.A.

R. R. WILLIAMS: Department of Agronomy and Soil Science, University of New England, Armidale, NSW 2351, Australia

E. A. ZANDVOORT: PhytoNova B.V., P.O. Box 98, 2230 AB Rijnsburg, The Netherlands

PROPAGATION AND TRANSPLANT PRODUCTION TECHNOLOGY OF NEW FLORAL CROPS

MARK S. ROH and ROGER H. LAWSON
United States Department of Agriculture
Agricultural Research Service
Florist and Nursery Crops Laboratory
10300 Baltimore Avenue, B-004, Room 208
Beltsville, Maryland 20705-2350 USA

ABSTRACT. In 1982 the new crops research program was initiated in the Florist and Nursery Crops Laboratory to increase the diversity of florist crop germplasm in the United States. Anigozanthos, Chamelaucium, Correa, a dwarf form of Eustoma grandiflorum 'Little Blue Belle', Chamelaucium, Cupressus, and other crops have been introduced to the industry with information that describes propagation, culture, and controlled flowering. Rooting of cutting or in vitro tissue culture propagation of new crops had included Achimenantha, Chamelaucium uncinatum, 'Purple Pride', and Eustoma grandiflorum, 'Little Belle Blue', Ornithogalum dubium, Lachenalia aloides, and Lilium. In addition, a new production technology using bulbils as propagules was developed for Lilium elegans, Asiatic hybrid lily. This production technology is discussed.

1. Introduction

New crops research programs in The Netherlands (Noordegraaf 1987) and Denmark (Christensen and Friis 1987) have expanded the range of new plant materials available in the European market. A similar program was initiated for the United States (Armitage 1986; Roh and Lawson 1987b, 1990). Introduction of new crops requires identification, exploration and collection of interesting new plants that have the potential of becoming economically important. Hybridization and selection from a breeding program and the introduction of new cultural technology are important steps in the introduction of a new crop.

To meet the increasing demand by growers and consumers for new crops in the U.S., the Florist and Nursery Crops Laboratory initiated a program of germplasm development in 1982. We define new crops as newly discovered genera or species, newly introduced cultivars of plants grown in earlier years, but forgotten or

K. Kurata and T. Kozai (eds.), Transplant Production Systems, 1–20.

without complete cultural information, plants that are cultivated in foreign countries but have not been introduced in the United States, or crops that can be produced with new production technologies that can enhance crop quality and shorten the total production time.

2. New Crops Program

The overall scheme of identification, collection, propagation, evaluation, distribution and introduction is shown in Fig. 1. From the initial selection of a plant that may have commercial potential to an official introduction, we project 4 to 7 years.

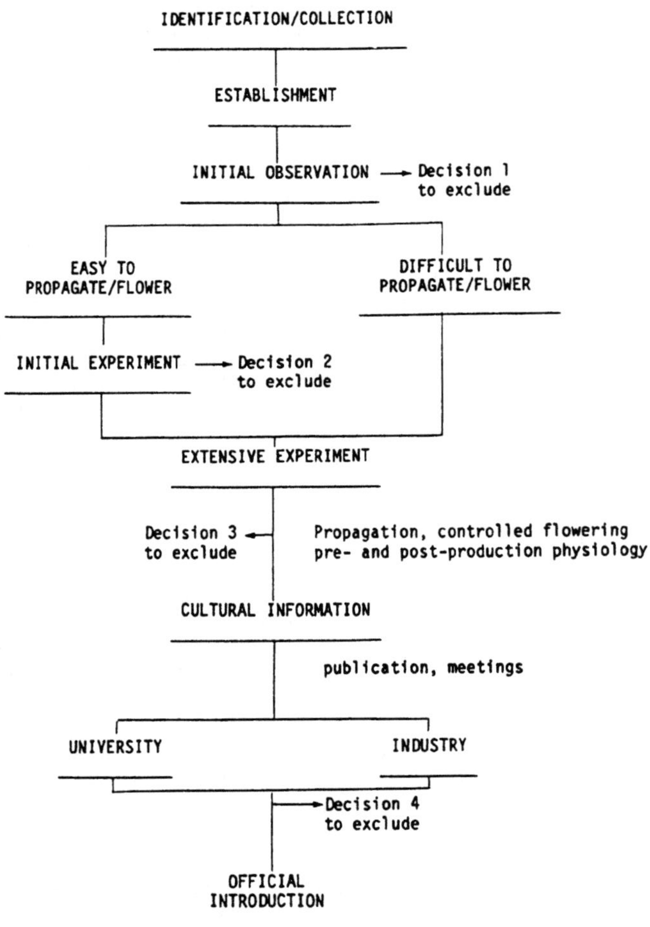

Figure 1. Overall scheme of the new crops evaluation system.

2.1. COLLECTION

The Laboratory collection includes many genera and species of native Australian plants including Correa, Chamelaucium, and Anigozanthos (kangaroo paw); Phalaenopsis orchid and Gesneriaceae such as Sinningia from specialty growers; Erica, Lachenalia and Ornithogalum from South Africa; and new cultivars of Eustoma and Clematis for pot plant culture from Japan.

2.2. PROPAGATION AND INITIAL OBSERVATION

When a limited number of propagules are collected, it takes approximately a year to propagate enough plants for the initial experimentation. Propagation of several crops will be discussed in detail later in this paper. During propagation and early growth and development, the rate and the ease of propagation as well as plant statue is evaluated. Observations are made on the compact and dwarf appearance of a plant, time of flowering, and susceptibility to diseases. Care must be taken not to eliminate potentially important germplasm. Flowering in a genus like Aeschynanthus may be erratic under natural greenhouse growing conditions. However, extensive evaluation of the requirements for controlled light and temperatures on flowering will make Aeschynanthus a new crop in the near future (Gertsson 1987; Roh et al., 1988; Welander 1984; Whitton, et al., 1990).

2.3. PRODUCTION

Following propagation and an initial period of observation, investigations on the types of growing medium and nutrition and light and temperature effects on plant quality are evaluated. A major factor to consider while assessing production capability is the forcing period. From the time the grower places the rooted plant in the greenhouse bench, a maximum production time of 18 weeks will allow a turnover of three to four crops a year. Should the plant be excluded if more than 18 weeks are required? The total production time of Eustoma grandiflorum Griesb. from seed germination to flowering requires about 7 to 8 months (Roh and Lawson 1984a; Tsukada et al., 1982). On the basis of cropping time, Eustoma might be excluded from the program, although post-production longevity is excellent. However, plug production technology has enabled greenhouse growers to shorten cropping time to 15 to 17 weeks (Halevy and Kofranek 1984; Roh and Lawson 1984a).

2.4. PRODUCTION - NATIONWIDE EVALUATION

After an initial production trial in the laboratory, cultural information is released (Griesbach 1985; Lawson 1990; Lee et al., 1988; Roh 1991a,b; Roh and Lawson 1984a; Roh and Lawson

1987a,b). Evaluation of the same production protocol is then tested at various locations under diverse environmental conditions. Development of a systematic and well coordinated national program on the new crop project is vital for the success of the total introduction program. The nationwide system described by the Florist and Nursery Crops Laboratory for new crop introduction requires support from all sectors of industry and from research scientists. Based on the criteria for selection and the nationwide evaluation program, recommendations are then made for production of specific cultivars in different areas of the country.

2.5. SCHEDULE FOR PRODUCTION TRIAL

The time required for a regional and national production trial may vary depending on the nature of the new crop. The time required also depends on the extent of the information to be developed. We project a maximum of 4 to 7 years if new plants like *Eustoma* or *Aeschynanthus* have not been investigated extensively in United States production systems. Two to 3 years may be required for plants like *Leptospermum* which are currently produced commercially in other countries.

2.6. PRE- AND POST-PRODUCTION TECHNOLOGY

Pre- and post-production physiology is an important factor in determining which plants should or should not be maintained in the program. Tiny flower buds of *Eustoma* will not open on stems after harvest nor will a full purple or pink color develop when cut stems or potted plants are placed in low-light intensity environments. This problem requires further study to extend useful post-production life of the crop.

2.7. RELEASE OF PLANT MATERIAL AND CULTURAL INFORMATION

Following the research phase, a potential new crop is distributed to selected growers so that plants can be tested and evaluated under commercial conditions. Plant material is distributed to the cooperators with complete instructions on how to produce the crop. Cultural information is then published in grower periodicals. Examples include articles on *Eustoma* (Roh and Lawson 1984a; Lawson 1990), *Anigozanthos* (Roh and Lawson 1987a), *Cyrtanthus* (Roh 1991a), *Eucrosia* (Roh 1991b) and *Phalaenopsis* orchid 'Toyland' (Griesbach 1985; Lee et al., 1988).

3. **RESEARCH RESULTS**

In this paper, we describe research results of propagation and present brief cultural information of several new crops following the initial propagation. When the choice of a propagation method is considered, efficiency of the methods, either _in vitro_ or _in vivo_, must be considered in terms of the number of propagules that can be produced most cost effectively. Further, uniform propagules must be produced so that plant growth immediately after transplanting will also be uniform. The final flowering responses must also be uniform.

New production technology to produce quality _L. longiflorum_ and _L. elegans_ lilies starting from bulblets after tissue culture or leaf cutting and bulbils that form naturally from the leaf axils, respectively, is presented.

3.1. _Eustoma grandiflorum_ Griesb.

Propagation by seeds (Tsukada et al., 1981) is the most common method to start a crop although tissue culture (Semeniuk and Griesbach 1987) and rooting of cuttings (Roh et al., 1986) are also possible. Seeds germinate in 15 days at 25 C. During the first 30 to 90 days after germination, temperatures higher than 25 C should be avoided to prevent rosette growth that prevents flowering. The optimum temperature for seedling growth is considered near 10 C to 15 C.

3.1.1. _Vegetative propagation._ Five-cm long stem tip cuttings root well even without hormone treatment. The best rooting occurred, however, when cuttings were treated with indole acetic acid (IAA) at 25 or 50 ppm (Table 1).

Table 1. Effects of IAA treatments on rooting of _Eustoma_ 'Kiri no Mine' stem tip cuttings

IAA (ppm)	Rooting (%)	Number of roots
0	73	5.8
12.5	80	5.6
25	83	6.3
50	90	9.4
100	63	8.8

Physiological disorders such as leaf scorch and die back of the growing point were observed in many plants propagated from cuttings. Plants from rooted cuttings flowered significantly later and produced fewer flowers than seed propagated plants (Table 2). Some rooted cuttings showed a reduction in growth vigor and failed to flower until all seeds propagated plants flowered. Since there was no difference in the amount of calcium in the leaves from plants with or without any apparent physiological disorder, physiological disorder may not be solely related to the calcium and other elements. When the leachate was analyzed, the leachate contained primarily glucose, protein, and glucan (Table 3). The reduction in growth vigor may be caused by leaching of calcium in conjunction with carbohydrates and protein when cuttings were propagated under intermittent mist (Roh et al., 1986).

Table 2. Growth and flowering of Eustoma, 'Kiri no Mine' as influenced by propagation methods

Propagation methods	Shoot position	Days to flower	No. of node	No. of total flower	Flowering %
Seed	main	228	10	22	100
	bottom break	-[z]	-	-	-
Rooted cuttings	main	244	9	13	83
	bottom break	256	7	6	66

[z] -: Plants from seeds did not form bottom break.

Table 3. Ethanol soluble and non-cellulosic carbohydrate contents (g/g.fw) in leachate from Eustoma, 'Kiri no Mine' shoot tip cuttings. Five mg of leachate was analyzed after trifluoroacetic acid hydrolysis

	Ethanol soluble			Non-cellulosic					
Tissue	Glc	Frc	Suc	Rha	Ara	Xyl	Man	Gal	Protein
Tip	33	1.1	3.7	0.9	3.4	2.2	2.3	3.7	33.5
Base	41	1.0	3.1	1.3	4.8	3.2	3.2	5.3	38.5

3.1.2 Seed propagation. Seeds germinate in 12 to 15 days when the temperature is maintained at 21 C. After germination, seedlings grow very slow and the slow growing seedling stage may last for more than 3 to 4 months (Roh and Lawson 1984; Tsukada, et al., 1981, 1982). During the seedling growth stage, the temperature must be maintained lower than 21 C to produce quality cut flowers (Roh and Lawson 1984; Roh and Lawson 1987b).

In the Florist and Nursery Crops Laboratory, genetically dwarf Eustoma selections were obtained from tissue culture and propagated sexually without losing the dwarf characteristics (Griesbach et al., 1988; Semeniuk and Griesbach 1987). If the temperature is maintained higher than 32/26 C during the first 30 days after sowing, seedlings become rosetted and fail to flower within a year from seeding (Table 4).

Table 4. Effect of temperature treatment on growth and flowering of Eustoma, 'Little Belle Blue'. Seeds were sown on December 21, 1989

Temperature (C) treatment days after sowing 0 - 30	31 - 60	Days to flower	Rosette %	No. of flower
32/27	23	153	39	8
27/21	23	154	11	9
21/15	23	155	0	10
32/27	15	172	22	9
27/21	15	175	0	9
21/15	15	173	0	9

3.2. Chamelaucium uncinatum

When stem tip cuttings are propagated between October and May, about 90% of cuttings root. When cuttings were obtained from June to September and treated with indole butyric acid (IBA) at concentrations of 1,600 - 3,200 ppm, rooting was about 50% (Whitton, et al., 1988). Rooting percentage is also controlled by the temperature conditions during the growth period of stock plants. If stock plants are grown at 26 C, and cuttings are treated with 8,000 ppm IBA, the rooting percentage is increased to up to 90 % (Table 5).

Generally one or two thick roots are formed in about 20 to 30 days after sticking cuttings. The root(s) is easily broken during transplanting. Therefore it is suggested that rooted cuttings should be left in the rooting trays and experience a wet - dry-wet cycle before transplanting. With this controlled watering practice, roots are firmly attached to the stem and secondary roots will form from the primary roots (Fig. 2). Rooted plants grow slowly and may require up to 4 to 5 months before an inductive treatment for flowering can be given. Liquid feeding at 200 to 400 ppm N or one gram of slow release fertilizer such as 14N-6P-8K Osmocote per 10 cm pot will ensure optimum vegetative growth.

Table 5. Rooting of Chamelaucium uncinatum, 'Purple Pride' cuttings as influenced by temperatures during stock plant culture and rooting hormone

Temperature (C)	IBA concn (ppm)	Rooting %	No. of root
21	0	14	0.3
26	0	14	0.3
32	0	18	0.2
21	4,000	46	0.7
26	4,000	73	1.1
32	4,000	35	0.5
21	8,000	77	1.4
26	8,000	88	1.9
32	8,000	66	1.1

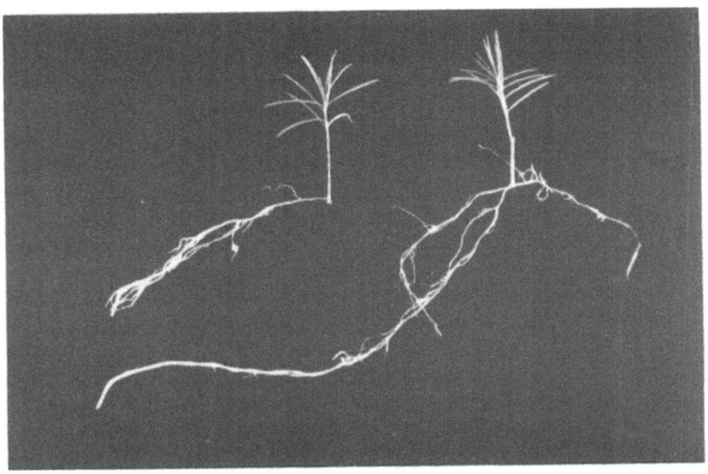

Figure 2. Rooted cuttings of Chamelaucium uncinatum, 'Purple Pride' after wet - dry - wet cycle.

3.3. Achimenantha, 'Inferno'

Selected Achimenes longiflora hybrids and Achimenantha, 'Inferno' were evaluated for forcing during winter. Due to dormancy that lasts more than 3 months in Achimenes and 5 months in Achimenantha, tissue culture propagation was employed to maintain stock plants in the laboratory under long day photoperiod environment so that plantlets could be brought to a greenhouse during winter. Scale pieces were successfully used as an explant source to produce shoots longer than 5 mm (Table 6) (Roh and Wocial 1989) and the regenerated plants showed no phenotypical variations at flowering regardless of the explant sources. When plants were transplanted into soil in February and March, plants from rhizome explants flowered in 89 and 66 days, respectively (Roh and Wocial 1989). To avoid dormancy, rooted cuttings should be forced with 16 hours of long days from high irradiance discharge lamps.

Table 6. Effect of explant source on in vitro propagation of Achimenantha 'Inferno'

Season	Explant source	Regene-ration %	No. of shoots per explant	Shoots longer than 5 mm	
				No.	%
December 1985	Rhizome	100	5.0	3.1	62
	Flower bud	100	33.3	12.7	38
	Stem	55	21.0	6.4	30
	Mid-rib	55	12.0	7.4	62
	Leaf blade	44	17.0	7.5	43
	Pedicel	22	17.0	7.5	44
January 1986	Rhizome	100	3.6	2.9	81
	Flower bud	100	22.6	8.1	36
	Stem	28	6.0	5.0	83
	Mid-rib	42	3.6	3.3	92
	Leaf blade	85	7.2	5.2	72
	Pedicel	25	5.5	5.0	91
February 1986	Rhizome	100	3.7	3.0	81
	Flower bud	100	19.9	6.9	35
	Stem	55	8.4	4.8	57
	Mid-rib	33	3.3	3.3	100
	Leaf blade	57	4.0	3.5	88
	Pedicel	0	0	0	0

3.4. INTRA-SPECIFIC HYBRID of Ornithogalum dubium and Lachenalia aloides var. pearsonii

Onithogalum dubium intra-specific hybrid (30-0) is noted for its bright yellow to deep yellow (R.H.S. 7A to 28A) colored flowers on a 10 to 20 cm long inflorescence. The effect of leaf age and position of explants from a single leaf on in vitro propagation was attempted. Young and newly expanding leaves gave the best results, regardless of the position of explants from these leaves. Old and mature leaves that were formed before the inflorescence formation gave the poorest multiplication results.

Up to 40 regenerated plantlets were produced from 1.5 cm^2 leaf explants when cultured in Murashige and Skoog (MS) media supplemented with 0.2 mg/l NAA and 2.0 mg/BA in 5 months (Fig. 3, Table 7). When leaf cuttings were obtained in the greenhouse at the same time the explants were taken, the cut surface of leaf showed soft-rot symptoms and failed to regenerate under the mist due to the high temperatures (23 - 28C).

Figure 3. Illustration of scoring of in vitro propagated Ornithogalum dubium intra-specific hybrid (30-0).

Leaves of Lachenalia were harvested at different stage of maturity in relation to inflorescence development. When leaves were harvested when the inflorescence became visible, leaf explants obtained from the proximal end of the leaves produced the highest number of bulblets with 100 % regeneration. If explants were obtained from leaves 30 days after the visible inflorescence first appeared, the number of bulblets was reduced and regeneration percentage became lower (Table 8). In both Ornithogalum and Lachenalia, the proximal end of the leaf regenerated better than the distal end. The physiological stage of the donor plants determined the rate of regeneration of Lachenalia species propagated in vitro (Niederwieser and Vcelar 1990).

3.5. Lilium longiflorum and L. elegans

Lilies are generally propagated by tissue culture taking explants from scales or scaling. In addition to tissue culture or scaling, Lilium longiflorum can be propagated by leaf cuttings and

Table 7. The effect of leaf position in plants and explant position in a leaf on in vitro propagation of Ornithogalum dubium intra-specific hybrid (30-0)

Leaf position in plant[z]	Explant position in leaf	Scoring 1	2	3	4	5
		distribution (%)				
Last leaf /prior	distal	90	10	0	0	0
	middle	58	14	28	0	0
	proximal	36	28	18	18	0
First leaf /after	distal	31	38	8	15	8
	middle	8	39	23	15	15
	proximal	22	14	7	43	14
Second leaf /after	distal	37	7	7	21	28
	middle	13	7	13	40	27
	proximal	26	7	7	40	27

[z] - Last leaf/prior : Last leaf formed prior to inflorescence formation. This leaf is the oldest or matured.
First leaf/after : First leaf formed after the inflorescence formation.
Second leaf/after : Second leaf formed after the inflorescence formation. This leaf is the youngest.

L. elegans can be propagated by harvesting bulbils that form naturally from the axil of the leaves. Generally, L. elegans is not propagated by leaf cuttings except with 'Inferno' or 'Enchantment'.

Leaf cutting propagation of L. longiflorum has been described (Roh 1982). When detached leaves of 'Georgia' lily were propagated at 21/16 C after treatment with IAA and 6-benzylamino purine (BA) alone or in combination, the number of bulblets was highest when leaves were treated with a combination of 25 mg/l IAA and 25 or 50 mg/l BA (Table 9) (Roh 1990).

Since bulblet formation is inhibited as the IAA concentration was increased, it is possible that lily leaves require a low but steadily available free IAA source for best bulblet formation. The optimum concentration of IAA could be provided during the critical stage of bulblet regeneration by applying IAA-amino acid conjugates. Leaves treated with 100 mg/l IAA-glycine (IAA-Gly) plus 50 mg/l BA gave the best results yielding 85% leaves retaining green color after 47 days and a total of 18 bulblets (Table 10) (Roh 1990). During this

Table 8. Effect of position of explants and the leaf harvest date at different stages of maturity on the number of bulblets and regeneration percentage of Lachenalia aloides var. pearsonii.

Leaf harvest stage	Leaf position					
	dis-tal	mid-dle	prox-imal	dis-tal	mid dle	prox-imal
	No. of bulblets			% of regeneration		
At visible inflorescence	8.0	8.3	11.3	10	40	100
30 days after visible inflorescence	5.7	6.4	6.8	10	33	90

Table 9. Effects of growth regulator treatments on leaf cutting propagation of L. longiflorum 'Georgia'

IAA (mg/l)	BA (mg/l) 0	25	50	100
	No. of leaves that regenerated bulblets			
0	6	6	9	7
25	5	10	11	6
50	5	19	9	9
100	2	8	7	8
	Mean no. of bulblets per leaf			
0	1.3	1.1	2.3	1.5
25	1.5	3.6	3.2	1.1
50	1.0	2.1	2.1	1.4
100	0.3	1.8	1.5	1.3

experiment, the leaves became chlorotic within 2 weeks after sticking in the rooting medium, possibly due to a high air temperature.

Leaf chlorosis and senescence leading to leaf death should be prevented as long as possible to promote bulblet formation. The base of IAA-Gly + BA treated leaves, which produced the highest number of bulblets at 50 mg/l IAA-Gly (Table 11), remained green, even 94 days after the start of the treatment (Roh 1990).

These results show that lily leaves clearly can utilize IAA-conjugates, such as IAA-Gly, for bulblet regeneration. Leaf

Table 10. Effect of free IAA and IAA amino acid conjugates on leaf cutting propagation of 'Georgia' lily. Leaves which retained green color at the base of the cutting after 47 days (green leaves) were recorded.

IAA or IAA-conjugate (100 mg/l)	BA (mg/l)	green leaves %	Total no. of bulblets regenerated	Total no. of bulblets forming
IAA	0	8	0	0
	50	23	2	0
IAA-L-Gly	0	39	8	0
	50	85	18	6
IAA-L-Phe	0	15	0	0
	50	23	2	0
IAA-L-Ala	0	0	0	0
	50	8	0	0

Table 11. Effect of free IAA and IAA-amino acid conjugates on leaf cutting propagation of 'Georgia' lily.

IAA and IAA conjugates (mg/l)	Treatment						
	IAA	BA (50 mg/l)	IAA + BA	IAA -Gly	IAA -Gly + BA	IAA -Phy	IAA -Phe + BA
	% green leaves - total no. of bulblets						
0	0-0	0-0	4-1	0-0	0- 1	0-0	0-1
25	0-0	4-1	4-3	0-1	63- 0	0-0	17-5
50	0-0	8-2	13-4	13-4	63-21	0-0	8-4
100	0-0	4-1	13-4	17-5	50-16	0-0	0-0

cutting propagation is a particularly important propagation method in lilies where seed propagation can not be employed to produce a homogenous population.

Commercially acceptable plants of short stemmed Easter lily, *Lilium longiflorum*, were produced either starting from seeds or bulblets in 14 months or 10 months, respectively. For seed propagated lilies attempts have been made to eliminate traits from *L. formosanum* since hybrids between *L. longiflorum* and *L. formosanum* are very susceptible to virus infection and do not flower uniformly.

Seedlings or bulblets either from scaling or tissue culture propagation are used to produce quality flowering plants in less than 10 to 14 months, respectively (Fig. 4). Attention should be given to the shoot apex size, not the size of the bulb. The shoot apex of L. longiflorum, 'Nellie White' from large (78 g) or small (6 g) bulbs becomes very similar in size when long day inductive treatments for flowering is given (Fig. 5). Therefore, it is possible to use seeds or bulblets to produce quality plants in 10 to 14 months, respectively. To make this technology feasible and applicable in the greenhouse forcing environments, temperature and photoperiod must be controlled.

Figure 4. Bulblet propagated L. longiflorum, 'Nellie White' in 10 cm pot.

Although seeds or bulblets from tissue culture and leaf cuttings can be used as starting propagules, L. elegans can not be started from seeds or leaf cuttings. Therefore, other ways of obtaining propagules must be developed. We have been breeding L. elegans lilies that produce bulbils from the axil of the leaves (Fig 6).

We have developed a production process method to produce L. longiflorum or L. elegans in less than a year starting from seeds,

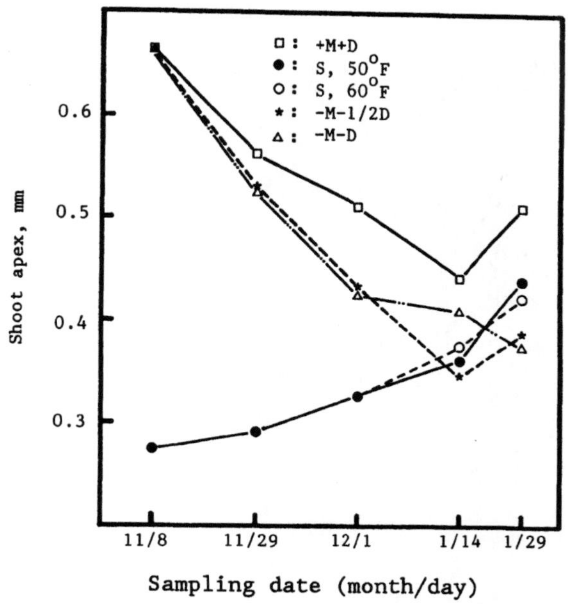

Fig. 5. Changes in shoot apex size of L. longiflorum, 'Nellie White' during development. From a 78 g bulb, all scales were intact (+M +D), all mother scales and one-half outer daughter scales were removed (-M -1/2D, 21 g in weight), or all scales were removed (-M -D, 11g in weight). Some plants from small bulbs (S : 6 g in weight) were grown at 10 C (50 F) or 15.5 C (60 F) from 1 December, 1990. Long day photoperiod was given starting from January 29, 1991.

bulblets, or bulbils as starting propagules. Thus the traditional bulb production phase in the field was completely eliminated and all phases of production are performed in the greenhouse or controlled environments. In this production system diseases can be more effectively monitored and controlled compared to field culture.

Using bulbils, quality plants for cut flower and pot plants can be produced (Roh, 1990). The general production procedures include the following steps.

1. Harvest 400 to 500 mg bulbils having a weight of 500 mg from the leaf axils of mother plants 45 to 60 days after flowering;

Figure 6. Bulbil formation in L. elegans hybrid lily.

2. Treat the harvested bulbils in a sequence of low (5 C for 20 days) - high (10 to 13 C for 14 days) - low (5 C for 20 days) temperatures;

3. Grow potted bulbils at temperatures of 26 to 32 C during the day and 21 to 26 C during the night to promote scaly leaf formation. Continue treatment until termination of scaly leaf formation;

4. Maintain the bulbils at temperatures of 16 to 21 C during the day and 10 to 13 C during the night to induce shoot emergence;

5. Force the plants to flower.

Using this protocol cut flowers with 2 to 4 flowers on a stem longer than 60 cm (Fig. 7) or pot plants with up to 5 flowers can be produced in 240 to 280 days.

3.6. Phalaenopsis 'Toyland' Orchid.

Phalaenopsis 'Toyland' orchid, developed in the Laboratory (Griesbach, 1985) has been evaluated to investigate the effect of temperature and light intensity on vegetative growth. Seed propagated 'Toyland' plants are generally uniform in vegetative growth and flowering response. When seedlings grown in growth chambers for 210 days at 25/23 C, day/night, under 7 W/m^2 were larger than plants grown at 20/18 C or 31/29 C at the same light intensity (Table 12) (Lee et al., 1988).

Fig. 7. Lilium elegans hybrid lily at anthesis. The plant was grown in 10 cm pot. Two flowers were formed on a 60 cm long stem.

4. CONCLUSION

New crops research is dynamic and the interest in new crops is rapidly expanding among scientists, growers, and the general public. Each of the steps described in this article from the identification of new plant material with potential commercial value to its release with cultural information is important for successful introduction of a new plant.

Table 12. Effect of temperature and light intensity on leaf growth of Phalaenopsis, 'Toyland' seedling grown for 210 days

Temp (C) Day/night	Light intensity (W/m^2)	Leaf length (cm)	Leaf width (cm)	Leaf no
20/18	1.5	1.9	1.2	2.8
	7.0	2.3	1.8	3.2
	22.0	2.3	1.8	2.9
25/23	1.5	2.5	1.6	2.7
	7.0	3.9	2.4	2.9
	22.0	3.3	2.2	2.9
31/29	1.5	2.2	1.3	2.5
	7.0	2.4	1.5	2.7
	22.0	1.9	1.2	2.7

The program at the United States Department of Agriculture is a joint venture between the floral industry and the government. Systematic evaluation procedures with university and industry sponsors are designed to produce complete cultural information that can be utilized nationwide. This research program is an example of how government and industry can work together and how production technology can be developed with close communication and cooperation among individuals representing various geographical areas and interests.

5. LITERATURE CITED

Armitage, A.M. (1986) Evaluation of new floricultural crops: A system approach. HortScience 21:9-11.

Christensen, O.V., and K. Friis (1987) Research and development of unknown pot plants. Acta Hort. 205:33-37.

Gertsson, U.E. (1987) Influence of light on flowering in Aeschynanthus speciosus. J. Hort. Sci. 62:71-74.

Griesbach, R.J. (1985) An orchid in every pot. Florists' Rev. 176(4548):26-27,29-30.

Griesbach, R. J. (1985) Phalaenopsis orchid as potential pot plants. HortScience 20:624.

Griesbach, R. J., and P. Semeniuk (1987) Use of somaclonal variation in the improvement of Eustoma. J. Heredity 78:114-116.

Griesbach, R.J., P. Semeniuk, M. Roh, and R.H. Lawson (1988) Tissue culture in the improvement of Eustoma. HortScience 23:658,791.

Halevy, A.H., and A.M.Kofranek (1984) Evaluation of lisianthus as a new flower crop. HortScience 19:845-847.

Lawson, R.H. (1990) Eustoma. Growers' Notebook. Greenhouse Manager. Dec. 11-12.

Lee, J.S., M.S. Roh, R. Griesbach, and K.C. Gross (1988) Growth responses of Phalaenopsis 'Toyland' seedlings affected by different light intensity and temperature. J. Kor. Soc. Hort. Sci. 29(1):58-62.

Niederwieser, J. G., and B. M. Vcelar (1990) Regeneration of Lachenalia species from leaf explants. HortScience 25:684-687.

Noordegraaf, C. vonk (1987) Development of new cutflower crops. Acta Hort. 205:25-31.

Roh, M. S. (1990a) A method for producing Lilium elegans. U. S. Patent Application Serial No. 07/501,638.

Roh, M. S. (1990b) The effects of growth regulators on bulblet formation from Easter lily leaves. Quar. Bull. Pl. Growth Regulator Soc. Amer. 18(3):140-146.

Roh, M. (1991a) Eucrosia for cuts. Greenhouse Grower 9(10):100, 102, 104.

Roh, M. (1991b) Vallota for pots. Greenhouse Grower 9(12):104-105.

Roh, M. S. (1982) Propagation of Lilium longiflorum Thunb. by leaf cuttings. HortScience 17:607-609.

Roh, M.S., J.D. Cohen, and K.C. Gross (1986) Vegetative propagation of Eustoma grandiflorum under mist. HortScience 21:799. (Abstract).

Roh, M.S., and R.H. Lawson (1984) The lure of lisianthus. Greenhouse Manager 2(11):103-104, 108, 110, 112-114, 116-121.

Roh, M.S., and R.H. Lawson (1987a) Kangaroo-paws, a classy new floral experience. Greenhouse Manager 5(9):86-87, 90, 92, 96-97, 97-100, 103, 105, 106-108.

Roh, M.S., and R.H. Lawson (1987b) Research and development on new crops in the United States Department of Agriculture. Acta Hort. 205:39-48.

Roh, M. S., and R.H. Lawson (1990) New Floricultural Crops. p. 448-453. In: Proc. 1st National Symposium. New Crops Research, Development, and Economics. Eds. J. Janik and J. E. Simon. Timber Press. Portland, OR.

Roh, M., B. Whitton, and W. Healy (1988) Interaction of propagation method and nutrition on flowering of Aeschynanthus speciosus Hook. HortScience 23:219. (Abstr.)

Roh, M.S., and M. Wocial (1988) In vitro production of Achimenantha 'Inferno' as influenced by season and the source of the explants. Acta Hort. 252:129-136.

Semeniuk, P., and R.J. Griesbach (1987) In vitro propagation of prairie gentian. Plant Cell, Tissue, Organ Culture 8:249-253.

Tsukada, T., T. Kobayashi, and Y. Nagase (1981) Studies on the physiological characters and cultivation of prairie gentian. I. Physiological characters of germination and raising of seedling. Bull. Nagano Veg. Ornam. Crops Expt. Sta. 1:39-46.

Tsukada, T., T. Kobayashi, and Y. Nagase (1982) Studies on the physiological characters and cultivation of Russel prairie gentian. II. Effect of temperature and photoperiod on growth and flowering. Bull. Nagano Veg. Ornam. Crops Expt. Sta. 2:77-88.

Welander, N.T. (1984) Influence of temperature and daylength on flowering in Aeschynanthus speciosus. Scientia Hort. 22: 157-161.

Whitton, B., M. Roh, and W. Healy (1988) The influence of plant growth regulators and cutting length on rooting of Chamelaucium uncinatum, 'Purple Pride'. HortScience 23:231 (Abstract).

Whitton, B., W. Healy, and M. Roh (1990) Flowering of Aeschynanthus, 'Koral' at fluctuating and constant temperatures. J. Amer. Soc. Hort. Sci. 115:906-909.

COMPARISONS OF DIFFERENT PLANT PRODUCTION METHODS FOR FOREST TREES

M.I. MENZIES
Forest Research Institute
Private Bag 3020
Rotorua
New Zealand

and

J.T. ARNOTT
Forestry Canada
506 W. Burnside Road
Victoria, B.C. V8Z 1M5
Canada

ABSTRACT. While natural regeneration and seeding have been used extensively to re-establish forests, it is now more common to plant, using seedlings or vegetative propagules. The main plant production methods are bare-root, transplant, and containerized systems. There are advantages and disadvantages for all systems. Bare-root plants are cheap to raise, and mechanized systems have been developed for most operations. The planting season is restricted to the winter months when the trees are dormant, although some species may be cool-stored successfully for early spring planting. Transplants are raised initially either bare-rooted or in containers and then transplanted into nursery soil to be raised as bare-rooted transplants. Transplants have been an alternative in harsher climates, where tree growth is slow and a longer time is required in the nursery. Containerized systems have been developed extensively in the 1970s and 1980s and are now widely used in harsher climates where bare-root production is more risky, and in the tropics where seedlings often do not become dormant, or planting must be done in the summer during the rainy season. When selecting a plant production method, the plant specifications necessary for successful establishment on the particular site can be used as a guide. Any cost-benefit analysis should include the total cost of restocking the area and not just the cost of the plants in the nursery.

1. Introduction

Forestry is an important industry throughout the world, based on natural and planted forests. Following logging, the forest areas need to be restocked, either by natural regeneration, seeding, or planting. Natural regeneration has been relied on extensively, but there can be a number of disadvantages, particularly a lack of control over whether there will be too few or

K. Kurata and T. Kozai (eds.), Transplant Production Systems, 21–44.

too many seedlings regenerating, the distribution of seedlings, the species, and the level of genetic improvement. Seed trees of the desired species can be left standing to shed seed, but this does depend on reliable seed fall from these remaining trees. Supplementation of natural seed has also been practised, by aerial seeding or manual seeding. However, this can be a very inefficient use of seed unless the seed is carefully placed in scarified spots and protected from predation by birds and mammals. While natural or supplemented seeding can be a cheap option, there is no guarantee that it will result in successfully established forest for future harvesting. Planting of areas is now more widely accepted as the most reliable and cost-effective method, using seedlings or vegetative propagules.

Most of the planting is done with seedlings, but cuttings have been used for centuries (Toda 1974), particularly where seed has been scarce or clones with particular characteristics have been desired. Poplars (*Populus*) and willows (*Salix*) would be two genera which have been widely planted in the field as clones, using either rooted or unrooted cuttings (Wilkinson *et al.* 1981). Even radiata pine (*Pinus radiata* D. Don) unrooted cuttings have been directly planted in the field successfully (NZ Ministry of Forestry 1992).

Other types of planting stock which will become more important in the future are based on micropropagation by organogenesis or embryogenesis. Organogenesis can be expensive for forestry (Smith 1986), although one forestry company in New Zealand is using this system commercially to establish clonal forests (Gleed 1991). Somatic embryogenesis offers the potential of higher multiplication rates and lower costs, and somatic seedlings have been planted for spruces (*Picea*), larch (*Larix*), Douglas fir (*Pseudotsuga menziesii* [Mirb.] Franco), loblolly pine (*Pinus taeda* L.), and radiata pine. The use of micropropagation techniques for forestry is discussed in more detail at this Symposium by Aitken-Christie and Connett (1992).

There are three different plant production methods used to produce seedlings, rooted cuttings, plantlets or somatic seedlings for forest planting, and these are bare-root, transplant and container systems. In a bare-root system, plants are raised in soil in nursery beds and when they are ready for planting, they are lifted out of the nursery beds and sent to the field sites with a minimum of soil protecting the roots. Plants for transplants are raised either as bare-root or containerized plants initially, and then transplanted into nursery soil to be raised as bare-rooted transplants. In a container system, the plants are raised in a potting mix in a container, and the plants are usually sent to the field rooted in their containers. The root mass holds the potting mix together, and the plants may be planted with or without the container, depending on whether the container is biodegradable and permeable to the root system. While there are distinctly different production methods, with various advantages and disadvantages for each, the principles for raising the plants are similar. This paper will discuss the specifications for forest planting stock and compare these production methods.

2. Planting Stock Specifications

Planting stock must be raised to suit the sites where it will be planted. It is important that the new crop not only survives after field planting, but also has vigorous and uniform growth in the first and subsequent years. The optimum time for planting is usually when the plants are most dormant in winter, but this is not always feasible, particularly if sites are not accessible because of snow, or where rainfall is predominantly in the summer months.

Seedling quality specifications have traditionally been based on certain morphological characteristics, including height, diameter at the root collar, sturdiness (height/diameter ratio), root/shoot ratio, and some root features, although there is no universal agreement on which variables are the most useful (Aldhous 1967, Cleary *et al.* 1978, Lavender and Cleary 1974, Schmidt-Vogt 1974). No single variable is sufficient. Tall trees are often preferred because they are thought to be better for weedy sites, or where animal damage can be expected, but tall trees can be produced by close spacing and high nitrogen fertilization, both of which can result in poor quality seedlings (Schmidt-Vogt 1974). Similarly root/shoot ratio is not reliable as a single indicator since trees with the same ratio may have a large heavy taproot and few laterals, or a mass of fibrous root; the latter is considered more desirable (Cleary *et al.* 1978). A combination of height, root collar diameter, sturdiness, and root quality would seem preferable.

The physiological quality of a seedling is also important for successful establishment, but is more difficult to assess. Poor physiological condition is often apparent only after outplanting, when seedlings perform poorly or die without any new root growth or before flushing (Lavender and Hermann 1976). When a seedling is planted in the field, the first problem is usually to restore and maintain a suitable water balance. Stomatal closure will help conserve water, but may be insufficient to prevent wilting. The seedling must establish a good contact between its roots and the water supply in the surrounding soil. A large root surface area will help, but often it is the fine water-absorbing fibrous roots that are damaged and killed before planting. New roots must be initiated and elongated into the surrounding soil, using stored or generated photosynthates. If the seedling is under water stress, with stomata closed, then the production of current photosynthates will be minimal, and any new root growth will be dependent on stored food reserves (Menzies *et al.* 1985b).

Plants of many species from cold or temperate climates have a definite dormancy cycle, with hard winter resting buds formed. Unless the full dormancy cycle is successfully completed, bud flushing in spring may not be vigorous or may not occur (Cleary *et al.* 1978). The essential physiological attributes therefore include an adequate water content and food reserves, a healthy root system, preferably with root initials already formed, and a suitable dormancy state.

3. Factors Affecting Raising Of Planting Stock

3.1. CLIMATE

Probably the three most important aspects of climate affecting plant growth are temperature, light, and rainfall. These factors interact, and can be manipulated to control plant growth and development.

3.1.1. *Temperature.* Temperature affects the rate of chemical reactions, such as photosynthesis and respiration. Below about 5°C, little metabolic activity occurs (Cleary *et al.* 1978), while the optimum varies with species and is usually in the range of 20-30°C. Both air and soil temperatures are important. Germination in spring will be slow until the soil temperatures have warmed above 10°C or more. The length of the growing season will depend on how long there is between conditions warming up in spring and cooling down in autumn. If a species is slow growing and the growing season is short, then more than one year may be required for the plant to reach the desired size. In autumn, shortening photoperiods followed by lower temperatures are important for inducing dormancy and frost tolerance (Weiser 1970, Lavender and Cleary 1974, Greer and Warington 1982, Glerum 1985).

3.1.2. *Light.* Both light intensity and photoperiod are important for plant growth and development. Light intensity is rarely a problem with plants raised outdoors, unless the density of the plants is too great, restricting light to the lower foliage. Long photoperiods are essential to maintain shoot growth of tree seedlings at higher latitudes, but short photoperiods are necessary to induce dormancy and frost tolerance in autumn (see reviews by Vince-Prue 1975 and Arnott & Mitchell 1982).

3.1.3. *Rainfall.* Sufficient moisture is critical during the growing season to avoid excessive plant water deficits affecting plant growth and development. If soil water becomes limiting during the growing season, then irrigation will be required to maintain optimum rates of photosynthesis. Once plants have reached the desired size, limiting water availability and inducing plant water deficits is a method for inducing the first stage of dormancy. However, when used on containerized seedlings, it can lead to reduced numbers of stem units and needle primordia (O'Reilly *et al.* 1989, Grossnickle *et al.* 1991). Furthermore, the timing of application, the intensity and the duration of the water deficit are all critical in reducing the potential for damage during subsequent cold storage and maintaining seedling vigour (Macey and Arnott 1986).

3.2. SOIL

The physical characteristics of the soil are important for long term bare-root nursery operations. Well-drained, deep, coarse-textured soils – loamy sands or sandy loams – with silt plus clay contents of between 10 and 25% have the best physical characteristics, although limited capacity to retain nutrients (Knight 1981). Heavy clay soils can give problems with cultivation access and drainage, and should be avoided if possible. Soil texture and structure affect aeration, water movement, storage capacity and penetrability to roots. While texture is fixed by soil type, soil structure is affected by nursery practice and it is important that soil structure be maintained (Knight 1981). This may be done by the addition of organic amendments or the regular use of green manure crops, such as grass and legumes for 2–3 years.

A wide range of materials has been used as potting mixes for containers, including various soil types, peat, pumice, perlite, and vermiculite. The medium requirements are similar to those for bare-root nursery soil, and include a slightly acid pH, high cationic exchange capacity, low inherent fertility, good porosity and freedom from pests (Landis 1990a). Roots must be able to penetrate the medium easily. Peat, or peat and vermiculite, have been the most popular in North America (Tinus 1982).

3.3. FERTILIZERS

Plants need an adequate and balanced supply of nutrients for optimum growth. For bare-root seedlings, soil analysis before cultivation and sowing will allow a base dressing to be applied to adjust soil pH to 5.3–5.4 and supply a balanced level of nutrients. Relatively low-cost agricultural fertilizers such as potassic serpentine superphosphate, dolomite, and muriate of potash can be used (Knight 1981, Chavasse 1981). Nitrogen is often better supplied during the growing season, because it is so mobile (Knight 1981), and can be supplied as side dressings or in liquid form as urea. Side dressings of granulated complete NPK are used to maintain active growth during the growing season. Trace elements such as Mg, Fe, Zn, and Cu can also be supplied as side dressings or in liquid form if required. Plants take significant quantities of nutrients out of the nursery when they are lifted for field planting, and these need to be replaced for subsequent crops.

Incorporation of slow-release fertilizers into the growing medium is sometimes used in container nurseries, particularly when liquid fertilizer injection equipment is not available, or when the seedlings are being grown outdoors in areas with heavy rainfall. However, the method has major drawbacks because of difficulties in obtaining even distribution throughout the growing medium and breakdown of the fertilizer pellets in the mixing process (Matthews 1982). Injection of nutrients into the irrigation supply is by far the most common method of fertilization technique used throughout North America. This method offers the best means of controlling plant growth throughout all phases of seedling development, especially during the critical period of dormancy induction prior to storage and/or outplanting (Landis 1989a). Fertilizer regimes have been devised for various species, and the proportions of various nutrients are changed to meet the requirements for optimum growth to reach the desired size and then induction of dormancy before outplanting.

3.4. IRRIGATION

An adequate supply of water to plants is essential for optimum growth. With bare-root nurseries, natural rainfall may be the only source of water, although irrigation can be an excellent insurance, particularly for critical periods such as seed germination, after application of certain herbicides and fertilizers and root conditioning, as well as for ensuring that growth is not limited by water availability. Irrigation equipment and reliable supply of clean water are necessary if irrigation is going to be used.

In container nurseries, plants are growing in a limited soil volume, and regular watering is necessary to ensure optimum growth and development. The individual containers are usually separated, preventing flow of water between containers, and so irrigation systems giving even distribution are essential. Travelling booms are often used as they provide the most uniform irrigation pattern. The boom passes several times over the crop for each watering cycle, to ensure even watering. Fixed sprinkler systems can be very inefficient, resulting in excess amounts of run-off into the waste water disposal system. With the possibility of nitrate and pesticide run-off, this can pose a problem with nurseries located near urban areas (Landis 1989b). A freely-draining container mix is necessary to avoid waterlogging. Potting mixes can be very difficult to rewet if they dry out completely, particularly if there is a high proportion of peat in the mix.

3.5. DISEASE AND INSECT CONTROL

Disease and insect damage may slow growth, and can even result in plant death if left untreated. Sterilization of nursery soil and potting mixes is sometimes done before sowing seed or setting cuttings to kill any soil-borne fungi, insects, and weed seeds. This is an expensive operation, especially for bare-root nurseries, and treating the seed or cuttings and spraying the soil with appropriate chemicals as necessary is often preferred. If the soil is sterilized, then reinoculation with mycorrhizal fungi may be necessary, particularly for conifer crops.

Diseases and pests are of particular concern in containerized nurseries that tend to use high seedling densities and enclosed growing facilities. The latter can, if not properly controlled, have high humidity levels and poor air movement, which are ideal for promoting plant disease outbreaks. However, excellent manuals are now available to aid workers in quickly identifying diseases and pests before they become a problem (Landis 1989c). Growers can use a variety of cultural and chemical measures to manage these problems and limit their losses.

3.6. SOWING AND PLANT DENSITY

The three important aspects of sowing are time of sowing, sowing depth, and sowing density.

3.6.1. *Time of Sowing.* Sowing time is one of the methods of controlling plant size, with smaller seedlings resulting from later sowings (Chavasse 1977). In bare-root nurseries, seed is usually sown in spring depending on climate, ground conditions, and final desired seedling size. After the ground is prepared, the sowing programme is completed as rapidly as possible so that germination occurs over a relatively short time. When the sowing period is prolonged, birds may follow seedling emergence over the nursery, causing damage over the whole area. If the emergence time is short, bird damage is usually restricted to edge areas, adjacent to shelter (Menzies *et al.* 1985b). Time of sowing for containerized crops will depend on the growing regime being used, but is timed to produce plants of a suitable size and dormancy, ready for the planting season. In container nurseries that produce multiple crops during a single growing season, up to three sowings can occur within a calendar year.

3.6.2. *Sowing Depth.* Sowing depth affects both germination percentage and crop uniformity. If seed is sown too shallow it is subject to drying and bird predation. If it is sown too deep, it may not have the germinative energy to emerge, or emergence may be delayed, resulting in an uneven crop. Sowing depth is more critical with small seed (Menzies *et al.* 1985b). Seed is usually sown in a continuous furrow in the soil for bare-root plants, with the soil pushed back to cover the seeds. Sowing depth is set by the depth of the furrow. Alternatively, seed may be covered by fine gravel (Balneaves and Fairhurst 1974), sawdust, or other covering, particularly where nursery soils are heavy, and soil crusting may inhibit seedling emergence.

With containerized nurseries, seed is covered by a medium such as fine gravel, crushed rock or styrofoam pellets, and the depth of cover can be adjusted to suit seed size (Nyborg and Schikaze 1974), but it must be as uniform as possible to facilitate uniform plant growth and development (Tinus and McDonald 1979).

3.6.3. *Plant Density.* The density at which plants are grown is one of the most important factors affecting plant quality (Menzies *et al.* 1985b, Menzies 1988). If plants are grown too closely, they tend to be spindly, with small diameters, poor sturdiness and with poorly developed root systems. Lower foliage may also be too shaded during the nursery phase, resulting in death of the lower foliage from the low light levels while in the nursery, or from sun scorch after planting out in the field.

As plant spacing in the nursery becomes wider, seedling height and sturdiness ratio decreases while root collar diameter and root volume increases (Menzies *et al.* 1985b). There is an optimum spacing beyond which there is no significant further improvement. This spacing varies with species and the climate of the nursery locality. Optimum seedling spacing is 5 cm x 12.5 cm (160 plants m^{-2}) for 1-year-old radiata pine seedlings at Rotorua, but at least 6 cm x 12.5 cm (133 plants m^{-2}) in the south of the South Island (Menzies *et al.* 1985b). Similar trends have been reported for Douglas fir containerized seedlings, with plants grown at 270 plants m^{-2} being significantly better quality than those grown from 540-1100 plants m^{-2} (Tanaka and Timmis 1974).

When selecting a density for growing plants in the nursery, there is a compromise between increased costs and improved quality from lowering plant density. Sowing in bare-root nurseries is often done with precision seeders, which sow seed at precise spacing. The sowing density is given by the within-row and between-row spacing, assuming that there is a single seed sown in each sowing spot, rather than doubles or trebles. If there are doubles and trebles, as well as many missed spots, the average density may be the same as the target density, based on spacing, but the quality of the seedlings raised in clumps from double or treble sowing will be lower. It is important that seed with a high germination percentage is used, otherwise there will be gaps in the stocking.

Similarly for containerized seedlings, it is important that as many containers as possible have a plant growing in them to avoid empty cells. Seed should be cleaned to obtain at least 75% germination (Tinus 1982) unless multiple seeding per container is going to be used. Multiple seeding involves wasting seed as well as extra costs for thinning to a single plant per cell (Belcher 1978), but this approach is used by a majority of seedling growers to ensure the maximum numbers of filled containers. Very few blank cells can be tolerated, because of expensive greenhouse growing space (Tinus and McDonald 1979). A wide array of container types, with many different combinations of cell volumes and densities, now enables nursery managers to select the most appropriate container for a particular plant size they wish to produce.

3.7. MYCORRHIZAE

Mycorrhizae are beneficial root fungi which can be important for establishment and growth of plants. In bare-root nurseries, mycorrhizal fungi are usually present from previous crops, unless the soil is sterilized, while they are not present in containerized soil mixes unless they are introduced. Different species of mycorrhizae may colonise a tree species, and they may have different levels of efficiency (Bowen 1965). *Rhizopogon spp.* are more efficient with radiata pine seedlings than other mycorrhizae (NZ Forest Service 1983). Introduction of mycorrhizae is usually achieved by use of forest duff, fungal mycelia, or fungal sporocarps (Chu-Chou and Grace 1990) and these methods are suitable for both bare-root and containerized nurseries. Although the importance of mycorrhizal inoculants is recognized by container nursery managers throughout North America, a recent survey of 78 nurseries indicated that only 6 percent actually had a fungal inoculation program (Castellano and Molina 1989).

3.8. CONDITIONING

Plants need to be conditioned before planting to withstand the stresses of transplanting and allow them to grow rapidly in their first growing season in the field. This also involves correct induction of dormancy and frost tolerance. Douglas fir seedlings must go through a cycle of four dormancy phases before they will initiate vigorous root growth and flush in time in the spring (Lavender and Cleary 1974). These phases are dormancy induction, from mid-summer to early autumn, dormancy deepening up to early winter, deep dormancy during winter, and post dormancy up until bud burst. The first phase is triggered by moisture stress in late summer (Lavender and Cleary 1974) and/or by short photoperiods followed by low temperatures (Tanaka 1974), and during this period terminal buds are initiated and develop.

Seedling conditioning has been largely an empirically based operation, subject to wide differences of opinion and controversy. Commonly undercutting, drought conditioning by withholding water, or nutrient limitation, or a combination of these have been used (Mullin 1966). All three methods can be used with bare-root plants, particularly undercutting and wrenching. For containerized plants, drought conditioning and nutrient limitation are used in conjunction with environmental control of photoperiod and temperature.

3.8.1. *Conventional Drought Conditioning.* The aim of drought conditioning has been to start the dormancy cycle in mid-summer to late summer, causing height growth to cease, and terminal bud initiation and development, so that the dormancy sequence can be completed on time without a fall flush of growth. This can be done by subjecting the seedlings to a plant water potential of about -1.2 to -1.5 Mpa predawn, or -1.8 Mpa during the day (Lavender and

Cleary 1974). The length of time water needs to be restricted varies from one to three months before the planting season (or the beginning of autumn rains) (Goor 1955, Cleary *et al.* 1978). If the weather is excessively dry and hot, light irrigation may be needed to cool the seedlings and maintain photosynthesis (Lavender and Cleary 1974). If there is excessive rainfall, root pruning by undercutting may be necessary to control height growth (Darby 1962). In container nurseries, the timing of short photoperiods and moisture stress and the intensity and duration of drought stresses are all critical in determining the ultimate degree of seedling hardiness (Macey and Arnott 1986). Furthermore, such treatments may not be particularly effective on indeterminate species such as western red cedar (*Thuja plicata* D. Don) (Krasowski and Owens 1991).

3.8.2. *Conditioning by Undercutting and Wrenching.* Undercutting is the process of passing a horizontal, flat, thin, sharpened blade beneath the seedbed, to sever taproots of seedlings, or sinker roots of cuttings, usually at a depth of 8–10 cm, although often up to 20–30 cm in North American nurseries.

Wrenching is the process of passing a horizontal, thicker, tilted, blunt blade beneath the seedbed to prevent deep rooting and height growth renewal, to aerate the root zone, and to promote fibrous rooting. This is usually done just below the undercutting depth, although in heavy soils it may need to be at the undercutting depth, and is usually done every 2–3 weeks once the trees have reached the desired height in the summer.

Lateral root pruning is the process of passing vertical knives or coulters between the rows of seedlings, severing any lateral roots down to the wrenching depth. This is usually done every 4–6 weeks once wrenching has commenced.

Undercutting or wrenching have been widely used usually as an empirical method of inducing dormancy and controlling height growth in late summer/early fall, to aerate the seed beds, to aid fibrous root formation, and to aid lifting of nursery stock (Davey 1964, Stoeckeler 1965, Mullin 1966, van den Driessche 1969, van Dorsser and Rook 1972, Eis and Long 1973). To get full benefit of an undercutting and wrenching program, it is necessary to combine the control of bed density, with undercutting, wrenching, and lateral root pruning (Rook 1969, 1971, van Dorsser and Rook 1972). The conditioning causes height growth to nearly cease, while diameter growth and root growth continues. This results in seedlings with better sturdiness, root/shoot ratio and with a compact fibrous root system that can withstand the stresses of transplanting (Rook 1971).

3.8.3. *Conditioning by Nutrient Limitation.* Plants will cease to grow when nutrients are limiting, and this can be used to induce dormancy, provided watering is continued, although plants require less water as they harden off. This has been very effective with containerized eucalypt (*Eucalyptus*) seedlings (Faulds and van Dorsser 1987).

3.9. LIFTING AND HANDLING AND PACKAGING

Bare-root plants need to be lifted and packed for planting in the field, while containerized plants can be sent to the field in their containers or removed from their containers and packed in cartons. Lifting and handling procedures can have a marked effect on plant establishment and growth (Trewin 1981, Tabbush and Ray 1989). Bare-root seedlings can be lifted by hand or by machine. Lifting by hand is currently preferred in New Zealand because there is less damage to the seedlings. It is important to remember that the goal of handling, storage and transportation is to ensure that the full growth potential of the seedling when it leaves the nursery is present when it is outplanted (Mitchell *et al.* 1990).

3.9.1. *Time of Lifting.* With bare-root seedlings, survival after planting is likely to be higher if seedlings are lifted later in the winter rather than at the beginning of winter. Seedlings can be conditioned to enable planting at any time of the year on an experimental basis, but careful handling is necessary for unseasonal planting, and drought following planting can be a problem in summer and autumn (Menzies *et al.* 1985b).

Containerized seedlings have a much broader range of planting seasons than bare-root stock (Arnott 1974), but the best choice of planting season will be one that minimizes stress during and after planting. In British Columbia, for example, this is usually early spring in coastal areas and during the summer in the interior regions of the province (Mitchell *et al.* 1990).

3.9.2. *Lifting and Handling Methods for Bare-Root Seedlings.* Before lifting, the soil must be loosened, so that seedlings can be easily removed without stripping off roots. This is done by passing a wrenching blade repeatedly under the bed at a steeper angle and at about double the normal wrenching speed, so that the soil is well broken up, and the seedling roots loosened. The seedlings can then be pulled up easily, either singly, or preferably in groups, to minimize loss of fibrous roots. Seedlings must be carefully separated, and the seedlings root trimmed in small bundles, so that they are all of even length. This allows planting without tangling or distorting root systems. Root trimming in large bundles gives uneven results, with some trees missed in the middle of bundles, and edge trees often overtrimmed (Trewin 1981).

Once seedlings have been lifted and root trimmed, they have a fixed morphological quality that is not affected by handling treatment. However, the physiological quality can be affected. For example, water content of seedlings can be replenished by wetting the roots. Food reserves will decline with time, due to losses by respiration, but storage conditions can minimize the losses. Therefore handling procedures must be designed to minimize any loss of these food and water reserves (Menzies *et al.* 1985b).

3.9.3. *Packaging.* Both bare-root and containerized plants need to be protected while in storage and during transport to the field to avoid desiccation and crushing. Plastic bags are often used as a container to prevent desiccation, but they can not prevent crushing, bruising and breakage of tops and roots, and plants are susceptible to overheating if bags are exposed to sunlight. Rigid cardboard or plastic cartons will provide protection against this type of damage, and a combination of both plastic bags and cartons are sometimes used to provide protection against desiccation and physical damage. Multi-walled kraft paper bags with a plastic lining are also used.

Cartons or bags can be stored and transported in crates for protection as well as ease of mechanized handling (Trewin and Cullen 1985).

4. Typical Regimes For Raising Stock

4.1. BARE-ROOT SEEDLINGS

The length of time required in the nursery to raise bare-root seedlings ranges from under 1 year to up to 4 years. If the seedlings are kept for more than 1 year, they may be raised at close spacing for 1 or 2 years, and then transplanted to wider spacing for a further 1 or 2 years. The age and transplant history of seedlings are commonly described by 2 numbers; the first giving the time (in years) that the seedlings have been in seedling beds, and the second number giving the time (in years), the seedlings have been in transplant beds. Thus

1/0 (or 1+0) seedlings are raised in nursery beds for up to a year and are not transplanted, while 2/2 (or 2+2) seedlings are raised for 2 years in seedling beds and 2 years in transplant beds for a total of 4 years in the nursery.

Production regimes for radiata pine seedlings and cuttings at FRI Nursery, Rotorua, New Zealand will be used as examples of 1/0 bare-root seedlings, and 1-year-old bare-root cuttings. More detailed information on radiata pine is available from Chavasse (1981) and Menzies *et al.* (1985a, b).

4.1.1. *Sowing.* Seed is sown in spring (September–November), usually October. A vacuum drum precision seeder is used to form raised seedbeds and sow the seed at the required spacing (Hiscock *et al.* 1981). Seed is sown in grooves at a depth of approximately 6 mm and then covered with soil.

4.1.2. *Weed Control.* The principle followed is to prevent weeds germinating in the crop, and so a pre-emergence spray of propazine at 1 kg active ingredient (a.i.) ha^{-1} and chlorthal at 7 kg a.i. ha^{-1} is applied after sowing but before emergence, and this spray is repeated two times during the growing season as post-emergent sprays (Brown 1991). Other herbicides, such as knock-down herbicides with or without residual properties may be used if necessary to control germinated weeds. Glyphosate is an example of a knock-down herbicide which can be applied before seedling emergence, while haloxyfop can be applied after emergence. Tebuthylazine is a herbicide which can be used for post-emergence knockdown with some residual control.

4.1.3. *Fertilization.* A base dressing is applied before sowing, depending on soil tests. Once seedlings are growing, then maintenance side dressings of a granulated fertilizer (12% N, 5% P, 14% K, 4% S, 1.2% Mg + micronutrients, or similar) are applied in early and late summer (late December and late February). Each is applied at 130 kg ha^{-1}, using a fertilizer distributor which drops the fertilizer between the rows, without leaving fertilizer on the foliage.

Foliar feeding is also used to prevent deficiencies of N, Mg, and Fe. N is applied as a 2% (w/v) urea spray, together with 2% (w/v) magnesium sulphate regularly during the growing season, but particularly before and after conditioning commences. Fe is applied as 0.05% (w/v) iron chelate as required.

4.1.4. *Conditioning.* Seedlings are conditioned by undercutting, wrenching and lateral root pruning, commencing in summer, when the seedlings have reached the desired size of about 20 cm, using a reciprocating undercutter/wrencher, and a lateral root pruner (Rook 1971, van Dorsser and Rook 1972, van Dorsser 1981).

Seedlings are undercut at a depth of 5-8 cm, and then wrenched at approximately 4–weekly intervals at a depth of 10-12 cm, to avoid the undercut taproot. Seedlings are laterally root pruned on two sides by rolling coulters passing down between the rows, and cutting down to the wrenching depth. This is done about 6 weeks after undercutting, and again about 6 weeks later.

4.1.5. *Lifting and Handling.* Seedlings are lifted in winter, from May to August. After lifting and root trimming, seedling roots are dipped in water to keep the roots moist and the seedlings are then placed on their side in waxed cardboard cartons on the seedbed to minimize handling time. The cartons will hold 100-150 seedlings, depending on seedling size, and act as the planting container in the field. Cartons are packed in plywood crates for transport to the field (Trewin and Cullen 1985). Storage time between lifting and planting is kept to less than 48 hours if possible to minimize desiccation and loss of food reserves by respiration.

4.2. BARE-ROOT CUTTINGS

4.2.1. *Setting.* Bare-root radiata pine cuttings are raised on a similar regime to 1/0 seedlings (Menzies *et al.* 1985a, Faulds and Dibley 1989). Raised beds are formed using the bed-former from the precision seed sower, and the spacing for setting (8 cm x 15 cm) is marked on the beds using a roller. With 15 cm between rows, there are only 6 rows per bed, rather than 8 as for seedlings. Cuttings are collected from stool beds in winter (May–July) and loosely packed in waxed cardboard cartons and watered. Cuttings should be 7–10 cm long, with a minimum basal diameter of 3 mm. They are set at a depth so that the tips of the basal needles are exposed (2–4 cm). There is no need for rooting hormones to be applied.

4.2.2. *Weed Control.* The same weedicide regime as for 1/0 seedlings is used, with the beds sealed with propazine before setting.

4.2.3. *Fertilization.* A slow-release fertilizer, 'Magamp' (7%N, 40%P, 6%K and 12%Mg), is added to the soil before bed forming at a rate of 100–300 kg/ha, depending on soil fertility and nursery location (Menzies and Faulds 1991).

Cuttings go pale in spring (October–November) as they form roots, and a granulated complete fertilizer is applied in December, as with 1/0 seedlings. A further application of the granulated fertilizer is applied in late summer–autumn (February–April) if the cuttings are growing too slowly. Foliar feeding is also done, as with 1/0 seedlings.

4.2.4. *Conditioning.* Conditioning by undercutting, wrenching, and lateral root pruning is done as for 1/0 seedlings. The initial undercutting is done earlier, in mid-summer (depending on plant size), at a depth of 10 cm, followed by lateral root pruning in late-summer. Wrenching and further lateral root pruning are done as for 1/0 seedlings (Faulds and Dibley 1989).

4.2.5. *Control of Plant Size.* Cuttings tend to be larger than seedlings, and an ideal size is 25–35 cm in height with a root collar diameter of 8-10 mm. Size can be restricted by a number of methods (Faulds and Dibley 1989), including rootpruning stock plants, reducing fertilizer applications or rates, setting smaller cuttings, setting later than July (although irrigation may be needed with later setting, and losses may be higher), and earlier and more intensive conditioning.

4.2.6. *Lifting and Handling.* These are the same as for 1/0 seedling crops.

4.3. TRANSPLANTS

Transplants are grown for a number or reasons including:

- maximizing use of scarce seed. Seed can be germinated under controlled conditions before plants are lined out, and there is no thinning of plants in seedbeds.
- maximizing use of nursery seedbeds, particularly when using seed with poor germination. Also seedlings can be raised initially at higher densities before transplanting, and this can save nursery bed space, particularly for 3– or 4–year crops.
- raising larger plants than can be grown in normal containers.
- minimizing time in the nursery. Seedlings can be raised quickly in containers under controlled conditions, before being lined out.

- controlling root development. Seedlings can be root pruned before lining out, to stimulate fibrous root development.
- avoiding weed control problems with small plants. Some species are sensitive to weedicides when they are small, and this means expensive hand-weeding may be necessary. Raising seedlings initially in containers avoids this problem.

Seedlings for transplanting can be raised by several methods, including broadcast sowing into trays or seedbeds, sowing in individual containers, and sowing in rows in seedbeds. The size of plants before transplanting can vary from 1–2 cm up to 30 cm or more. Bare-root seedlings are usually transplanted in winter, while they are dormant, unless they have been hardened off especially by conditioning or nutrient starvation. Eucalypt seedlings in seed trays can be hardened off very successfully at any desired height by withholding nutrients but not water so that the seedlings become starved, with woody stems, and thicker leaves (Faulds and van Dorsser 1987). If bare-rooted seedlings are transplanted in spring or early summer, then shading or irrigation may be needed to ensure survival. Seedlings can be lifted in winter when they are most dormant and cool-stored until they can be transplanted in spring. This avoids transplanting in winter during the busy lifting season, and also ensures beds are available for transplanting following the lifting season. Transplanting can be done manually or by machine. With bare-root seedlings, it is important that the roots are trimmed before transplanting and that the seedlings are transplanted without distorting the root systems. Containerized seedlings can be transplanted without root distortion if the roots have bound the soil mix together, but are not extending beyond the soil plug. Containerized seedlings can be planted out at any time of the year, as long as their dormancy requirements are met.

4.4. CONTAINERIZED SEEDLINGS

Two excellent reviews of containerized seedling practice are the Proceedings of the North American Containerized Forest Tree Seedling Symposium, held at Denver, Colorado, USA in 1974 (Tinus *et al.* 1974) and the Proceedings of the Canadian Containerized Tree Seedling Symposium, held at Toronto, Ontario, Canada, 1981 (Scarratt *et al.* 1982). Furthermore, an updated version of the American Container Tree Nursery Manual (Tinus and McDonald, 1979) is presently being published in an expanded series of separate volumes dealing with specific topics. Three volumes have been published to date (Landis *et al.* 1989, 1990a, 1990b) and another four are in preparation. There are many different systems used around the world depending on the species being grown, the facilities available, the container type, and the planting season.

4.4.1. *Facilities.* Growing facilities can range from just growing the plants outdoors with no extra facilities, through simple shadehouses, to computer-controlled greenhouses, where the environment is closely monitored and controlled. More sophisticated facilities are required in harsher climates or where growth is being manipulated out of phase with natural conditions. Control of temperature, light intensity, photoperiod, moisture, and mineral nutrition is necessary to optimize growth and induce dormancy, particularly if two crops are raised each year (Owston 1974). In British Columbia, container production methods have emphasized single-crop, low-risk rotations using the normal growing season (van Eerden 1974, 1982). While open outdoor compounds are still extensively used, recent trends show an expansion of fully-controlled greenhouses to produce single, high-quality seedling crops (Van Eerden 1990).

4.4.2. *Containers.* Containers can be divided into two functional categories: those planted with the tree seedling and those removed before the seedling is planted (Tinus and McDonald 1979).

Some of the earliest container designs were developed to grow an acceptable seedling which could then be out-planted directly in its container. These containers included plastic bullets, paper pots, tubes and jiffy pots. However, solid plastic-walled containers caused some cases of root strangulation (van Eerden 1982) and some concerns are still being expressed in Canada about poor root egress from paper pots (Sims 1988).

The most popular container systems used in North America use the plug seedling concept; tree seedling roots hold the growing medium together to form a firm mass, or plug, which is then extracted from the container prior to shipment to the field. These containers include rootrainers, styrofoam blocks, multipots and ecopots. An excellent listing of the many designs currently used throughout North America is found in Landis (1990a). Plug seedlings appear to have gained the widest acceptance because of the root systems they develop and because the containers in which they are grown can be easily re-used several times. However, the trend to develop the ideal container will undoubtedly continue because there is no single container that is perfect for all applications (Landis 1990a). For example, miniature containers have been recently introduced to produce small, young seedlings for container/bare-root transplants (Hee *et al.* 1988, Klapprat 1988).

Containers are commonly tapered, with a drainage hole at the base, and with vertical ribbing to prevent root spiralling. The containers are raised off the ground to allow air pruning of any roots growing through the drainage hole. The container size must have sufficient volume and spacing between containers to allow quality seedlings to be grown to the size required for the site. Cavity volumes of 40 to 60 cm^3 and a spacing of 700 to 1100 cavities per m^2 are generally satisfactory for producing planting stock for most sites (van Eerden 1982). As discussed earlier, the most commonly used potting mixes are peat, or peat plus vermiculite (Tinus 1982). The proportion of peat moss to vermiculite by volume ranges from 1:1 to 3:1 throughout North America (Landis 1990b). Proper filling procedures are essential to avoid overcompaction (> 0.14 g ml^{-1}) of the growing medium (Matthews 1982).

4.4.3. *Sowing and Germination.* Seed is usually sown with a vacuum precision seeder, and models that can sow over 90,000 cavities/hour are available (Arnott 1986). Seed is covered with coarse, industrial washed sand to a depth of 3 to 6 mm to provide a favourable germination environment, prevent build up of moss and algae on the surface, and provide physical protection. Sand particle size of 2-4 mm is recommended (Matthews 1982).

Germination can occur in the controlled environment of a germinator or directly in the growing environment. If a germinator is used, styroblocks are stacked 12 high on pallets for 7 days before being placed in the growing environment. Greenhouses are currently used as the germinating unit but they are less efficient than germinators in promoting fast germination and can be more expensive to heat in the early part of the year (Matthews 1982). Misting 2-3 times daily has been sufficient in British Columbia for air temperatures of 20-30°C (van Eerden 1974).

4.4.4. *Mycorrhizae.* Mycorrhizae should be encouraged on containerized seedlings (Marx and Barnett 1974), and inoculation with fungal mycelia or sporocarps should be done where possible (Chu-Chou and Grace 1990).

4.4.5. *Cultural Conditions.* General procedures followed for raising containerized seedlings in British Columbia will be used as an example (Matthews 1982).

With a wide range of species and climate, growers use a mixture of open outdoor compounds, shelterhouses and greenhouses to produce their crops. Following germination in greenhouses, day temperatures are maintained near 20°C for optimum growth. For crops started outdoors, warmer locations in southern regions are preferred. Wherever possible, high light levels are used to maximize plant growth and when excessively high temperatures (> 30°C) are encountered, cooling is achieved by venting the greenhouses, evaporative irrigation or by the use of shadecloth.

Species originating in northern and/or high elevations are likely to initiate bud set soon after germination, particularly when grown in southern locations. To avoid this, natural photoperiods are either extended, or the night interrupted by low intensity lighting (1.6 micromols m^{-2} s^{-1}) from high pressure sodium vapour lamps (Arnott 1989).

Seedlings are irrigated to field capacity two or three times a week using fixed sprinklers or irrigation booms. Growers are particularly careful to avoid drying out the peat substrate which can be very difficult to re-saturate. Nutrient prescriptions vary with the species, container size, growing conditions and crop schedules. Calcium and magnesium are usually supplied by the dolomite lime incorporated into the growing medium although this procedure is being discouraged by some North American growers for several reasons (Landis 1990b). Nutrients are supplied following germination, the ratios varying with the growing season. A typical regime would be fertilizing twice a week with 10-52-10 (N-P-K) at a concentration of 625 mg l^{-1} for 2 weeks following germination. Thereafter, fertilizer (28-14-14) is applied twice a week at a concentration of 310 mg l^{-1} during summer, and then changed to 10-52-10 at a concentration of 625 mg l^{-1} every two weeks during autumn (Arnott and White 1991). Seedlings can be fertilized with 10-52-10 (625 mg l^{-1}) from one to three times a month during winter until planting.

A few weeks before the desired seedling height has been reached (usually in mid-summer), a conditioning process begins. For spruce, the photoperiod lighting is turned off, the fertilizer changed to 10-52-10, and drought stressing is applied for two to three weeks. This is done by repeatedly drying down the styroblocks to about 75% of their saturated weight. Seedling pre-dawn xylem water potentials fall to approximately -1.5 MPa. Depending on the weather, and the facility, this usually takes from 3 to 4 days, after which the seedlings are re-saturated with water only and the dry-down cycle is repeated for the remainder of the three weeks. While effectively arresting shoot growth of most species, this treatment can have detrimental effects on seedling morphology and vigour (Macey and Arnott 1986, O'Reilly *et al.* 1989, Krasowski and Owens 1991). A much superior method of seedling conditioning is using black out facilities to provide artificially-shortened days over a two week period (Hawkins and Draper 1988, Grossnickle *et al.* 1991). This immediately triggers bud set and initiates the dormancy phases leading to full winter hardiness with no adverse morphological or phenological effects.

In late fall, after the shoots have reached frost hardiness levels of at least -18°C, the seedlings are extracted from the containers, packaged in bundles of 25 with the roots wrapped in plastic film, and placed in waxed cartons. These are stored in sheds at temperatures of +2°C to -2°C, depending on species, for 3 to 5 months.

4.5. CONTAINERIZED CUTTINGS

Containerized cuttings are raised the same way as seedlings, except that cuttings are set in the containers. Smaller cuttings can be used than for bare-root setting, as they are being raised in a more protected environment, and there are not the problems of soil splash and frost heave that are usually encountered outdoors.

Misting is essential during the first weeks after setting and bottom heat is usually necessary to promote the development of callus and roots. In Western Canada, this system is used to produce seedlings from genetically-improved spruce seed (Russell and Ferguson 1990, Russell *et al.* 1990).

5. Field Comparisons Of Stock Types

5.1. BARE-ROOT AND CONTAINERIZED

It is now recognized that many of the first field trials comparing containerized seedlings with bare-root seedlings were compromised by use of container stock which was too small and of poor quality (Gutzwiler and Winjum 1974, Kinghorn 1982). There have been many trial results reported where either bare-root or containerized seedlings have had better survival and early growth rate, depending on the stock and site conditions. In general, containerized seedlings have grown as well as or better than bare-root seedlings (Arnott 1974, Hite 1974, Scarratt 1974, Rasanen 1982, Vyse 1982). Containerized stock has usually performed well on lower quality, easier, drier sites, where there is only light to moderate competing vegetation (Hobbs *et al.* 1982, Scarratt 1974, 1982) and where there is a minimal animal population (Gutzwiler and Winjum 1974). They have not always done well on fertile sites with competing woody vegetation (Scarratt 1974, 1982). Bare-root seedlings are often taller initially than containerized seedlings, and this can cause a 1 or 2 year height growth advantage compared with containerized seedlings (Scarratt 1974, Krause 1982).

Longer term trial results were recently reported by Arnott and White (1991). For Douglas fir and western hemlock (*Tsuga heterophylla* [Raf.] Sarg.) on coastal sites in British Columbia, there were no significant differences in growth after 11 years between containerized and bare-root seedlings, although the bare-root seedlings were taller initially and after 5 years. Containerized western hemlock seedlings had significantly better survival than bare-root seedlings at both 5 and 11 years after planting. On the cooler, drier sites of interior British Columbia, containerized seedlings of white spruce (*Picea glauca* [Moench] Voss) and lodgepole pine (*Pinus contorta* Dougl.) had significantly better survival than bare-root seedlings, 10 years after planting. Bare-root seedlings were taller initially but the differences soon disappeared.

5.2 SEEDLINGS AND VEGETATIVE PROPAGULES

Cuttings from young donor trees survive and grow as well as or better than seedlings of the same genetic quality (Menzies and Klomp 1988, Menzies *et al.* 1991, Mason 1991). Although there is some loss of early diameter growth with cuttings from older donor trees (4 years old or older) of radiata pine, there are indications that these cuttings have better form, with less malformation, especially on fertile farm sites (Menzies *et al.* 1991), and these cuttings can yield greater volumes of higher quality products (Eldridge and Spencer 1988).

The field performance of micropropagated plantlets has been reviewed by Gupta *et al.* (1991). Plantlets regenerated from adventitious buds tend to show early maturation traits (Smith 1986, Frampton and Amerson 1989) and slower initial growth compared with seedlings (Ritchie and Long 1986, Menzies, M.I. unpublished data). If a juvenile growth pattern is desirable, then axillary bud micropropagation is preferred (Gupta *et al.* 1991).

6. Costs Of Planting Stock

Cost comparisons of various planting systems are scarce (Kinghorn 1982). Production cost of container-grown stock is higher than that of bare-root stock (Arnott and White 1991) and can be up to double that of bare-root stock (Rasanen 1982). However, production costs are only part of the establishment costs. Container systems can save about US$100/ha compared with bare-root systems in Sweden, Finland and Norway because of easier outplanting as a result of mechanical site preparation and improved planter productivity from using the 'Potti-putki' planting tube (Rasanen 1982). In British Columbia, it is still slightly cheaper to plant bare-root seedlings than containerized seedlings on a per hectare basis, but when costs are adjusted for seedling survival, containerized seedlings are cheaper (US$90–165 ha^{-1}) for western hemlock, white spruce, and lodgepole pine, but still more expensive (US$55 ha^{-1}) for Douglas fir (Arnott and White 1991).

Vegetative propagules are more expensive to produce than seedlings, if the cost of seed is excluded (Menzies *et al.* 1985a, Mason 1991), and this cost can be 2–3x for cuttings and 7–9x for micropropagated plantlets (Menzies *et al.* 1985a, Smith 1986). The best genetic quality seed is usually scarce and very expensive, and currently in New Zealand it is cheaper to produce bare-root cuttings than seedlings (Arnold 1990, Dibley and Faulds 1991). Control-pollinated seed prices would need to decrease by about 30% before seedlings would be cheaper to produce than cuttings. Use of vegetative propagation will also allow a greater area to be established with the best genetic quality seed, and this potential extra production should be accounted for when cost-benefit analyses are done. In British Columbia, cost-benefit analyses have demonstrated the advantages of using this method for maximizing the numbers of shippable propagules from genetically improved seed (Russell and Ferguson 1990).

7. Advantages And Disadvantages Of Bare-Root And Containerized Production Systems

Bare-root systems have been successfully producing millions of seedlings economically in many parts of the world, and programmes in the southern part of the USA, Britain, and the Southern Hemisphere (especially New Zealand and Australia) are proof of this. However, in the northern latitudes of the Northern Hemisphere, the climate is harsher, and bare-root production is more risky and unpredictable, and can take 3 or 4 growing seasons to produce stock of a suitable size (Kinghorn 1982). In tropical areas plants often do not become dormant in winter, so planting must be done in summer, during the rainy season. Containerized systems are often preferred in these areas. The advantages and disadvantages of bare-root and containerized production systems have been discussed in many papers (e.g. Gutzwiler and Winjum 1974, Kinghorn 1974, 1982, Cleary *et al.* 1978, Faulds and van Dorsser 1987, Ball and Brace 1982, Mason and Jinks 1990) and are summarised below.

7.1 ADVANTAGES OF BARE-ROOT PRODUCTION SYSTEMS

1. Production costs are lower.
2. Larger seedlings can be produced, and so there are fewer problems with competing vegetation and animal and insect damage.
3. The root system can be trimmed to increase planting efficiency and minimize root distortion.

7.2 DISADVANTAGES OF BARE-ROOT PRODUCTION SYSTEMS

1. It is more difficult to control the growing environment, which can cause problems.
2. In colder climates, it can take 3 or 4 years to produce suitable sized plants.
3. Bare-root plants can be prone to damage, especially desiccation, during lifting, packing, storage, and transport to the field.

7.3 ADVANTAGES OF CONTAINERIZED PRODUCTION SYSTEMS

1. Faster production of planting stock is possible, with greater control and more flexibility over growing schedules. This can shorten the lead time for production from 2–4 years down to one growing season.
2. Higher quality plants can be produced for some species, such as some eucalypt species, western hemlock, and the true firs.
3. Container systems maximize seed use for scarce seeds, since germination conditions can be closely controlled (although very small seedlings can be pricked out as transplants in a bare-root nursery).
4. Container nurseries do not need an extensive area for a nursery with a suitable climate and soil.
5. There are fewer weed problems.
6. There is less transplanting shock with an intact root system bound with a potting mix.
7. The planting season can be extended beyond the winter months.
8. Containerized seedlings are easier and faster to plant, especially in rocky soil or debris-covered sites.
9. Container systems are more adaptable for mechanization, both in the nursery and in the field.

7.4 DISADVANTAGES OF CONTAINERIZED PRODUCTION SYSTEMS

1. There is a large capital cost in setting up suitable facilities, especially in harsher climates.
2. With small containers, there can be a rapid moisture loss.
3. There can be problems with poor root systems, such as spiralling, which can lead to tree toppling problems.
4. There can be problems in conditioning crops for field conditions and in overwintering stock.
5. There is a higher disease/pest risk in container nurseries with their enclosed growing areas, high humidities, and high seedling densities.
6. Frost heaving can be a problem, especially with small plugs.
7. There can be water deficit problems if there is not good contact between the root plug and the soil after planting, or if the rooting depth is too shallow after planting.
8. Small containerized plants can be vulnerable to browsing animals.

8. Conclusion

Forest areas are now being established more commonly by planting seedlings or vegetative propagules, rather than relying on natural regeneration or seeding. While seedlings are the most commonly used planting stock, more cuttings, tissue-cultured plantlets, and somatic seedlings from embryogenesis will be used in the future, particularly for clonal forestry and expansion of the use of genetically-modified plants.

The main plant production methods are bare-root, transplants, and containerized systems. With bare-root systems, plants are raised in open nursery beds. The plants are cheap to raise, and mechanized systems have been developed for most operations. The planting season is restricted to the winter months when the trees are dormant, although some species may be cool-stored successfully for early spring planting.

Containerized systems are now widely used in harsher climates where bare-root production is more risky, and in the tropics where seedlings often do not become dormant, or planting must be done in the summer during the rainy season. Mechanized systems have now been developed for all operations from sowing to tree planting.

With the transplant system, plants are raised either bare-rooted or in containers and then transplanted into nursery soil to be raised as bare-rooted transplants.

There are advantages and disadvantages with all plant production methods. Bare-root systems have been used extensively in the temperate regions of the world, while container systems have been used more in the tropics and in the more northern parts of the northern hemisphere. Transplants have been an alternative in harsher climates, where tree growth is slow and a longer time is required in the nursery.

When selecting a plant production method, the plant specifications necessary for successful establishment on the particular site can be used as a guide. Any cost-benefit analysis should include the total cost of restocking the area and not just the cost of the plants in the nursery.

9. References

Aitken-Christie, J. and Connett, M. 'Micropropagation of forest trees', this conference.

Aldhous, J.R. (1967) 'Review of research and development in forest nursery techniques in Great Britain, 1949-1966', Paper for FAO Symposium on Man-Made Forests, 1967, For. Comm. Res. and Devel. Paper No. 46, 11 pp.

Arnold, R.J. (1990) 'Control pollinated radiata pine seed - a comparison of seedling and cutting options for large-scale deployment', N.Z. For. 35 (3), 12-17.

Arnott, J.T. (1974) 'Performance in British Columbia', in R.W. Tinus, W.I. Stein, and W.E. Balmer (eds.), Proc. North American Containerized Forest Tree Seedling Symposium', Denver, Colorado, USA, August 1974, Great Plains Ag. Council Pub. 68, pp. 283-290.

Arnott, J.T. (1986) 'Douglas fir stand establishment overview: Coastal British Columbia', in C.D. Oliver, D.P. Hanley and J.A. Johnson (eds.), Douglas fir stand management for the future, Coll. of For. Resources, Univ. of Wash., Inst. For. Resources Contrib. No. 55, pp. 219-229.

Arnott, J.T. (1989) 'Regulation of white spruce, Engelmann spruce and mountain hemlock seedling growth by controlling photoperiod', Forestry Suppl. 62, 157-168.

Arnott, J.T. and Mitchell, A. (1982) 'Influence of extended photoperiod on growth of white and Engelmann spruce seedlings in Coastal British Columbia', in J.B. Scarratt, C. Glerum and C.A. Plexman (eds.), Proc. Canadian Containerized Tree Seedling Symposium, Toronto, Ontario, Canada, September 1981, COJFRC Symp. Proc., O-P-10, pp. 139-152.

Arnott, J.T. and White, W. (1991) 'The influence of nursery systems on plantation establishment in British Columbia', in M.I. Menzies, G. Parrott, and L.J. Whitehouse (eds.), Proc. IUFRO Symp. Efficiency of Stand Establishment Operations, Rotorua, September 1989, N.Z. Ministry of Forestry, FRI Bulletin No. 156 (in press).

Ball, W.J. and Brace, L.G. (1982) 'Production, use, and field performance of container seedlings in the prairie provinces', in J.B. Scarratt, C. Glerum and C.A. Plexman (eds.), Proc. Canadian Containerized Tree Seedling Symp., Toronto, Ontario, Canada, September 1981, COJFRC Symp. Proc. O-P-10, pp. 313-320.

Balneaves, J.M. and Fairhurst, M.J. (1974) 'Effects of seedbed covering on emergence of radiata pine (Rangiora Nursery)', N.Z. J. For. 19(1), 132-133.

Belcher, E.W. (1978) 'Aspects of seed quality', in Proc. Intermountain Nurseryman's Conf., USDA For. Serv., Pacific Southwest Region, San Francisco, CA, Section D, pp. 54-60.

Bowen, G.D. (1965) 'Mycorrhiza inoculation in forestry practice', Austr. For. 29, 231-237.

Brown, W.D. (1991) 'The cost of chemical weed control in *Pinus radiata:* a study of three Central North Island New Zealand Nurseries', in M.I. Menzies, G. Parrott, and L.J. Whitehouse (eds.), Proc. IUFRO Symp. Efficiency of Stand Establishment Operations, Rotorua, September 1989, NZ Ministry of Forestry, FRI Bulletin No. 156 (in press).

Castellano, M.A. and Molina, A. (1989) 'Mychorrhizae', in T.D. Landis, R.W. Tinus, S.E. McDonald and J.P. Barnett (eds.), The Container Tree Nursery Manual, Volume 5, Agric. Handbook 674 Washington, D.C., USDA For. Serv., pp. 101-167.

Chavasse, C.G.R. (1977) 'The significance of planting height as an indicator of subsequent seedling growth', N.Z. J. For. 22(2), 283-296.

Chavasse, C.G.R. (ed.) (1981) Forest Nursery and Establishment Practice in New Zealand, New Zealand Forest Service, Forest Research Institute Symposium No. 22.

Chu-Chou, M. and Grace, L.J. (1990) 'Mycorrhizal fungi of radiata pine seedlings in nurseries and trees in forests', Soil Biol. Biochem. 22(7), 959-966.

Cleary, B.D., Greaves, R.D., and Owston, P.W. (1978) 'Seedlings', in B.D. Cleary, R.D. Greaves, and R.K. Hermann (eds.), Regenerating Oregon's Forests, Oregon State University Extension Service, Corvallis, Oregon, pp. 63-97.

Darby, S.P. Jr. (1962) 'Intensified nursery practices - the key to high-quality, custom-grown, field-graded forest tree seedlings', U.S. For. Serv., Tree Planters' Notes No. 52, 7-11.

Davey, C.B. (1964), 'Root pruning in forest nursery management', in Proc. Region 8 Forest Nurseryman's Conf., Morganton, NC., 1964, pp. 38-42.

Dibley, M. and Faulds, T. (1991) 'Production and costs of juvenile radiata pine cuttings', in M.I. Menzies, G. Parrott, and L.J. Whitehouse (eds.), Proc. IUFRO Symp. Efficiency of Stand Establishment Operations, Rotorua, September 1989, NZ Ministry of Forestry, FRI Bulletin No. 156 (in press).

Eis, S. and Long, J.R. (1973) 'Root pruning in the nursery', U.S. For. Serv., Tree Planters' Notes 24(1), 20-22.

Eldridge, K.G. and Spencer, D.J. (1988) 'Field performance of cuttings in Australia', in M.I. Menzies, J.P. Aimers, and L.J. Whitehouse (eds.), Workshop on Growing Radiata Pine from Cuttings, Rotorua, New Zealand, May 1986, N.Z. Ministry of Forestry, Forest Research Institute Bulletin No. 135, pp. 42-55.

Faulds, T. and van Dorsser, J.C. (1987) 'Growing eucalypts in containers', N.Z. Forest Research Institute, What's New in Forest Research No. 80.

Faulds, T. and Dibley, M.J. (1989) 'Growing radiata pine from juvenile cuttings', N.Z. Forest Research Institute, What's New in Forest Research No. 176.

Frampton, L.J. Jr. and Amerson, H.V. (1989) 'Influence of tissue culture method on field growth of loblolly pine', *In Vitro* Cell. Dev. Biol. 25(3), Part II, 13A.

Gleed, J.A. (1991) 'Towards clonal afforestation', in J.T. Miller (ed.), Proc. FRI/NZFP Forests Ltd Clonal Forestry Workshop, Ministry of Forestry, Forest Research Institute Bulletin No. 160, p. 61.

Glerum, C. (1985) 'Frost hardiness of coniferous seedlings: principles and applications', in Duryea, M.L. (ed.), Proc. workshop held at Forest Research Laboratory, Oregon State University, Corvallis, Oregon, October 16-18, 1984, pp. 107-123.

Goor, A.Y. (1955), 'Tree planting practices for arid zones', F.A.O. For. Development Paper No. 6, 126 pp.

Greer, D.H. and Warrington, I.J. (1982) 'Effects of photoperiod, night temperature, and frost incidence on development of frost hardiness in *Pinus radiata*, Austr. J. Pl. Physiol. 9, 333-342.

Grossnickle, S.C., Arnott, J.T., Major, J.E., Tschaplinski, T.J. (1991) 'Influence of dormancy induction treatments on western hemlock seedlings. I. Seedling development and stock quality assessment', Can. J. For. Res. 21, 164-174.

Gupta, P.K., Timmis, R. and Mascarenhas, A.F. (1991) 'Field performance of micropropagated forestry species', *In Vitro* Cell. Dev. Biol. (in press).

Gutzwiler, J.R. and Winjum, J.K. (1974) 'Performance of containerized coniferous seedlings in recent forest regeneration trials in Oregon and Washington', in R.W. Tinus, W.I. Stein, and W.E. Balmer (eds.), Proc. North American Containerized Forest Tree Seedling Symposium', Denver, Colorado, USA, August 1974, Great Plains Ag. Council Pub. No. 68, pp. 291-297.

Hawkins, C.D.B. and Draper, D.A. (1988) 'Height control of interior spruce by means of photoperiodic induction', in Landis, T.D. (Tech. Coordinator), Proc. Combined meeting of the Western Forest Nursery Associations, August 8-11, 1988, Vernon, British Columbia, USDA For. Serv., Gen. Tech. Rep. RM-167, pp. 45-49.

Hee, S.M., Stevens, T.S., Walch, D.C. (1988) 'Production aspects of mini-plug transplants', in Landis, T.D. (Tech. Coordinator), Proc. Combined meeting of the Western Forest Nursery Associations, August 8-11, 1988, Vernon, British Columbia, USDA For. Serv., Gen. Tech. Rep. RM-167, pp. 168-171.

Hiscock, D., van Dorsser, J. and Mason, E. (1981) 'Precision seed sowing becomes a reality', New Zealand Forest Research Institute, What's New in Forest Research No. 96.

Hite, W.A. (1974) 'Container field performance in the Rockies and Plains', in R.W. Tinus, W.I. Stein, and W.E. Balmer (eds.), Proc. North American Containerized Forest Tree Seedling Symposium', Denver, Colorado, USA, August 1974, Great Plains Ag. Council Pub. No. 68, pp. 306-309.

Hobbs, S.D., Lavender, D.P. and Wearstler, K.A. (1982) 'Performance of container-grown Douglas fir on droughty sites in Southwest Oregon', in J.B. Scarratt, C. Glerum and C.A. Plexman (eds.), Proc. of the Canadian Containerized Tree Seedling Symp., Toronto, Ontario, Canada, September 1981, COJFRC Symp. Proc. O-P-10, pp. 373-388.

Kinghorn, J.M. (1974) 'Principles and concepts in container planting', in R.W. Tinus, W.I. Stein, and W.E. Balmer (eds.), Proc. North American Containerized Forest Tree Seedling Symp., Denver, Colorado, USA, August 1974, Great Plains Ag. Council Pub. No. 68, pp. 8-18.

Kinghorn, J.M. (1982) 'Summation: Containerisation - boon or boondoggle?', in J.B. Scarratt, C. Glerum, and C.A. Plexman (eds.), Proc. Canadian Containerized Tree Seedling Symp., Toronto, Ontario, Canada, September 1981, COJFRC Symp. Proc. O-P-10, pp. 425-427.

Klapprat, R.A. (1988) 'Greenhouse transplants for bareroot stock production', in Landis, T.D. (Tech. Coordinator), Proc. Combined meeting of the Western Forest Nursery Associations, August 8-11, 1988, Vernon, British Columbia, USDA For. Serv., Gen. Tech. Rep. RM-167, pp. 165-167.

Knight, P.J. (1981) 'The maintenance of productivity in forest nurseries', in C.G.R. Chavasse (ed.), Forest Nursery and Establishment Practice in New Zealand, Part 1 - Nursery Practice in New Zealand, N.Z. Forest Service, Forest Research Institute Symposium No. 22, pp. 48-69.

Krasowski, M.J. and Owens, J.N. (1991) 'Growth and morphology of western red cedar as affected by photoperiod and moisture stress', Can. J. For. Res. 21, 340-352.

Krause, H.H. (1982) 'Early growth of bare-root and paperpot plantations at various locations in New Brunswick', in J.B. Scarratt, C. Glerum and C.A. Plexman (eds.), Proc. Canadian Containerized Tree Seedling Symp., Toronto, Ontario, Canada, September 1981, COJFRC Symp. Proc. O-P-10, pp. 355-366.

Landis, T.D. (1989a) 'Mineral nutrients and fertilization', in T.D. Landis, R.W. Tinus, S.E. McDonald and J.P. Barnett (eds.), The Container Tree Nursery Manual, Volume 4, Agric. Handbook 674, Washington, D.C., USDA For. Serv., pp. 1-67.

Landis, T.D. (1989b) 'Irrigation and water management', in T.D. Landis, R.W. Tinus, S.E. McDonald and J.P. Barnett (eds.), The Container Tree Nursery Manual, Volume 4, Agric. Handbook 674, Washington, D.C., USDA For. Serv., pp. 69-118.

Landis, T.D. (1989c) 'Disease and pest management', in T.D. Landis, R.W. Tinus, S.E. McDonald and J.P. Barnett (eds.), The Container Tree Nursery Manual, Volume 5, Agric. Handbook 674, Washington, D.C., USDA For. Serv., pp. 1-99.

Landis, T.D. (1990a) 'Containers: types and functions', in T.D. Landis, R.W. Tinus, S.E. McDonald and J.P. Barnett (eds.), The Container Tree Nursery Manual, Volume 2, Agric. Handbook 674, Washington, D.C., USDA For. Serv., pp. 1-39.

Landis, T.D. (1990b) 'Containers and growing media', in T.D. Landis, R.W. Tinus, S.E. McDonald and J.P. Barnett (eds.), The Container Tree Nursery Manual, Volume 2, Agric. Handbook 674, Washington, D.C., USDA For. Serv., pp. 41-85.

Landis, T.D., Tinus, R.W., McDonald, S.E., Barnett, J.P. (1989) The Container Tree Nursery Manual, Volume 4, Seedling Nutrition and Irrigation, Agric. Handbook 674, Washington, D.C., USDA For. Serv., 119 pp.

Landis, T.D., Tinus, R.W., McDonald, S.E., Barnett, J.P. (1990a) The Container Tree Nursery Manual, Volume 2, Containers and the Growing Media, Agric. Handbook 674, Washington, D.C., USDA For. Serv., 88 pp.

Landis, T.D., Tinus, R.W., McDonald, S.E., Barnett, J.P. (1990b) The Container Tree Nursery Manual, Volume 5, The Biological Component: Nursery Pests and Mycorrhizae, Agric. Handbook 674, Washington, D.C., USDA For. Serv., 171 pp.

Lavender, D.P. and Cleary, B.D. (1974) 'Coniferous seedling production techniques to improve seedling establishment', in R.W. Tinus, W.I. Stein, and W.E. Balmer (eds.), Proc. North American Containerized Forest Tree Seedling Symp., Denver, Colorado, August 26-29, 1974, Great Plains Ag. Council Pub., No. 68, pp. 177-180.

Lavender, D.P. and Hermann, R.K. (1976) 'Role of forest tree physiology in producing planting stock and establishing plantations', paper presented to IUFRO Conference, 1976, Oslo, Sweden, 12 pp.

Macey, D.E. and Arnott, J.T. (1986) 'The effect of moderate moisture and nutrient stress on bed formation and growth of container-grown white spruce seedlings', Can. J. For. Res. 16, 949-954.

Matthews, R.G. (1982) 'Contrasting approaches to containerized seedling production, I. British Columbia', in J.B. Scarratt, C. Glerum and C.A. Plexman (eds.), Proc. Canadian Containerized Tree Seedling Symp., Toronto, Ontario, Canada, September 1981, COJFRC Symp. Proc. O-P-10, pp. 115-122.

Marx, D.H. and Barnett, J.P. (1974) 'Mycorrhizae and containerized forest tree seedlings', in R.W. Tinus, W.I. Stein, and W.E. Balmer (eds.), Proc. North American Containerized Forest Tree Seedling Symp.', Denver, Colorado, USA, August 1974, Great Plains Ag. Council Pub. No. 68, pp. 85-92.

Mason, W.I. (1991) 'Commercial development of vegetative propagation of genetically improved Sitka spruce (Picea sitchensis [Bong.] Car.) in Great Britain', in M.I. Menzies, G. Parrott, and L.J. Whitehouse (eds.), Proc. IUFRO Symp. Efficiency of Stand Establishment Operations, Rotorua, September 1989, NZ Ministry of Forestry, FRI Bulletin No. 156 (in press).

Mason, W.L. and Jinks, R.L. (1990) 'The use of containers as a method for raising tree seedlings', For. Comm., Research Information Note 179.

Menzies, M.I. (1988) 'Seedling quality and seedling specifications of radiata pine', What's New in Forest Research No. 171, N.Z. Forest Research Institute.

Menzies, M.I. and Faulds, T. (1991) 'Fertilization of vegetative propagules', in R. van den Driessche (ed.), Mineral Nutrition of Conifer Seedlings, CRC Press, Boca Raton, Florida, pp. 169-182.

Menzies, M.I. and Klomp, B.K. (1988) 'Effect of parent age on growth and form of cuttings, and comparison with seedlings', in M.I. Menzies, J.P. Aimers, and L.J. Whitehouse (eds.), Workshop on Growing Radiata Pine from Cuttings, Rotorua, New Zealand, May 1986, N.Z. Ministry of Forestry, Forest Research Institute Bulletin No. 135, pp. 18-41.

Menzies, M.I., Faulds, T., Dibley, M., and Aitken-Christie, J. (1985a) 'Vegetative propagation of radiata pine in New Zealand', in D.B. South (ed.), Proc. Int. Symp. Nursery Management Practices for the Southern Pines, Montgomery, Alabama, USA, 1985, Alabama Agricultural Experiment Station, Auburn University, pp. 167-190.

Menzies, M.I., Klomp, B.K., and Holden, D.G. (1991) 'Optimal physiological age of propagules for use in clonal forestry', in J.T. Miller (ed.), Proc. FRI/NZFP Forests Ltd Clonal Forestry Workshop, Rotorua, New Zealand, May 1989, Forest Research Institute Bulletin No. 160, pp. 142-146.

Menzies, M.I., van Dorsser, J.C., and Balneaves, J.M. (1985b) 'Seedling quality - radiata pine as a case study', in D.B. South (ed.), Proc. Int. Symp. Nursery Management Practices for the Southern Pines, Montgomery, Alabama, USA, 1985, Alabama Agricultural Experiment Station, Auburn University, pp. 384-415.

Mitchell, W.K., Dunsworth, G., Simpson, D.G., Vyse, A. (1990) 'Planting and seeding', in D.P. Lavender, R. Parish, C.M. Johnson, G. Montgomery, A. Vyse, R.A. Willis and D. Winston (eds.), Regenerating British Columbia's Forests, University of British Columbia Press, Vancouver, B.C., pp. 235-253.

Mullin, R.E. (1966) 'Root pruning of nursery stock', For. Chron. 42, 256-264.

N.Z. Forest Service (1983) Report of Forest Research Institute, pp. 17-18.

N.Z. Ministry of Forestry (1992) Forest Research Institute Annual Report' (in press).

Nyborg, E.O. and Shikaze, G. (1974) 'Development of high-capacity precision seedling, loading, and handling equipment for container nurseries', in R.W. Tinus, W.I. Stein, and W.E. Balmer (eds.), Proc. North American Containerized Forest Tree Seedling Symp., Denver, Colorado, August 1974, Great Plains Ag. Council Pub., 68, pp. 146-157.

O'Reilly, C.O., Arnott, J.T., Owens, J.N. (1989) 'Effects of photoperiod and moisture availability on shoot growth, seedling morphology, and cuticle and epicuticular wax features of container-grown western hemlock seedlings', Can. J. For. Res. 19, 122-131.

Owston, P.W. (1974) 'Two-crop production of western conifers', in R.W. Tinus, W.I. Stein, and W.E. Balmer (eds.), Proc. North American Containerized Forest Tree Seedling Symp., Denver, Colorado, USA, August 1974, Great Plains Ag. Council Pub. 68, pp. 104-111.

Rasanen, P.K. (1982) 'Containerized forest tree seedling production and development prospects in Finland and Scandinavia', in J.B. Scarratt, C. Glerum and C.A. Plexman (eds.), Proc. Canadian Containerized Tree Seedling Symp., Toronto, Ontario, Canada, September 1981, COJFRC Symp. Proc. O-P-10, pp. 9-17.

Ritchie, G.A. and Long, A.J. (1986) 'Field performance of micropropagated Douglas fir', N.Z. J. For. Sci. 16(3), 343-356.

Rook, D.A. (1969) 'Water relations of wrenched and unwrenched *Pinus radiata* on being transplanted into conditions of water stress', N.Z. J. For. 14(1), 50-58.

Rook, D.A. (1971) 'Effect of undercutting and wrenching on growth of *Pinus radiata* D. Don seedlings', J. Appl. Ecol. 8, 477-490.

Russell, J. and Ferguson, C. (1990) 'Production of genetically improved stecklings of Interior spruce, A Growers Manual', Ministry of Forests, Research Branch, Victoria, B.C., FRDA Report No. 110, 15 pp.

Russell, J., Grossnickle, S.C., Ferguson, C., Carson, D.W. (1990) 'Yellow-cedar stecklings: nursery production and field performance', Ministry of Forests, Research Branch, Victoria, B.C. FRDA Rep. No. 148, 20 pp.

Scarratt, J.B. (1974) 'Performance of tubed seedlings in Ontario', in R.W. Tinus, W.I. Stein, and W.E. Balmer (eds.), Proc. North American Containerized Forest Tree Seedling Symp., Denver, Colorado USA, August 1974, Great Plains Ag. Council Pub. 68, pp. 310-320.

Scarratt, J.B. (1982) 'Container stock specifications for Northern Ontario' in J.B. Scarratt, C. Glerum and C.A. Plexman (eds.), Proc. Canadian Containerized Tree Seedling Symp., Toronto, Ontario, Canada, September 1981, COJFRC Symp. Proc. O-P-10, pp. 343-354.

Scarratt, J.B., Glerum, C., and Plexman, C.A. (1982) Proc. Canadian Containerized Tree Seedling Symp., Toronto, Ontario, Canada, September 1981, COJFRC Symp. Proc. O-P-10.

Schmidt-Vogt, H. (1974) 'Planting material', in Proc. Symp. Stand Establishment, IUFRO Joint Meeting Divs. 1 and 3. Wageningen, The Netherlands, October 1974, pp. 69-107.

Sims, J. (1988) 'Shifting from paperpots to ecopots', Silviculture Magazine 3 (2), 10-12.

Smith, D.R. (1986) 'Radiata pine (*Pinus radiata* D. Don)', in Y.P.S. Bajaj (ed.), Biotechnology in Agriculture and Forestry, Vol.1: Trees, Springer-Verlag, Berlin, Heidelberg, pp. 274-291.

Stoeckeler, J.H. (1965) 'Conifer Nursery Practice in the Prairie Plains', USDA. For. Serv. Agricultural Handbook No. 279, 93 pp.

Tabbush, P.M. and Ray, D. (1989) 'Effects of rough handling and microsite on the establishment of Sitka spruce on clearfelled sites in upland Britain', Forestry Suppl. 62, 289-296.

Tanaka, Y. (1974) 'Increasing cold hardiness of container-grown Douglas fir seedlings', J. For. 72(6), 349-352.

Tanaka, Y. and Timmis, R. (1974) 'Effects of container density on growth and cold hardiness of Douglas fir seedlings', in R.W. Tinus, W.I. Stein, and W.E. Balmer (eds.), Proc. North American Containerized Forest Tree Seedling Symp., Denver, Colorado, August 1974, Great Plains Ag. Council Pub., 68, pp. 181-186.

Tinus, R.W. (1982) 'Environmental control of seedling physiology', in J.B. Scarratt, C. Glerum, and C.A. Plexman (eds.), Proc. Canadian Containerized Tree Seedling Symp., Toronto, Ontario, Canada, 1981, COJFRC Symp. Proc. 0-P-10, pp. 75-82.

Tinus, R.W. and MacDonald, S.E. (1979) 'How to grow tree seedlings in containers in greenhouses', USDA For. Serv., Ft. Collins, Colorado, Gen. Tech. Rep. RM-60, 256 pp.

Tinus, R.W., Stein, W.I., and Balmer, W.E. (1974) Proc. North American Containerized Forest Tree Seedling Symp., Denver, Colorado, USA, August 1974, Great Plains Ag. Council Pub. No. 68.

Toda, R. (1974) 'Vegetative propagation in relation to Japanese forest tree improvement', N.Z. J. For. Sci. 4(2), 410-417.

Trewin, A.R.D. (1981) 'The importance of a good outplanting system for the establishment of bare-root pine seedlings', in C.G.R. Chavasse (ed.), Forest Nursery and Establishment Practice in New Zealand, Part 2 - Forest Establishment Practice in New Zealand, N.Z. Forest Service, Forest Research Institute Symposium No. 22, pp. 226-244.

Trewin, A.R.D. and Cullen, A.W.J. (1985) 'A fully integrated system for planting bare-root seedlings of radiata pine in New Zealand', in D.B. South (ed.), Proc. Int. Symp. on Nursery Management Practices for the Southern Pines, Montgomery, Alabama, USA, 1985, Alabama Agricultural Experiment Station, Auburn University, pp. 524-548.

van den Driessche, R. (1969) Forest Nursery Handbook, Research Notes, Brit. Columbia For. Serv., Victoria, B.C., Canada, No.8, 44 pp.

van Dorsser, J.C. (1981) 'Seedling Conditioning' in C.G.R. Chavasse (ed.), Forest Nursery and Establishment Practice in New Zealand, Part 1 - Nursery Practice in New Zealand, N.Z. Forest Service, Forest Research Institute Symposium No. 22, pp. 128-141.

van Dorsser, J.C. and Rook, D.A. (1972) 'Conditioning of radiata pine seedlings by undercutting and wrenching: descriptions of methods, equipment, and seedling response', N.Z. J. For. 17(1), 61-73.

van Eerden, E. (1974) 'Growing season production of western conifers', in R.W. Tinus, W.I. Stein, and W.E. Balmer (eds.), Proc. North American Containerized Forest Tree Seedling Symp., Denver, Colorado, August 26-29, 1974, Great Plains Ag. Council Pub. 68, pp. 93-103.

van Eerden, E. (1982) 'The fundamentals of container seedling production', in J.B. Scarratt, C. Glerum, and C.A. Plexman, (eds.), Proc. Canadian Containerized Tree Seedling Symp., Toronto, Ontario, Canada, September 1981, COJFRC Symp. Proc. O-P-10, pp. 83-90.

van Eerden, E. (1990) 'Seedling production and processing; Container', in D.P. Lavender, R. Parish, C.M. Johnson, G. Montgomery, A. Vyse, R.A. Willis and D. Winston (eds.), Regenerating British Columbia's Forests, University of British Columbia Press, Vancouver, B.C., pp. 226-234.

Vince-Prue, D. (1975) Photoperiodism in Plants, McGraw-Hill Inc., London, 444 pp.

Vyse, A. (1982) 'Field performance of small-volume container-grown seedlings in the Central Interior of British Columbia', in J.B. Scarratt, C. Glerum and C.A. Plexman (eds.), Proc. Canadian Containerized Tree Seedling Symp., Toronto, Ontario, Canada, September 1981, COJFRC Symp. Proc. O-P-10, pp. 291-298.

Weiser, C.J. (1970) 'Cold resistance and injury in woody plants', Science 189, 1269-1278.

Wilkinson, A.G., Hathaway, R.L., and van Kraayenoord, C.W.S. (1981) 'Establishment and management of poplar and willow nurseries', in C.G.R. Chavasse (ed.), Forest Nursery and Establishment Practice in New Zealand, Part 1 - Nursery Practice, N.Z. Forest Service, Forest Research Institute Symposium No. 22, pp. 227-233.

LOW-TEMPERATURE STORAGE OF BEDDING-PLANT PLUGS

ROYAL D. HEINS, NATHAN LANGE, AND
THOMAS F. WALLACE, JR.
Department of Horticulture
Michigan State University
East Lansing, Michigan 48824-1325

ABSTRACT. Ageratum, begonia, marigold, and salvia seedlings growing in individual cells were stored to determine the effects of temperature, irradiance, and storage time on growth and forcing time after transplanting. Photosynthetic photon flux densities of 0, 1, and 5 $\mu mol \cdot m^{-2} \cdot s^{-1}$ were combined with temperatures of 0.0, 2.5, 5.0, 7.5, 10.0, and 12.5C to create 18 unique storage environments. Sample plants were removed at 1-week intervals for a period of at least 6 weeks, and were forced into flower. In all four species, chilling injury increased as storage at 0.0 and 2.5C was extended, and surviving plants flowered late relative to controls. Ageratum was most sensitive to chilling injury, followed in order by salvia, marigold, and begonia. Plant deterioration and ultimately death occurred as dark storage duration increased at 10 and 12.5C. Greatest percent survival and highest quality of seedlings stored in the dark occurred at temperatures just above those that caused chilling injury. Exposure to 1 $\mu mol \cdot m^{-2} \cdot s^{-1}$ during storage greatly increased quality of plants held at warmer-than-optimal temperatures. No substantial differences were observed among plants stored with 1 versus 5 $\mu mol \cdot m^{-2} \cdot s^{-1}$. All four species could be stored for 6 weeks in the light, but their responses to storage temperatures and durations differed. Optimal temperatures were 5-7.5C for begonia, 5C for marigold, and 7.5C for salvia and ageratum.

1. Introduction

Formerly, most seed-propagated plants were germinated in flats and transplanted bare root. Today, the majority of plants propagated from seed in the United States are germinated in individual cells called plugs. When the seedling is the proper size, it is transplanted with the entire root ball to a finish container (pot or flat).

Plants are sometimes ready for transplanting before greenhouse space or labor is available, or the market is ready to accept them. The grower is forced to restrict growth under such circumstances to avoid overgrowth of the closely spaced seedlings (up to 5,000 seedlings m^{-2}). Holding techniques include water stress, nutrient stress, or treatment with growth retardants. Stress and growth retardants can harden seedlings and, after transplanting, adversely affect growth rate, time until flower, and/or size at flower.

An alternative to holding seedling-plug flats in the greenhouse is low-temperature storage of the seedlings in a cooler. Low-temperature storage is common for many fruit,

K. Kurata and T. Kozai (eds.), Transplant Production Systems, 45–64.

vegetable, and nursery crops. Extensive research of these crops has provided data pertaining to temperature range, relative humidity, and storage diseases and durations (Hardenburg, et.al., 1986). In contrast, research of seedling storage is very limited. The only research we are aware of is that of Kumpf et. al. (1966), who found that seedlings growing in flats could be stored at 1.1 to 4.4C for 2 or more weeks if watered, placed in polyethylene bags, and kept under light 14 h per day. Ageratum, cineraria, cosmos, pepper, and tomato could be stored for 1 to 2 weeks; aster, browallia, dianthus, geranium, lobelia, and marigold for 2 weeks; and alyssum, petunia, salvia, and snapdragon for up to 6 weeks. Zinnia and coleus could not be stored at all.

Specific information about the effects of a range of temperatures and irradiances during storage is unavailable for seedlings of most bedding plants grown as plugs. We undertook this research to provide this information for ageratum, begonia, marigold, and salvia.

2. Materials and Methods

Size 406 plug sheets (2,570 plants m^{-2}, 406 plants per flat) of ageratum (*Ageratum houstonianum* 'Blue Danube'), begonia (*Begonia* x *semperflorens-cultorum* 'Vodka'), marigold (*Tagetes patula* 'Hero Yellow'), and salvia (*Salvia splendens* 'Red Hot Sally') were obtained from a commercial grower when the plugs had reached a transplantable size. One plug sheet from each species was placed in storage at each of 18 different temperature and irradiance combinations. Temperatures were 0, 2.5, 5, 7.5, 10, and 12C in the coolers and 20C in the greenhouse. Photosynthetic photon flux density (PPF) levels generated by cool white fluorescent bulbs were 0, 1, and 5 μmol m^{-2} s^{-1}. Darkness (0 μmol m^{-2} s^{-1}) was obtained by placing plug sheets in closed, corrugated-cardboard plug-shipping boxes. Relative humidity for seedlings stored in the light varied from 50 to 70%, but approached 100% for seedlings held in the dark.

All plants were subirrigated with water as needed during storage. Irrigation frequency varied from 3 to 20 days, depending on the temperature and humidity of each treatment environment. Contact between foliage and water was minimized to avoid fungal infection.

Ten representative plants were removed from each plug sheet after 1, 2, 3, 4, 5, and 6 weeks. One plug sheet, left in the greenhouse as a non-stored control, was used as a source of control seedlings. One plant per treatment was photographed immediately after removal, and then all the plants were measured before being potted in 10-cm pots. Initial plant height was measured from the root ball to the apical meristem. The plants were forced into flower in a glass greenhouse with a minimum temperature of 20C.

The date of first flower was recorded for each plant that survived storage. Plants were considered in flower when the buds reached anthesis. The average number of days from the start of forcing until first flower and the percentage of plant survival for each treatment was determined. Plant height for salvia, plant diameter for begonia, and plant diameter and height for ageratum and marigold were recorded at first flower. Plant area for begonia from diameter measurements and plant volume for ageratum and marigold from height and diameter was calculated by assuming that the plant was cylindrical.

3. Results

3.1 Ageratum

Height of ageratum seedlings stored in the dark increased as storage temperature and duration increased (Fig. 1). Plant height increased approximately 35 mm during the experiment on seedlings stored at 10 and 12.5C, but only 5 mm at 5C. In comparison, height of control seedlings (initial height 42 mm) increased 105 mm during the same 6-week period.

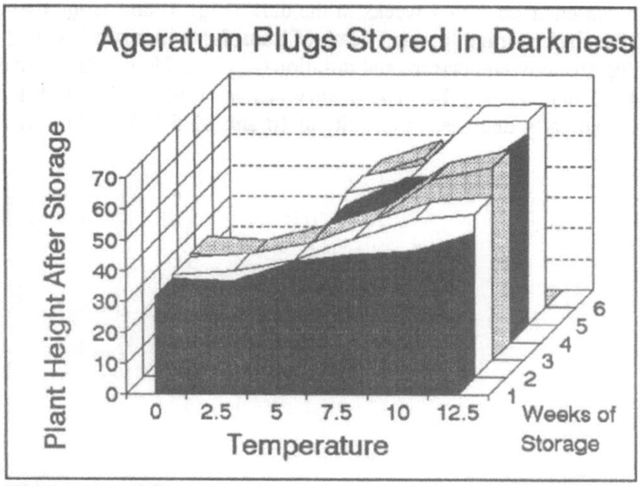

Figure 1: Influence of temperature on height (mm) of ageratum seedlings stored in the dark for 1 to 6 weeks.

The addition of 1 μmol m^{-2} s^{-1} of light during storage slowed the rate of seedling elongation (Fig. 2). For example, seedlings 62 mm tall after 3 weeks' dark storage at 12.5C were only 46 mm tall after 3 weeks' light storage. Final heights after 6 weeks at 12.5C, however, were similar in both the dark and light. Plant height did not increase significantly during 6 weeks' storage at 5C.

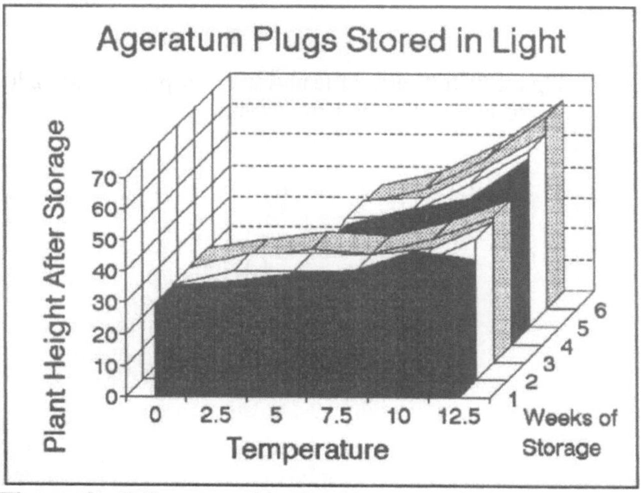

Figure 2: Influence of temperature on height (mm) of ageratum seedlings stored in the light for 1 to 6 weeks.

Plants stored at 0 and 2.5C showed signs of chilling injury after 2 weeks of storage. Percent mortality as a result of chilling injury at 0 and 2.5C increased from 0 to 100 as storage increased from 1 to 4 weeks in the dark (Fig. 3) and from 1 to 3 weeks in the light (Fig. 4). After dark storage, percent death of transplants increased correspondingly with the increase in plug-storage temperature and duration (Fig. 3). No plants survived after 6 weeks' dark storage at 10 and 12.5C, but 100% survived at 7.5C. The addition of 1 μmol m^{-2} s^{-1} of light dramatically decreased mortality at 10 and 12.5C (Fig. 4). All control plants survived.

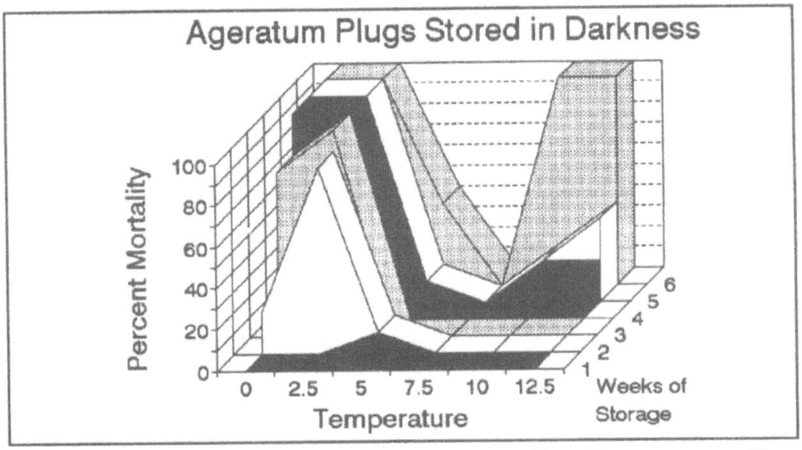

Figure 3: Influence of temperature on percent mortality of ageratum seedlings stored in the dark for 1 to 6 weeks.

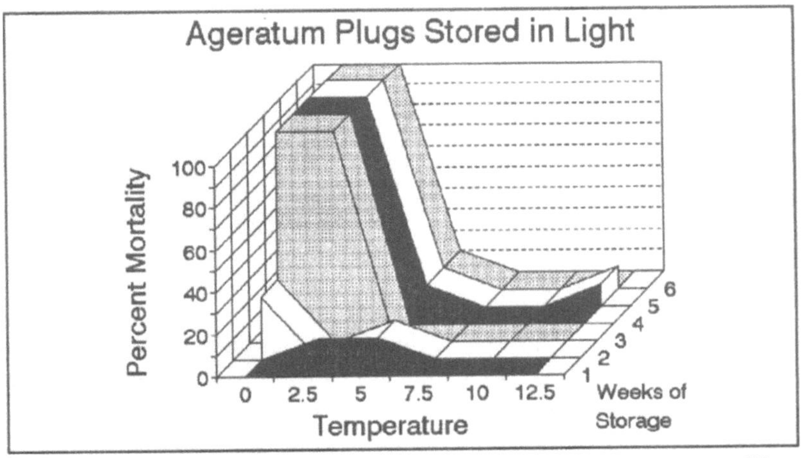

Figure 4: Influence of temperature on percent mortality of ageratum seedlings stored in the light for 1 to 6 weeks.

The time from transplant to flower was affected by storage in two ways. First, flowering was delayed on plants from treatments which caused chilling injury, but not death on all plants. For example, plants flowered in 86 and 65 days after 2 weeks' storage at 2.5C (Fig. 5) and 0C (Fig. 6), respectively, versus 46 days for unstored control plants. Second, flowering was also delayed on plants from dark storage at 10 and 12.5C after 5 weeks' storage, the longest duration plants survived at these two temperatures. The addition of 1 μmol m^{-2} s^{-1} did not prevent flowering delay on plants exposed to chilling-injury-inducing temperatures, but did prevent delayed flowering of plants held at 10 and 12.5C for up to 6 weeks (Fig. 6).

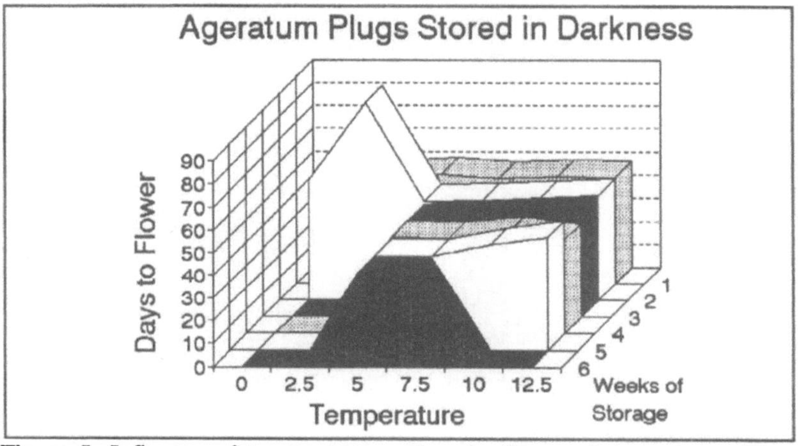

Figure 5: Influence of temperature on days to flower of ageratum seedlings stored in the dark for 1 to 6 weeks. Treatments with 0 days to flower are treatments where all plants died.

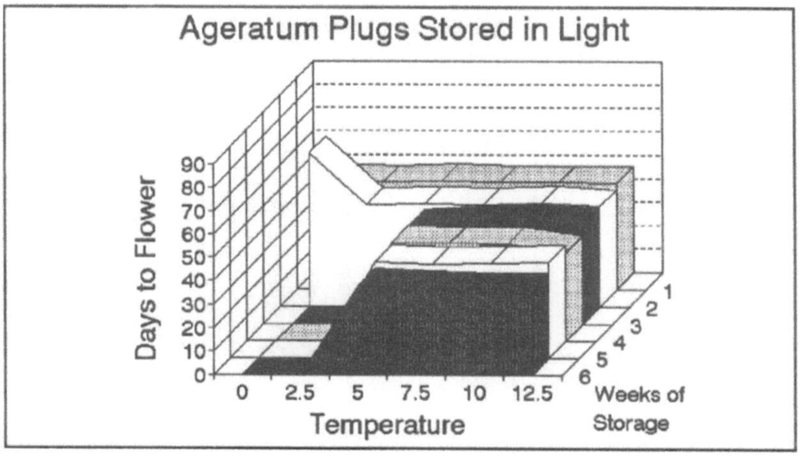

Figure 6: Influence of temperature on days to flower of ageratum seedlings stored in the light for 1 to 6 weeks.

There was no consistent storage-treatment effect on the size of plants when they flowered (Fig. 7 and 8). Flowering-plant size was more consistent across treatments on plants stored in the light (Fig. 8) than in the dark (Fig. 7). Size at flower tended to decrease as storage duration increased. Plants stored for only 1 week were significantly larger than those stored for 2 to 6 weeks. Size of control plants at flower decreased from 4050 cm^3 to 660 cm^3 as holding duration in the greenhouse increased from 0 to 6 weeks.

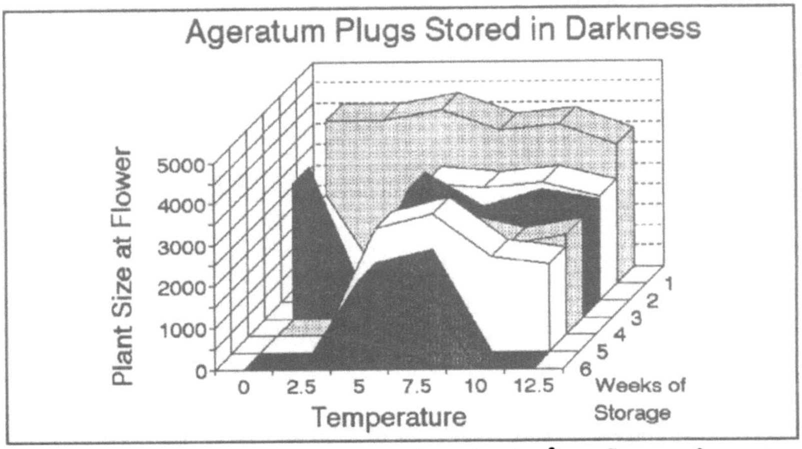

Figure 7: Influence of temperature on plant size (cm^3) at flower of ageratum seedlings stored in the dark for 1 to 6 weeks.

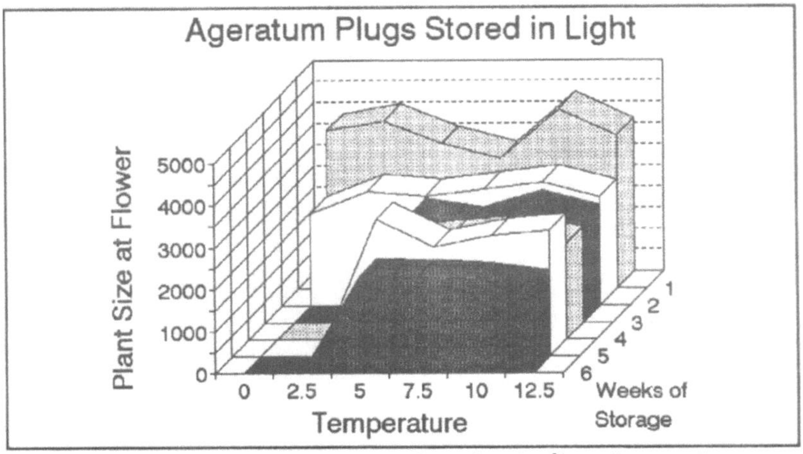

Figure 8: Influence of temperature on plant size (cm^3) at flower of ageratum seedlings stored in the light for 1 to 6 weeks.

3.2 Begonia

Height of begonia seedlings stored in the dark did not increase at 5C or lower, but increased as storage temperature and duration increased (Fig. 9); however, the gain in height was not great. More obvious than increased height was the etiolated appearance of plants stored at 10 and 12.5C. The addition of 1 μmol m^{-2} s^{-1} reduced, but did not prevent, petiole elongation during storage. Height of seedlings held in the plug sheet and maintained in the greenhouse increased from an initial height of 13 mm to 58 mm during the 6-week experiment period.

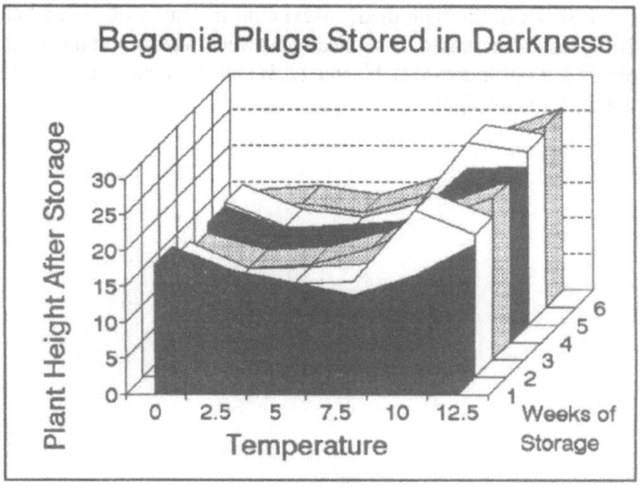

Figure 9: Influence of temperature on height of begonia seedlings stored in the dark for 1 to 6 weeks.

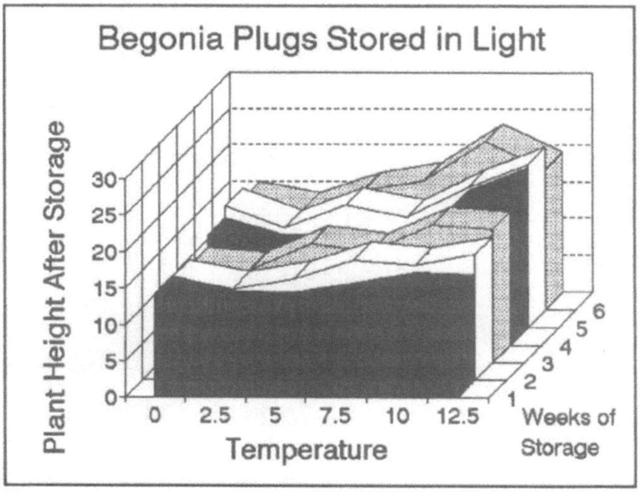

Figure 10: Influence of temperature on height of begonia seedlings stored in the light for 1 to 6 weeks.

Plants showed signs of chilling injury after 3 weeks' storage at 0C. Damage was not as severe on plants held in the dark (Fig. 11) as in the light (Fig. 12). All plants held at 0C for 4 weeks in the light died. Maximum mortality of plants held for 5 weeks in darkness was 20%; losses increased dramatically between weeks 5 and 6 (Fig. 11). The percentage of dead plants did not increase at 10 and 12.5C until plants had been stored at least 4 weeks, either in the dark or the light.

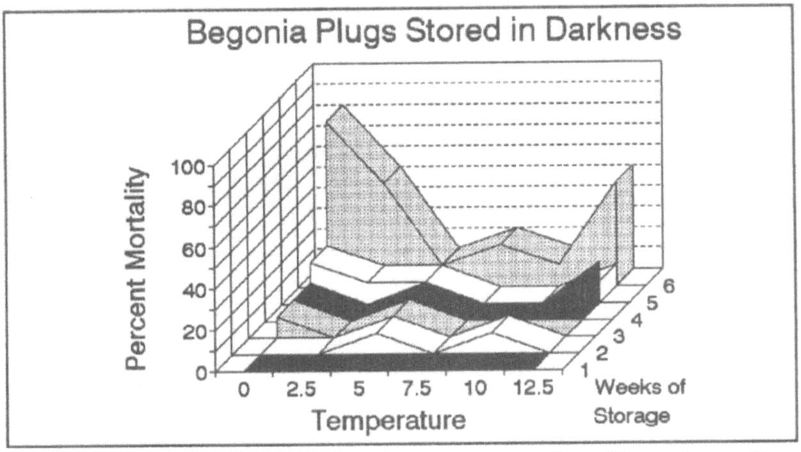

Figure 11: Influence of temperature on percent mortality of begonia seedlings stored in the dark for 1 to 6 weeks.

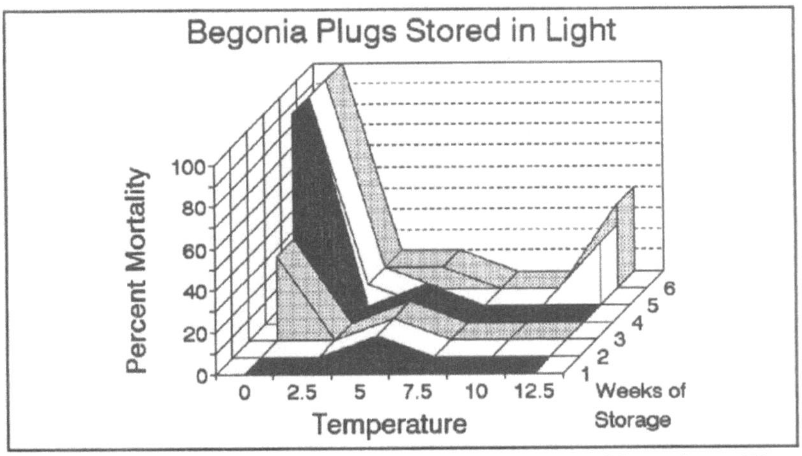

Figure 12: Influence of temperature on percent mortality of begonia seedlings stored in the light for 1 to 6 weeks.

Flowering was delayed in plants from treatments that caused chilling injury, but not death (Fig. 13 and 14). Compared to unstored control plants which flowered in 41 days, flowering was delayed 13 days for plants stored for 6 weeks in the dark at 0C (Fig 13), and 17 days for plants stored 3 weeks in the light (Fig 14). Flowering was also delayed 13 days on plants stored for 6 weeks in the dark at 12.5C (Fig. 13). There was no consistent pattern or delay in flowering of plants from other storage treatments.

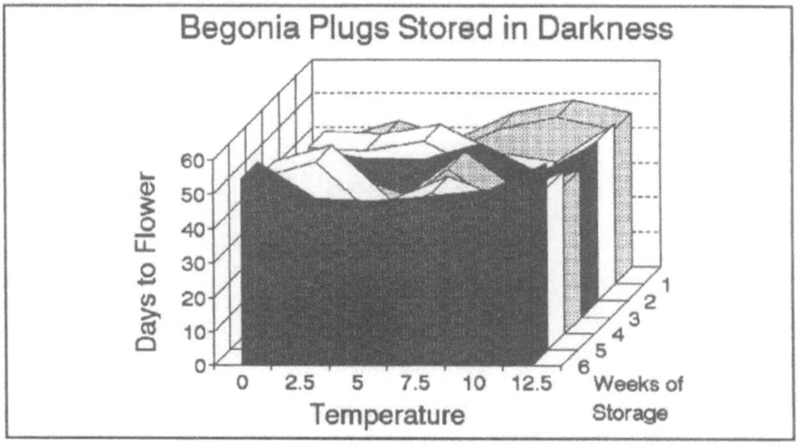

Figure 13: Influence of temperature on days to flower of begonia seedlings stored in the dark for 1 to 6 weeks.

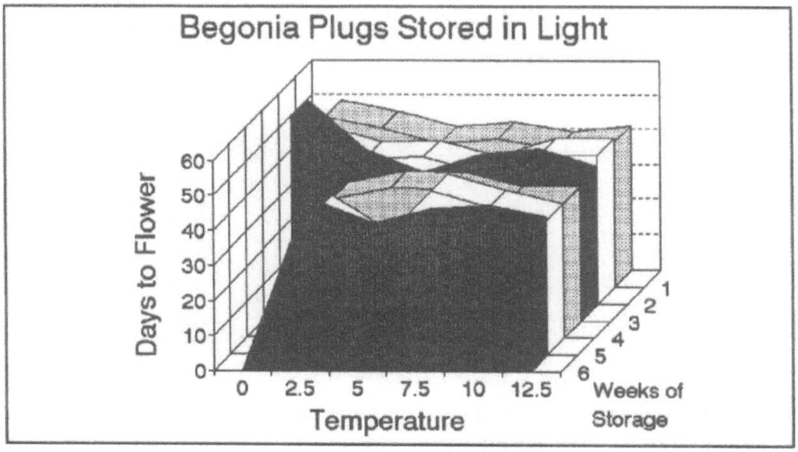

Figure 14: Influence of temperature on days to flower of begonia seedlings stored in the light for 1 to 6 weeks.

Plant size at flower followed no consistent pattern (Fig. 15 and 16), with the exception of plants stored for 6 weeks in the dark; plant size increased as storage temperature decreased. Size of control plants at flower decreased from 155 cm^2 to 115 cm^2 as holding duration in the greenhouse increased from 0 to 6 weeks.

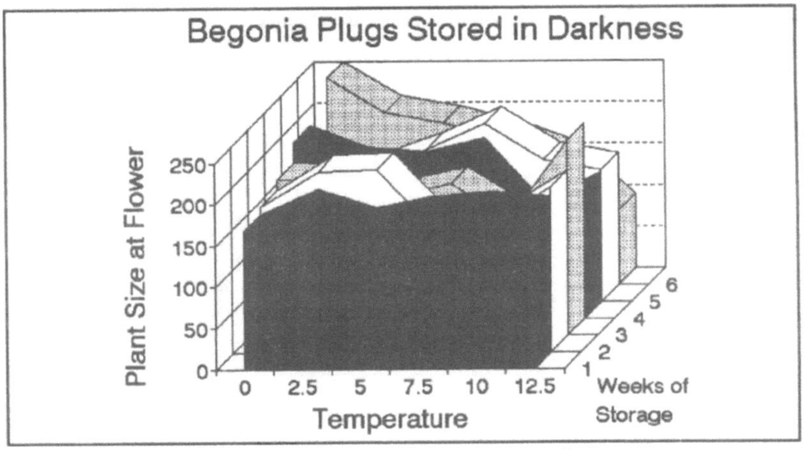

Figure 15: Influence of temperature on plant size (cm^2) at flower of begonia seedlings stored in the dark for 1 to 6 weeks.

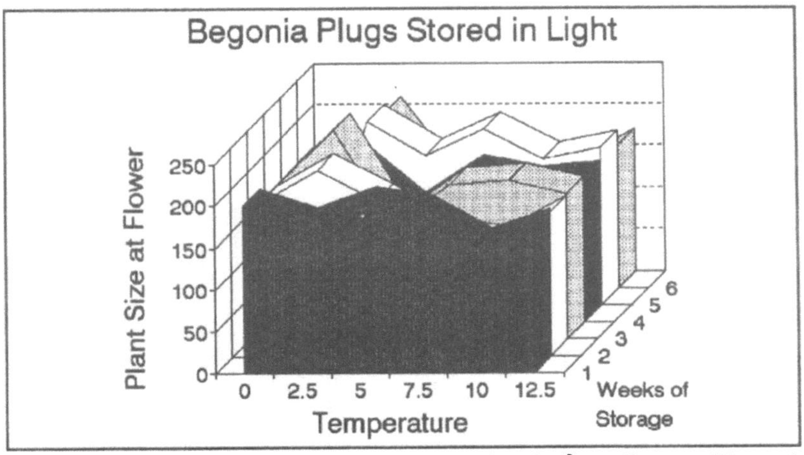

Figure 16: Influence of temperature on plant size (cm^2) at flower of begonia seedlings stored in the light for 1 to 6 weeks.

3.3 Marigold

Height of marigold seedlings increased slightly during storage in the dark (Fig. 17) (maximum 5 mm, or about 10% of the initial height). In the light (Fig. 18), height increased only at temperatures of 10C or greater. Height of control plants increased from 50 mm to 120 mm as holding duration in the greenhouse increased from 0 to 6 weeks.

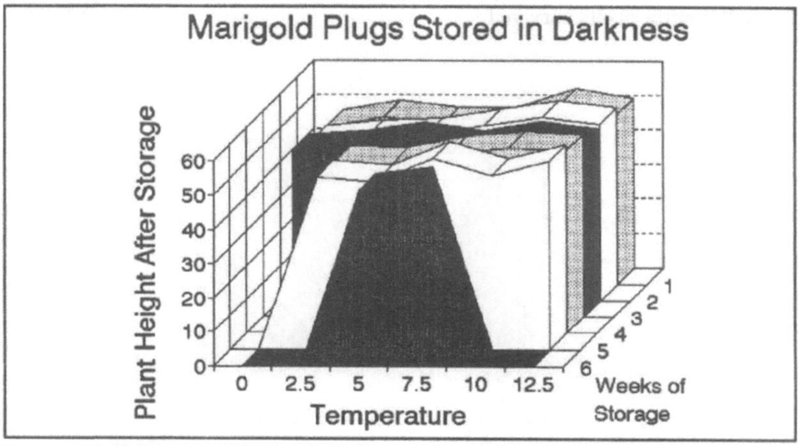

Figure 17: Influence of temperature on height of marigold seedlings stored in the dark for 1 to 6 weeks.

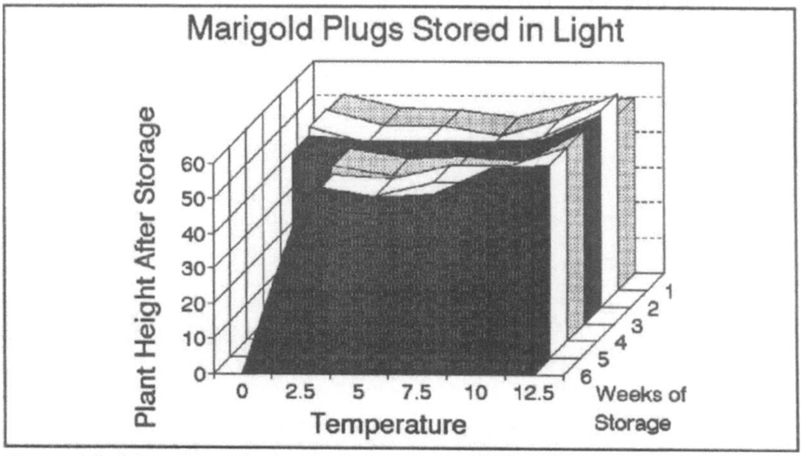

Figure 18: Influence of temperature on height of marigold seedlings stored in the light for 1 to 6 weeks.

Chilling injury was evident within 2 weeks of storage at 0C. Percent mortality increased from 0 or 10 after 2 weeks' light or dark storage at 0C to 100% after 4 weeks' storage at 0C (Fig. 19 and 20). At 2.5C, 4 or 5 weeks of storage were required to induce chilling injury in the dark (Fig. 19) or light (Fig. 20), respectively. Percent mortality at 2.5C was much greater in the dark than in the light, and increased as storage temperature and duration increased above 5C and 3 weeks' storage. In the dark, 5C was the only temperature at which substantial numbers of plants were able to survive for 6 weeks (Fig. 19). The addition of 1 $\mu mol \cdot m^{-2} \cdot s^{-1}$ dramatically decreased percent mortality at temperatures above 5C (Fig. 20). All control plants survived.

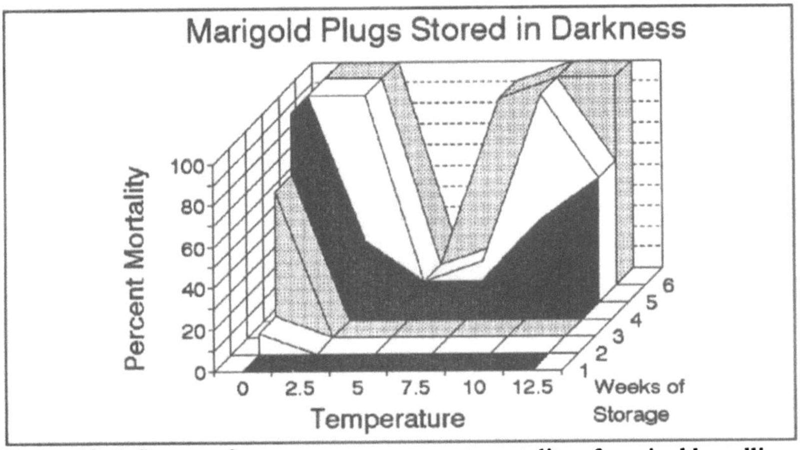

Figure 19: Influence of temperature on percent mortality of marigold seedlings stored in the dark for 1 to 6 weeks.

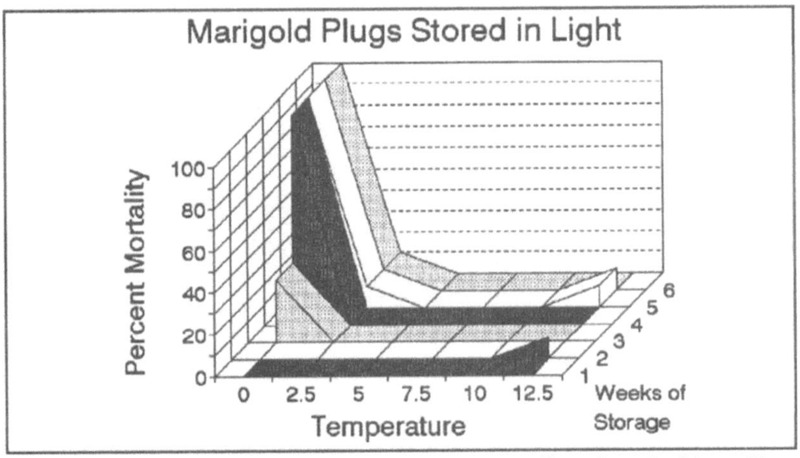

Figure 20: Influence of temperature on percent mortality of marigold seedlings stored in the light for 1 to 6 weeks.

The time from transplanting to flower was affected by storage in two ways. Treatments that induced chilling injury, but not death, caused flower-bud abortion and a subsequent flowering delay. Nonstored control plants flowered in 20 days. Flowering was delayed 28 or 24 days after 1 week's storage at 0C in the dark (Fig. 21) or light (Fig. 22), respectively. Plants stored in the dark at 10 and 12.5C also exhibited delayed flowering as storage duration increased. Flowering was delayed 25 days on surviving plants stored at 12.5C for 4 weeks (Fig. 21). Light did not prevent delayed flowering caused by chilling injury, but did reduce flowering delay on plants held at 10 and 12.5C (Fig. 22). Unstored plants in the greenhouse flowered in the plug sheets prior to transplanting within 4 weeks.

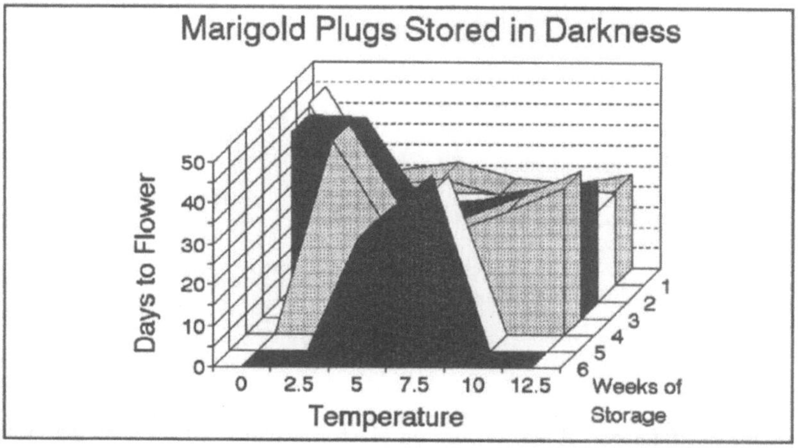

Figure 21 Influence of temperature on days to flower of marigold seedlings stored in the dark for 1 to 6 weeks.

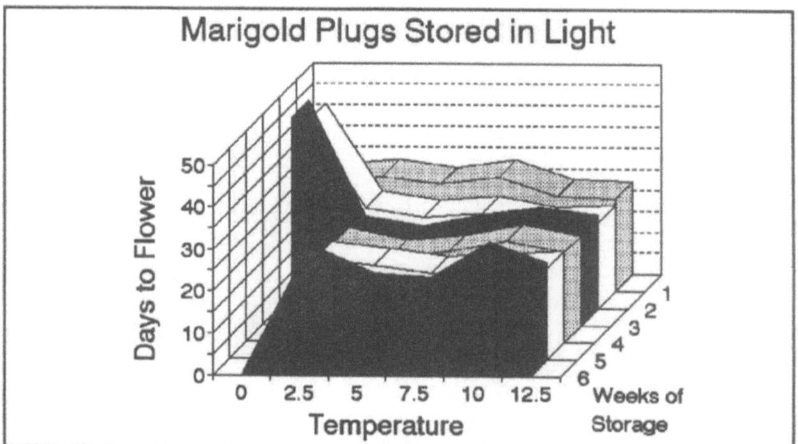

Figure 22: Influence of temperature on days to flower of marigold seedlings stored in the light for 1 to 6 weeks.

Plant size at flower was primarily related to the survival of the flower bud during storage. Plants in which flower buds aborted either because of chilling injury (e.g., 0C for 2 weeks) or because of extended storage (e.g., 7.5C for 5 and 6 weeks for plants in the dark [Fig. 23], 10 and 12.5C for plants in the light [Fig. 24]) were significantly larger at flower. The increased size was a result of the additional vegetative growth as lateral buds developed prior to flower. Plants held in the greenhouse flowered in the plug sheets and averaged only 20 cm^3 in volume at flower, whereas stored plants averaged 320 cm^3, and those that flowered after flower-bud abortion, up to 785 cm^3.

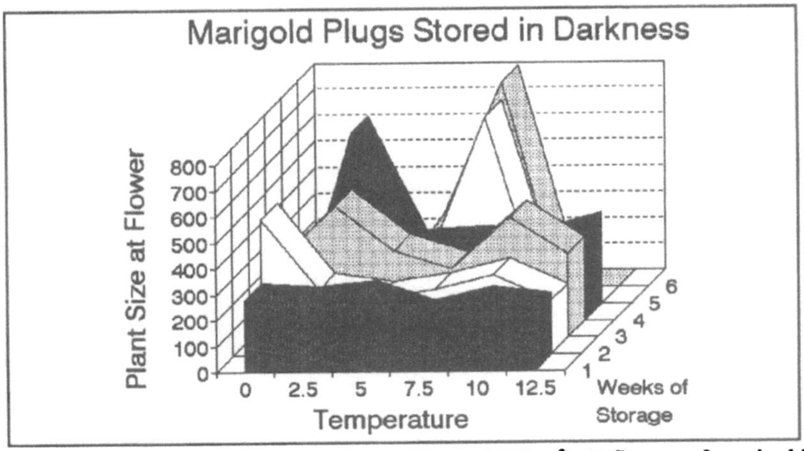

Figure 23: Influence of temperature on plant size (cm^3) at flower of marigold seedlings stored in the dark for 1 to 6 weeks.

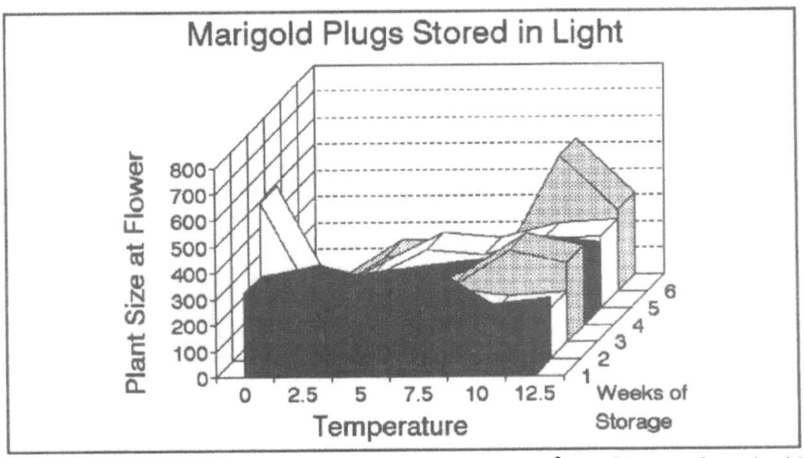

Figure 24: Influence of temperature on plant size (cm^3) at flower of marigold seedlings stored in the light for 1 to 6 weeks.

3.4 Salvia

Height of salvia seedlings stored in the dark increased as storage temperature and duration increased (Fig. 25). Seedling height increased approximately 15 mm at 10C, but at 5C, it increased only 3 mm. In comparison, height of seedlings held in the greenhouse increased 100 mm from an initial height of 54 mm during the same 6-week period.

Exposure to 1 μmol m^{-2} s^{-1} of light during storage slowed the rate of seedling elongation (Fig. 26). During 3 weeks' storage at 12.5C, seedlings grew 12 mm in the dark, but only 1 mm in the light.
No height increase occurred during 6 weeks' storage at 5C.

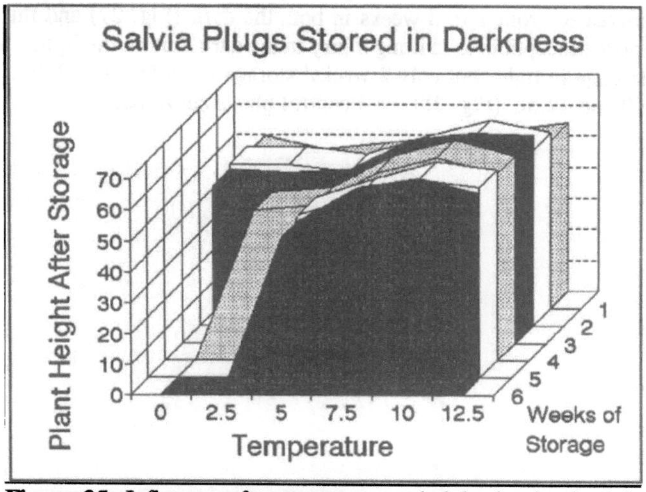

Figure 25: Influence of temperature on height (mm) of salvia seedlings stored in the dark for 1 to 6 weeks

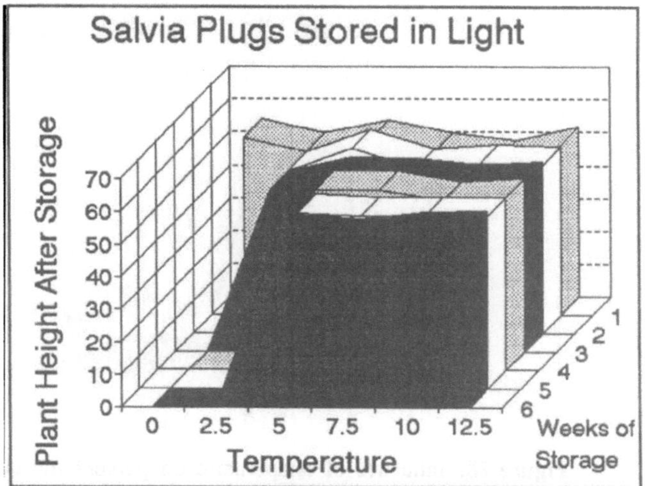

Figure 26: Influence of temperature on height (mm) of salvia seedlings stored in the light for 1 to 6 weeks.

Plants stored at 0 and 2.5C showed signs of chilling injury after 2 weeks of storage. Mortality as a result of chilling injury at 0 and 2.5C increased from 0 to 100% as storage time increased from 1 to 3 weeks in both the dark (Fig. 27) and the light (Fig. 28). Plants were more susceptible to chilling injury in the dark than in the light. Most plants survived 5 weeks' storage in light, but only 2 weeks' storage in darkness at 2.5C. Light prevented mortality at 10 and 12.5C (Fig. 28). All control plants survived.

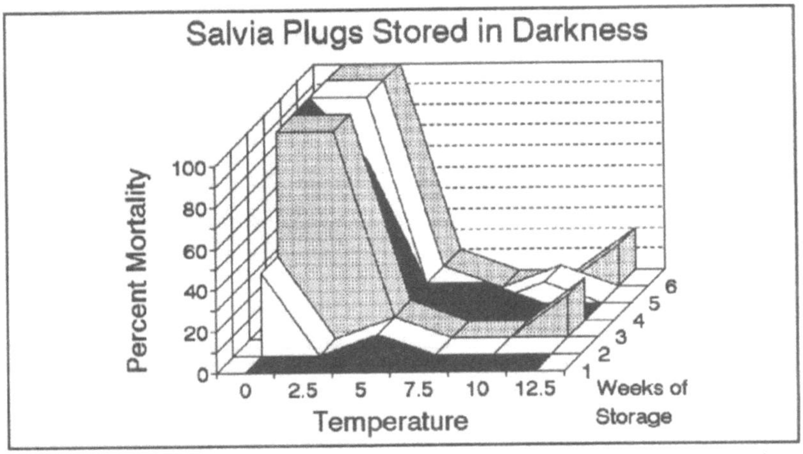

Figure 27: Influence of temperature on percent mortality of salvia seedlings stored in the dark for 1 to 6 weeks.

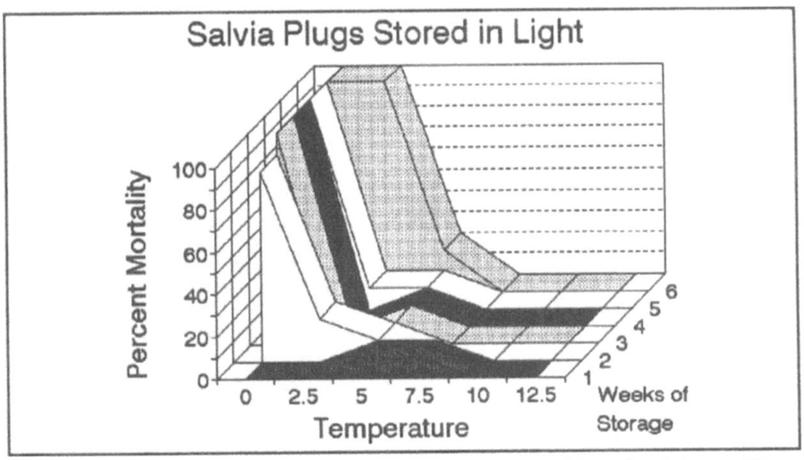

Figure 28: Influence of temperature on percent mortality of salvia seedlings stored in the light for 1 to 6 weeks.

The time from transplant to flower was affected by storage in two ways. First, flowering was delayed on plants that survived chilling injury from storage at 0 and 2.5C (Fig. 29 and 30). Compared to unstored control plants which flowed in 50 days, flowering was delayed up to 29 days on injured plants because of the abortion of the primary shoot axis. Second, flowering of plants held in the dark at 12.5C was delayed 18 days after 6 weeks' storage. Exposure to light during storage did not reduce the effect of chilling injury in delaying bloom (Fig. 30), but did prevent delayed flowering on plants held at 12.5C for 6 weeks.

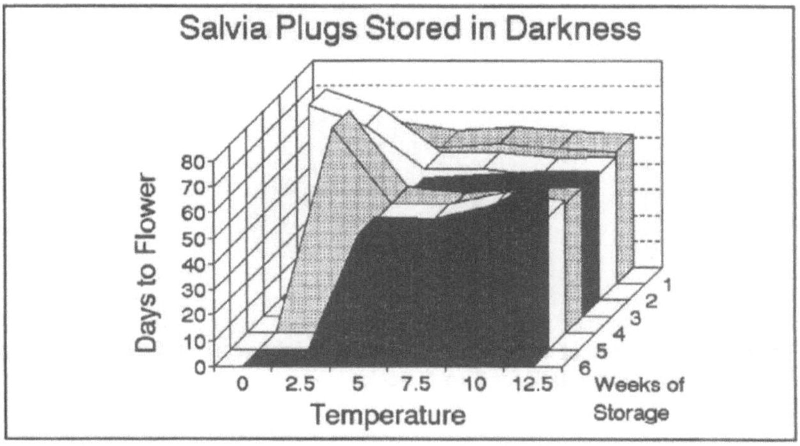

Figure 29: Influence of temperature on days to flower of salvia seedlings stored in the dark for 1 to 6 weeks.

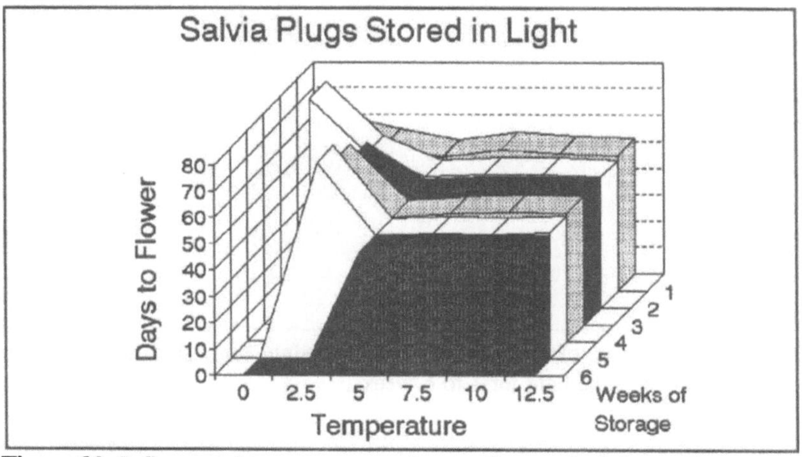

Figure 30: Influence of temperature on days to flower of salvia seedlings stored in the light for 1 to 6 weeks.

Chilling injury reduced plant height at flowering in both the dark and light (Fig. 31 and 32) because flowering lateral shoots were shorter than the primary shoot. In general, height at flower decreased as storage duration increased. For example, plants stored 1 week flowered at a height of approximately 35 cm, while those stored 6 weeks flowered at about 26 cm.

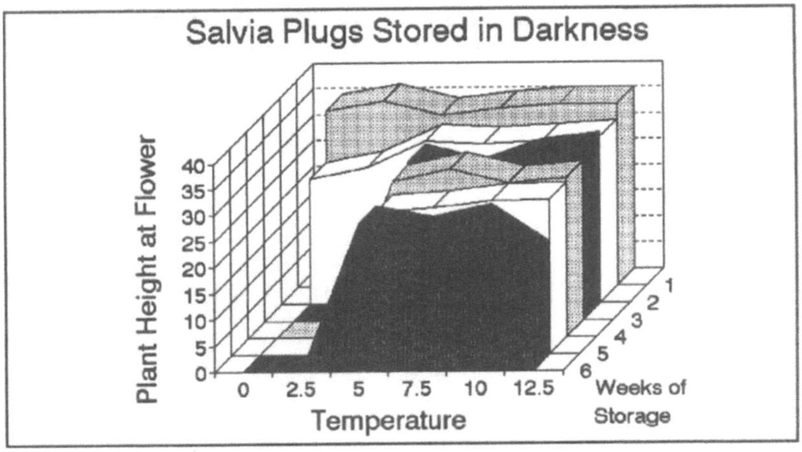

Figure 31: Influence of temperature on plant height (cm) at flower of salvia seedlings stored in the dark for 1 to 6 weeks.

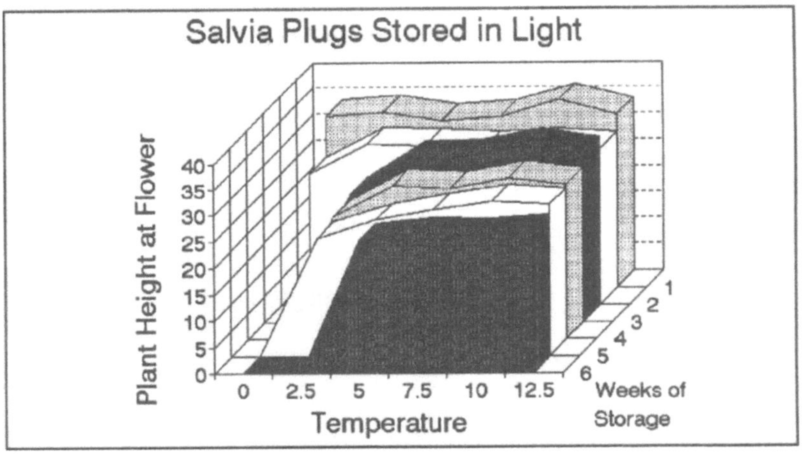

Figure 32: Influence of temperature on plant height (cm) at flower of salvia seedlings stored in the light for 1 to 6 weeks.

4. Discussion

Seedling survival and quality after storage depended on three storage parameters: temperature, presence or absence of light, and duration. The lowest treatment temperatures caused chilling injury, and the highest, plant deterioration and ultimately death as storage duration increased. Seedlings stored at 0 and 2.5C initially exhibited typical chilling-injury symptoms (Herner, 1990; Saltveit and Morris, 1990) of water-soaked leaf spotting, wilting, and death. In contrast, as storage temperature increased above 5C in the dark, deterioration occurred progressively faster and was exhibited by elongation, etiolation, leaf abscission, and flower and apical meristem abortion. Greatest percent survival and highest quality after dark storage occurred at temperatures just above those that caused chilling injury.

Exposure to light during storage had a dramatic effect on the quality of plants held at warmer-than-optimal temperatures. The amount of light necessary to improve quality was no more than 1 $\mu mol \cdot m^{-2} \cdot s^{-1}$. No substantial differences were observed among plants stored with 1 versus 5 $\mu mol \cdot m^{-2} \cdot s^{-1}$. In previous studies on geranium, pansy, petunia, and impatiens (Lange and Heins, 1991a,b), no significant differences in plant storability were evident among plants stored under 1, 5, and 25$\mu mol \cdot m^{-2} \cdot s^{-1}$, indicating that very low light levels are adequate to maintain plant quality during low-temperature storage.

All four species in this study were sensitive to chilling injury at temperatures less than 5C. Ageratum was most sensitive (Fig. 3 and 4,), followed by salvia (Fig. 27 and 28), marigold (Fig. 19 and 20), and begonia (Fig. 11 and 12). Kumpf, et. al. (1966) indicated salvia could be successfully stored at 1C for 2 weeks in the dark and 6 weeks at 4.5C in the light. Results from this study (Fig. 27 and 28) showed chilling injury on salvia seedlings after 2 weeks' storage at 0 and 2.5C, and therefore, storage is not recommended at temperatures less than 5C. Our results also suggest that 7.5C is superior to 5C for salvia storage in the dark or light (Fig. 27 and 28). Kumpf, et. al. (1966) also did not recommend marigold storage at 1C but did indicate marigold could be stored at 4.5C for 6 weeks in the light; both conclusions are in agreement with our observations.

Chilling injury occurs in many tropical plants (Morris, 1982) and in some temperate species (Bramlage, 1982) when the plants are exposed to temperatures above freezing, but below about 12.5C. The chilling injury on the species in this study is therefore not unexpected since ageratum originated in Mexico (Reilly, 1978), begonia and salvia in southern Brazil (Thompson and Thompson, 1981, and Reilly, 1978, respectively), and marigold in temperate and tropical Central and South America (Towner, 1961).

In salvia and marigold, percent mortality resulting from chilling injury was greater in the dark than in the light. The opposite was true for begonia and, to a lesser extent, ageratum (Fig. 12 and 13). While light interacts with temperature in some species to initiate or enhance chilling injury (van Hasselt, 1990), the interaction is dependent on light intensity. Little or no reduction in quantum yield occurred under conditions of 10% or less of full sunlight under chilling temperatures (Powles, et.al., 1983). Therefore, the lack of a strong response to light under chilling temperatures in this study is not unexpected.

Holding plants under 100% relative humidity during exposure to chilling temperatures reportedly prevents chilling injury in cucumber (Wright and Simon, 1973). However, in this experiment, exposure of plants to near 100% RH in the dark did not prevent chilling injury.

Apical meristems or flower buds aborted in some plants, leading to delayed flowering. These plants were typically larger at first flower than plants in which the apical meristem or flower buds did not abort. The increased size resulted from lateral shoot growth prior to flowering, a response which also occurs when apical meristems are pinched.

The value of storing seedling plugs at low temperatures is evident when greenhouse performance of optimally stored seedlings is compared to that of seedlings held in the greenhouse. In the most extreme example, marigold seedlings flowered in the plug sheet prior to transplanting. These plants were unacceptably small for commercial sale at time of flower. Similarly, size of ageratum, begonia, and salvia at flower decreased as duration of holding in the greenhouse increased. Quality of the flowering plant was maintained by proper storage.

When plant death and time to flower were taken into consideration, the optimum temperature range for storage of both ageratum and salvia was 7.5C. The acceptable storage-temperature range widened to the 7.5 to 12.5C range when the storage duration was short and when light was added during storage. The optimum temperature range for begonia storage was 5 to 7.5C. The acceptable storage-temperature range widened to the 2.5 to 10C range when the storage duration was 4 weeks or less, and when light was added during storage. The optimum temperature for marigold storage was 5C. Even at 5C, however, marigold stored poorly in the dark. Although all 4 species could be stored successfully for 1 week at 0C, storage at temperatures less than 5C is not recommended for any of these species because of the potential for chilling injury.

References

Bramlage, W. J. (1982) 'Chilling injury of crops of temperate origin', HortScience 17, 165-168.

Hardenburg, R. E., Watada, A. E., Wang, C. Y. (1986) 'The commercial storage of fruits, vegetables, and florist and nursery stocks', U.S. Department of Agriculture, Agriculture Handbook No. 66 (revised), 136p.

King, A. E., Wilson, J. M., Wright, M., and Simon, E. W. (1990) 'High humidity during chilling exposure', in C. Y. Wang, (ed.), Chilling Injury of Horticultural Crops, CRC Press, Inc., Boca Raton, Florida, p. 62.

Kumpf, J., Horton, F., and Langhans, R. W. (1966) 'Seedling storage', New York State Flower Growers Bul. 244, 30p.

Lange, N., Heins, R., and Carlson, W. (1991a) 'Store plugs at low temperatures', Greenhouse Grower 9(1), 22-28.

Lange, N., Heins, R., and Carlson, W. (1991b) 'Another look at storing plugs', Greenhouse Grower 9(2), 18-24.

Morris, L. L. (1982) 'Chilling injury of horticultural crops: an overview', HortScience 17, 161-162.

Reilly, A. (1978) Park's Success with Seeds, Geo. W. Park Seed Co., Inc., Greenwood, South Carolina.

Thompson, M. L., and Thompson, E. J. (1981) Begonias: the Complete Reference Guide, Times Books, New York, N.Y.

Towner, J. W. (1961) 'Cytogenetic studies on the origin of *Tagetes patula*. 1. Meiosis and morphology of diploid and allotetraploid *T. erecta* x *T. tenuifolia*', Amer. J. Bot. 48, 743-751.

PRESENT STATE OF TRANSPLANT PRODUCTION PRACTICES IN JAPANESE HORTICULTURAL INDUSTRY

T. ITO
Faculty of Horticulture
Chiba University
648, Matsudo, Chiba,
JAPAN

ABSTRACT. Transplant production in Japan has increased dramatically since 1985. This paper presents an in-depth, state-of-the-art review of plug transplant production practices in the horticultural industry of Japan. The current problems like labour shortage and the advantages such as labour saving, reduced time to produce a crop, uniformity of harvesting time and end-products plus other beneficial effects are narrated. Growth requirements and maintenance in the raising of plug transplants as well as environmental conditions, nutrition, irrigation, transportation and delivery systems are presented. New technology or modified methods of grafting techniques in transplants, its uses, and the necessary grafting conditions to be met are also enumerated. For instance, robotization and an upgraded version of the whip grafting method manually, are illustrated.

1. Horticultural Industry

1.1. INTRODUCTION

Total number of commercial farm household in Japan was about 3 million in 1990. A 10.4% of the total farm number has decreased in the last five years . Commercial farmers engaged in horticultural productions are 1.23 million, accounting for 41.5% of total farm household (Table 1).

TABLE 1. Total number of commercial farm household mainly engaged in horticultural industry.
(Ministry of Agr., Fores. and Fish., 1991)

Section	No. of Farm h. ($\times 10^3$) 1985	1990	'90/'85 (%)
Vegetables	72.5	60.6	−16.3
Fruits	61.6	52.7	−14.5
Flowers	11.9	9.9	−16.8
Sub−Total	146.0	123.2	−15.6
Others	185.5	173.9	-6.3
Total	331.5	297.1	−10.4

K. Kurata and T. Kozai (eds.), Transplant Production Systems, 65–82.

The cultivated land in this country is now estimated to be 5.27 million hectares (ha), resulting in yearly decrease of approximately 0.8 per cent. The planted areas for fruits, vegetable crops and ornamental plants, however, have been kept in stable state at 350, 630 and 11 ha respectively in recent years.

The production of ornamental plants has resulted in a steady progress year by year, although vegetable crops have shown little change in the last 10 years (Figure 1). The total value of horticultural plants and crops was 26.5 billion US$ in 1989. Vegetable crops accounted for two thirds of the total value (Figure 2).

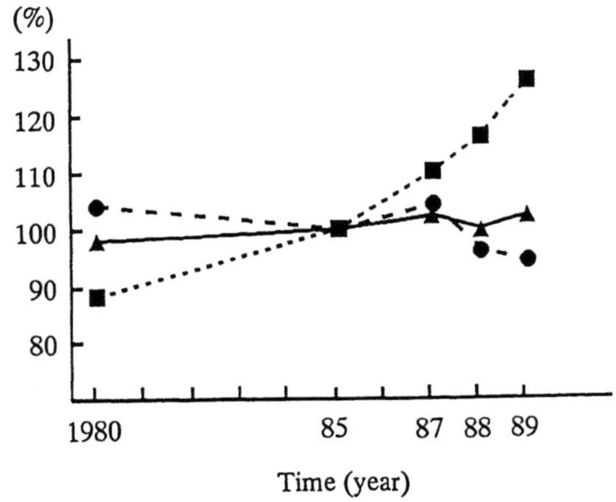

Figure 1. Changes in yearly productions of vegetables(▲), fruits(●) and flowers(■).
Relative values based on the amounts of 1985.
(Ministry of Agr., Fores. and Fish., 1991)

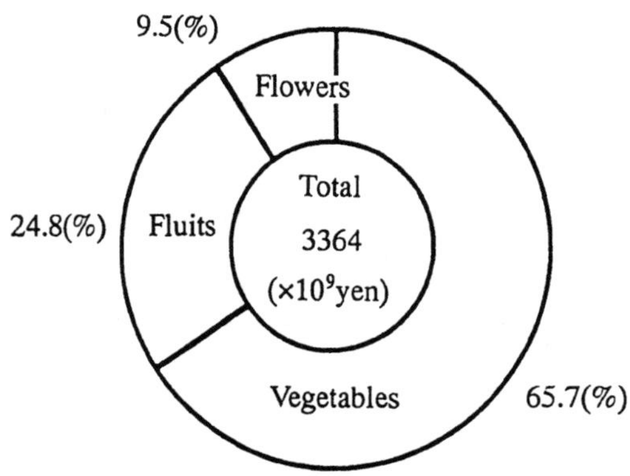

Figure 2. Annual values of horticultural products(1989).
(Ministry of Agr., Fores. and Fish., 1991)

Table 2 shows the planted area and the annual production of main horticultural plants and crops in 1989.

TABLE 2. Planted area and annual production of main horticultural plants and crops (1989). (Ministry of Agr., Fores. and Fish., 1991)

		Planted area (ha)	Annual production
Vegetables	Radish	62300	2449.0 (10^3tons)
	Cabbage	41400	1623.0
	Chinese-cabbage	29900	1334.0
	Onion	28200	1269.0
	Cucumber	20900	975.0
	Tomato	14500	773.1
	Watermelon	23200	763.6
	Carrot	24000	685.2
	Eggplant	17700	567.4
	Welsh onion	24000	541.9
Ornamental plants			
Cut flower	Chrysanthemum	5160	1748.0 (10^6pieces)
	Carnation	580	688.6
	Rose	420	373.6
	Lily	390	113.3
	Gentian	570	103.6
Potting plant	Foliage plant	300	42.8 (10^6pots)
	Flowering tree	220	24.8
	Succulent plant and Cactus	50	17.2
	Primrose	70	15.5
	Cyclamen	190	14.9
Bedding plant	Garden pansy	80	33.3 (10^6plants)
	Marigold	20	10.9
	Sage	20	10.8
Fruit trees	Mandarin	78800	2015.0 (10^3tons)
	Apple	49900	1045.0
	Japanese-pear	18600	439.1
	Grape	24600	274.7
	Persimmon	26300	265.7

1.2. CURRENT PROBLEMS AND TRANSPLANTS

Labour saving has been recognized as one of the important solution measures to overcome the shortage of farm labour population. New technology of transplant production has dramatically changed the conventional production technique of the vegetable and flower growers who are fascinated with plug transplants, since the labour for raising seedlings is estimated to be 5 to 15 per cent of total labour as shown in Table 3. In addition, the plug transplants have beneficial advantages for the users to be able to reduce the time for growing the final products with better uniformity and higher quality.

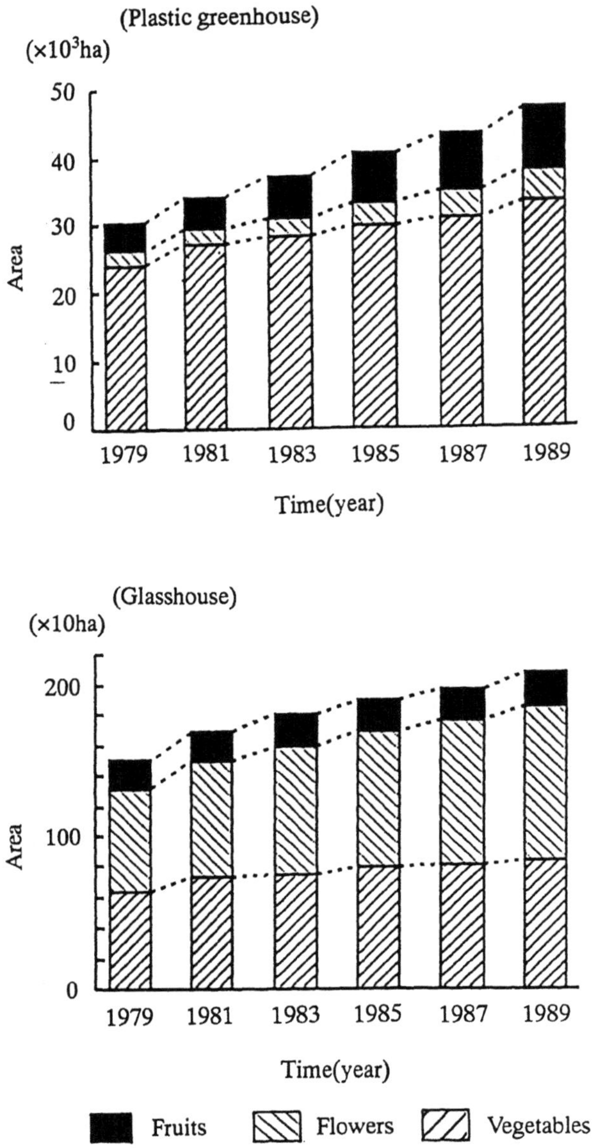

Figure 3. Yearly increase of plastic greenhouse and glasshouse areas in the last 10 years. (Ministry of Agr., Fores. and Fish., 1991)

Figure 3 shows yearly increase of greenhouse area, indicating that plastic greenhouses used for fruits and vegetables have increased markedly in the last ten years, while glasshouses have increased slowly in flower and vegetable production. Such intensive cultivations have resulted in high demand of transplants so as to use the facilities more efficiently. The best example is that of the carnation industry, where growers can purchase their cutting transplants anytime of the year hence can produce pot–plants or cut–flowers in commercial quantities all year round in their greenhouse.

2. New Technology of Transplant Production

2.1. PLUG TRANSPLANT PRODUCTION

2.1.1. *Introduction and development of the techniques.* Plug production technology in this country is far behind that of the United States and most European countries, because it was originally introduced for use in seedling transplants of bedding flowers by some private companies in 1985.

Dramatic impact occurred in the flower industry, especially for cutting transplants of ornamental flower plants. Consequently, plug transplant production has covered whole ornamental plants in the last few years, not only in the cutting transplants for potting–plants and cut–flowers but also in seedling transplants for bedding– and potting–plants. As for vegetable crops, Nagano Prefecture, one of the leading districts of open field vegetable crops, has developed "N K plug system" for vegetable seedlings, with several modifications of the European technique since 1988.

TABLE 3. Labour hour required for raising seedling. (Inayama, 1990)

Crops	Cropping programme	Days for raising (day)	Labour hour for culture (10^3hr/ha)	Labour hour for rasing (hr/ha)
Tomato	Forcing culture	70	14.9	940
	Semi–forcing c.	60	17.9	920
Eggplant	Semi–forcing c.	100	19.5	4010
	Low tunnel c.	90	22.2	1070
	Open field	90	28.7	1980
Cucumber	Forcing c.	30	18.4	1690
	Semi–forcing c.	35	14.3	2000
	Low tunnel c.	30	13.2	1960
	Late forcing c.	25	8.0	850

According to the 1991 Ministry of Agriculture, Forestry and Fishery reported on transplants, 100 millions of plug transplants were commercially produced in 1990, accounting for 72% increase of the previous year. Table 4 shows that lettuce (mainly iceberg lettuce), Italian broccoli, cabbage, tomato and onion are the predominant vegetables for plug seedlings, whereas cut–flowers, bedding– and potting–plants are being adapted for plug transplants.

Approximately, two thirds of the total plug transplants is produced by agricultural co-operative or individual producers, while the remaining transplants are provided from several nurseries as shown in Table 5. Major sources and suppliers of plug transplants in this country are listed in Table 6.

TABLE 4. Plug transplants production.
(Ministry of Agr., Fores. and Fish., 1991)

		No.of transplants produced ($\times 10^6$)		'90/'89 (%)
		1989	1990	
Vege-tables	Lettuce	9.2	21.7	230
	Broccoli	3.0	7.6	250
	Cabbage	3.1	5.8	191
	Tomato	2.5	4.4	177
	Onion	1.6	3.2	200
	Miscellaneous	9.2	15.0	163
	Subtotal	28.6	57.7	202
Orna-mental plants	Bedding plants	15.2	20.3	133
	Cut flowers	11.8	16.8	142
	Potting plants	4.2	8.0	190
	Subtotal	31.3	45.1	144
	Total	59.9	102.7	172

TABLE 5. Producer of plug transplants.
(Ministry of Agr., Fores. and Fish., 1991)

Crop	Year	No.of transplants produced ($\times 10^6$)		Total (%)
		Farmers[z]	Nursery	
Vege-tables	1989	20.8	7.7	28.6
	1990	41.5	17.2	57.7
	Increase (%)	198	223	202
Orna-mental plants	1989	15.7	15.3	31.3
	1990	23.8	21.3	45.1
	Increase (%)	152	139	144
Total	1989	36.5	23.0	59.5
	1990	65.0	38.5	103.5
	Increase (%)	178	167	174

[z]Farmers include agricultural co-operative and individual producers.

2.1.2. *Technology of plug transplant production.* Since few experiment on plug production has been carried out in the last five years, no beneficial information has been presented through experimental data to develop the original production techniques. Thus, the current technology of plug production is mainly based on techniques developed in the States or in European countries, with slight modifications from some conventional propagation techniques.

TABLE 6. Sources and suppliers of plug transplants.
(Natn'l. Feder. of Agri. Co–op. Ass'n.)

Trade name SOUR. & SUPPL. (CO.Ltd)	Item	Vege- tables	Flowers
Cell–Top Seedling SAKATA SEED	Plug transplants, Growing media, tray	●	●
Kubota Plug Seedling Sys. KUBOTA	Seeder, Nursery house, etc.	●	
Material for Plug System DAI–ICHI SEED	Growing media, tray, Planting machine		●
Naueru–System ISEKI	Plug transplants	●	
NK Plug Seedling NAGANO PREF. ECONO–FED OF AGRICULTURAL CO–OP.	Plug transplants, Growing & Planting Sys.	●	
PeSP Seedling YAMAGUCHI NURSERY TAKII & COMPANY	Plug transplants	●	
Sunny Plug Seedling DIA–TOPY	Plug transplants	●	●
T.M.B. Plug Seedling MIKADO SEED GROWERS MIYOSHI & COMPANY	Plug transplants	●	●
YANMAR Seedling YANMAR AGR. EQUIPT.	Growing media, tray, Seeder, Planting machine	●	

a. Tray size. The tray size of 600 × 300 mm is generally used in this country, because the same size of outer tray which has been commonly introduced to rice seedling production is often used instead of bench system in the small scale plug production.

b. Plug size. Plug size is one of the most important factors in producing transplants, not only for seedling performance but also for an efficient use of the facilities. Tables 7 and 8 show some examples of plug tray adapted for the vegetable and flower industries.

c. Growing medium. Peat moss imported from Scandinavian countries and Canada is widely accepted as the main growing medium in combination with perlite, vermiculite, granulate rockwool, compost, carbonized husk and so on. Many kinds of growing medium are available in the commercial market as shown in Table 9. Proper materials and precise mixing ratio will be needed, since a very small amount of growing medium is applied to each cell (Table 10).

d. Mixing and distribution. Domestic mixing machine is generally used even in the small scale plug production. More precise and accurate distribution of the growing medium is one of the most important technique for plug production, and a multi–purpose distributor is often used successfully for this purpose.

e. Seed quality. Germination and seedling establishment are markedly dependent on seed quality. Up to the present state, no refined seed is prepared for plug seedling production. Seed treatment such as priming is occasionally required to enhance the germination ability. Coated seeds are also necessary for some mechanical seeders.

TABLE 7. Plug sizes used in vegetable crops.

Plug tray[z]	Size of plug (mm)			Vegetables
	Top	Bottom	Depth	
253	20 × 20	10 × 10	40	Lettuce, Pak–choi, Tomato, Sweet pepper, Asparagus
171	25 × 25	10 × 10	50	Eggplant (scion)
136–4	30 × 30	20 × 20	40	Chinese cabbage, Cabbage
136–5	30 × 30	15 × 15	50	Broccoli, Cauliflower,

[z]Number of cells in a tray

TABLE 8. Plug sizes used in ornamental plants.

Plug tray[z]	Ornamental plants
406	Begonia, Petunia, Marigold, Impatiens, Cosmos, Madagascar periwinkle
288	Pansy, Petunia, Marigold, Primrose, Impatiens, Begonia, Russel prairie gentian, Dianthus, Madagascar periwinkle, Poppy, Persian buttercup,
200	Limonium, Pansy
135	Gypsophilla
128	Limonium
96	Carnation, Limonium
73	Anemone
51	Anemone

[z]Number of cells in a tray

f. Seeding. Vacuum rotary seeder should be adapted for a large nursery in order to improve seeding efficiency. Automatic seeder of feeder type is often convenient for accurate seeding of small scale production. Seeding plate, with definite hole so as to pass a pelletted seed separately, is prepared for a manual seeding by individual producers.

g. Germination. Seedling development is directly related to the quality of germination. Successful germination can be attained under controlled environmental conditions. Such favourable conditions will be easily achieved in the germination facility or in the greenhouse where light, temperature, moisture and oxygen levels are able to be controlled artificially. Light sometimes may play an essential role to prevent an unfavourable elongated hypocotyl.

h. Greenhouse facilities. Greenhouse covered with polyvinyl chloride (PVC) film is commonly used for the plug production. Since room temperature in the summertime in this country is often unfavorably high for the growth of plants, PVC films are generally removed and shading materials installed as needed. Heating system is required for the winter production. Overhead fog or mist system is adapted for precise watering to younger seedlings, although irrigation booms may be advantageous at the latter growing stage of plug seedlings. Bench is necessary for better development of plug roots, while the small individual producers may choose the outer tray instead of bench system.

TABLE 9. Sources and suppliers of growing medium.
(Natn'l. Feder. of Agri. Co–op Ass'n.)

Trade name	Property			
SOUR. & SUPPL. (CO.Ltd)	W.C.[z]	A.S.G.[y]	E.C.[x]	pH
Genki–kun No.1 CO–OP CHEMICAL	15–22	0.8	0.5–0.8	6.2–6.8
Grand–sol No.11 SUMITOMO CHEMICAL	27–33	0.9–1.1	-	6.2
Kureha Engei Baido KUREHA CHEMI. INDUST.	15	0.8–0.9	-	6.5
Pot Ace No.1 KATAKURA CHIKKARIN	30	0.7	-	6.0–6.5
Sansan–jirusi Engei Baido TAKII & COMPANY	35–38	0.7–0.8	0.4–0.7	6.0–6.4
Sukoyaka Yasai Youdo YANMER AGR. EQUIPMENT	–	–	–	6.0–6.5
Sukusuku Baido MIKADO SOIL & DEVELOP.	–	–	0.6	6.8
Super Soil No.1 MITSUI TOATSU CHEMI.	35	0.8	-	6.0–6.4
Topy G Soil TOPY GREEN	30	0.7	0.6–1.0	6.0–6.5
Yosaku V No.1 CHISSO–ASAHI FERTIL.	30	0.6–0.7	1.0–1.5	6.0–7.0

[z]W.C. :Water Content (%)
[y]A.S.G. :Apparent Specific Gravity
[x]E.C. :Electric Conductivity (mS/cm)

i. Environmental control. Light intensity should be properly controlled depending on the kinds and developmental stages of transplants. Shading may be applied to the cuttings or to younger plug transplants. Temperature control is also an essential factor to the plug plant performance, lower temperature may result in shorter and harder transplants. Carbon dioxide enrichment may have a beneficial effect on plant growth, however, no commercial application has been carried out yet in the plug transplant production.

j. Nutrition and irrigation. The volume of plug cell is so extremely restricted that proper amounts of watering should be applied every day with a uniform distribution. Manual watering may be necessary for the individual plug grower. An automatic irrigation system is beneficial for more intensive plug producers and nurseries. Mineral nutrients are generally applied by dissolving them into irrigation water as shown in Table 11. Since there are many kinds of mixed

TABLE 10. Effect of growing medium on seedling performance of cabbage plants. (Ohtsuka & Ito, 1992)

Growing medium[z]		Plant	No.of	Fresh weight		T/R
Component	Density[y]	height (cm)	leaves	Top (g)	Root (g)	ratio
Peatmoss:Verm	Low	11.4	3.3	0.93	0.31	33.3
	High	11.6	3.6	1.00	0.28	28.0
Rockwool:Verm	Low	13.0	3.5	1.26	0.22	17.5
	High	12.4	3.2	0.99	0.23	23.2

[z]Peatmoss:Vermiculite=1:1 in volume
Rockwool:Vermiculite=1:1 in volume
[y]High density is 120% of low density in weight in each cell

TABLE 11. "Enshishoho" nutrient solution. (Hori, 1966)

$Ca(NO_3)_2 \cdot 4H_2O$	950 g/ton
KNO_3	810
$MgSO_4 \cdot 7H_2O$	500
$NH_4H_2PO_4$	155
NaFe–EDTA	15–25
H_3BO_3	3
$MnSO_4 \cdot 4H_2O$	2
$ZnSO_4 \cdot 7H_2O$	0.22
$CuSO_4 \cdot 5H_2O$	0.05
Na_2MoO_4	0.02

TABLE 12. Effect of leaf pruning on the growth and root activity of lettuce seedlings. (Shiraishi & Ito, 1992)

Treatment	Irrigation frequency (times/day)	Dry weight[z] Top (mg)	Root (mg)	Root activ. Oxicid.Esc.[y] (μg/g/h)
Control	1	44	8	28.8
	2	53	9	51.6
	3	54	10	55.6
Pruning[x]	1	23	10	50.9
	2	22	7	66.1
	3	19	7	91.4

[z]Dry weight was examined 5 days after pruning.
[y]Respiratory activity of root was measured by the amounts of oxidated Esculin.
[x]Leaf was pruned at half in length 25 days after sowing. Dry weights of top and root at the leaf pruning stage were 37.3 and 7.3 mg respectively.

fertilizers provided with hydroponics in this country, the plug producers are able to select a fertilizer in accordance with their growing medium. If the grower uses a growing medium containing suitable fertilizers, he can grow seedlings with an occasional application of nutrient solution.

k. Timing and scheduling. Growth of plug transplants is often controlled mainly by irrigation frequency, nutrient concentration and temperature management, or partly by application of plant growth regulators. Leaf pruning will be carried out rarely when plug transplants grow too fast to apply for planting machines. Leaf pruning at proper timing will result in increasing root activity, when necessary, as shown in Table 12.

l. Hardening. In order to adapt to new environmental conditions, plug transplants should be hardened by exposing them to higher light intensities, lower temperatures or less amounts of irrigated water approximately one week before shipping. The strength of hardening will vary with environmental conditions where the transplants are planted, better circumstances may require milder hardening.

m. Shipping. Overnight delivery service is available for transportation of plug transplants in small scale industries. Private truck transportation by the individual nursery is more reliable and much easier than the next day service. Plug transplants for long distance shipment should be raised less succulent so as to endure water stress. The paper carton with small side vents is commonly used for the shipping of plug transplants.

n. Transplanting. One of the most important advantages for plug users must be labour saving provided with mechanical transplanting, though few reliable machines are available at this moment. It is very important for growers to transplant the single plug seedling mechanically into plastic–mulched planting hole, since mulching has prevailed as the usual practice in the open field vegetable production.

TABLE 13. Comparison of seedling performance of lettuce plants raised by two methods. (Nagano Agric. Res. Cent., 1987)

Days after sowing	Raising method	Plant height (cm)	No.of leaves	Fresh weight Top (g)	Fresh weight Root (g)	Root length (cm)
10	Plug system	2.6	1.5	.19	.11	3.6
	Traditional[z]	3.0	1.7	.23	.04	1.4
15	Plug system	3.7	2.3	.38	.19	5.8
	Traditional	4.3	2.3	.52	.08	3.9
20	Plug system	4.5	3.5	.56	.27	9.6
	Traditional	5.8	3.2	.73	.11	5.4
25	Plug system	4.7	3.8	.59	.30	10.3
	Traditional	6.5	3.7	.93	.18	5.9
30	Plug system	5.0	4.2	.62	.36	11.4
	Traditional	6.8	4.2	.99	.20	6.5

[z]The traditional raising was carried out with soil blocks.

o. Storage. The long term storage of plug seedlings will be beneficial for the plug producer in order to increase the efficient use of facilities and equipment. Refrigerator is often utilized in vegetable seedlings for the short term storage.

p. Quality of transplants. Quality of transplants may be evaluated with their adaptation to new environmental conditions. Early rooting to the new medium will promise better growth of the plants. Hence the physiological properties of plug transplants are considered to be important in this country, which makes it difficult to produce the transplants with mass productions (Table 13).

TABLE 14. Comparison of labour hour and production cost of broccoli seedling raised by two methods. (Kohno, 1990)

Raising method	Item	Labour hour (hr/ha)	Prod.cost[z] (10^3yen/ha)	Cost (yen/seedl.)
Plug system		0	504.3	12.60
Traditional		440	801.2	20.03
	Seed		63.9	
	Plastic pot		28.8	
	Soil sterilization		10.9	
	Growing medium		85.7	
	Heating system		61.1	
	Covering materials		30.8	
	Greenhouse		50.5	
	Charge of electricity		29.5	
	Labour cost		440.0	

[z]The cost was estimated with 40,000 seedlings required for planting of one hectare cultivation.

q. Cost and economical value. The production cost of vegetable seedlings is much more expensive in the traditional raising method than that of the plug system (Table 14). The price of plug transplants, however, is commercially accepted to be less than 10% of the end-product costs. Table 15 shows the cost of plug seedlings in some vegetable crops, indicating approximately 10 cents per plant.

TABLE 15. Cost of plug seedlings in vegetable crops.(Fukazawa, 1991)

Crops	Cost of each plug (yen/plug)	Cost of total plug (10^3yen/ha)	Prod.cost of total cultv. (10^3yen/ha)	Net income (10^3yen/ha)
Process.tomato	8	96	1,410	960
Fresh tomato s.c.[z]	9–12	234	8,060	7,550
Eggplant s.c.[z]	9–12	90	3,040	10,310
Lettuce a.c.[z]	10–13	730	1,660	2,500
Broccoli	10–13	420	1,420	1,930
Cabbage	10–13	550	1,690	2,430

[z]s.c. and a.c. mean summer and autumn crop respectively.

2.2. GRAFTED SEEDLING PRODUCTION

2.2.1. *Commercial values of grafted seedling.* It has been widely accepted that, the grafting cultivation is a useful and harmless technique to prevent soil–borne diseases in some fruit vegetables such as watermelon, cucumber, melon, tomato and eggplant. Table 16 presents the purpose of grafting and grafting method in each vegetable. Table 17 shows crops and areas of grafted plants in vegetables, showing that the leading grafted fruit vegetable crops are watermelon and cucumber in this country.

TABLE 16. Grafting culture in fruit vegetables.

Scion	Rootstock	Grafting method	Merit
Citrullus lanatus Matsum. et Nakai	*Lagenaria siceraria* Standl.	Hole	1.Fusarium wilt free 2.Encourage vigor
	Benincasa hispida Cogn.	Hole	1.Fusarium wilt free 2.Encourage vigor
	Cucurbita pepo L.	Tongue approach Wedge	1.Fusarium wilt free 2.Encourage vigor 3.Chilling tolerance
Cucumis sativus L.	*Cucurbita ficifolia*	Tongue approach	1.Fusarium wilt free 2.Chilling tolerance
	Cucurbita max. × mosch.		1.Fusarium wilt free 2.Encourage vigor 3.Extend growing term
Cucumis melo L. (var.cantalup.)	*Cucurbita max. × mosch.*	Tongue approach	1.Fusarium wilt free 2.Encourage vigor 3.Chilling tolerance
Cucumis melo L. (var. reticulatus)	*C. melo* L.	Tongue approach Wedge	1.Fusarium wilt free
Cucumis melo L. (var. conomon)	*Cucurbita max. × mosch.*	Tongue approach	1.Fusarium wilt free 2.Encourage vigor
Lycopersicum escul. Mill.	*L. pimpinel.* *L. hirsutum* *L. esculentum*	Tongue approch	1.Nematode resistance
Solanum melong. L.	*S. integr.* Poir. *S. torvum*	Tongue approach Wedge	1.Bacterial wilt free 2.Extend growing term

In general, grafted seedlings are mainly raised by individual growers for their own use. Several millions of grafted seedlings, however, are produced by the intensive nurseries or by agricultural co–operatives every year.

The grafting technique and acclimatization procedures require considerable skill and labour. Much more progress had been made in grafting cultivation, if the grafted seedlings could be provided timely at lower costs.

Intensive studies have been carried out to produce grafted seedlings either by mechanical procedure or by improving manual grafting techniques in the last five years.

TABLE 17. Planted areas of grafted fruit vegetables.
(Ministry of Agr. Fores. and Fish.,1981)

Crops	Cultivation	Planted area		Percentage
		Total(ha)	Grafted(ha)	(%)
Watermelon	Open field	24,933	23,812	95.3
	Greenhouse	3,244	3,213	99.0
	Total	28,237	27,025	95.7
Cucumber	Open field	14,765	4,575	31.0
	Greenhouse	7,035	6,003	85.3
	Total	21,809	10,578	48.5
Musk melon	Glasshouse	1,005	481	47.9
	Plastichouse	197	48	24.4
	Total	1,202	529	44.0
Cantaloupe	Open field	5,493	3,744	68.2
	Greenhouse	2,250	1,961	87.2
	Total	7,743	5,705	73.7
Tomato	Open field	6,888	355	5.2
	Plastichouse	4,731	763	16.1
	Glasshouse	377	58	15.3
	Total	11,996	1,176	9.8
Eggplant	Open field	13,352	2,673	20.0
	Greenhouse	1,694	1,634	96.4
	Total	15,046	4,307	28.6
Total		88,325	49,942	56.5

2.2.2. *Technology of grafted transplant production.*

a. Robotization of grafting. Fundamental Technology Department of IAM BRAIN developed a prototype grafting robot in 1987 and further improved it in 1989. Mechanization of grafting procedures was achieved by three wheels of scion, rootstock and clips in connection with cutters and programable controller. Single grafted plant requires only 3 seconds in all procedure. Whip grafting with single cotyledon of rootstock plant was adopted in this mechanical grafting as illustrated in Figure 4.

b. New manual technique for grafting. Agricultural technology Center of National Federation of Agri. Co–op. Ass'n. has developed a new grafting technique for Solanaceous plants. The grafting is performed much more speedy and at the 2.5–true–leaf stage of both the scion and the rootstock plants as contrasted with the 5–leaf stage of the traditional grafting(Table 18). In the first step, the stem of rootstock plant is cut at an angle of 30° at the internode 4 mm above the cotyledon. The next step is an insert of an elastic supporting tube at the same angle in the rootstock stem. The scion stem is also cut just the same as rootstock plant and is joined within the tube as illustrated in Figure 5. The rootstock plants are generally raised in the plug tray with 51 cells, and no transplanting is needed at the time of grafting. After carrying out the grafting of whole plant in the tray, the grafted plants are transferred to the growth chamber named "Nae-pit" in which the curing and acclimatization are performed under artificial environmental conditions as shown in Table 19 and Figure 6.

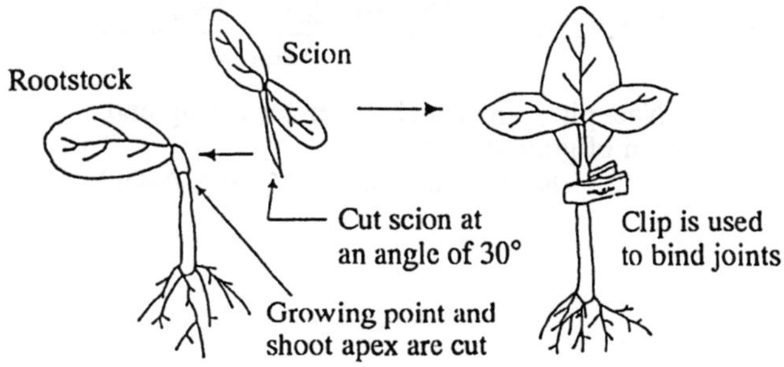

Figure 4. Whip grafting with single cotyledon.
(IAM BRAIN)

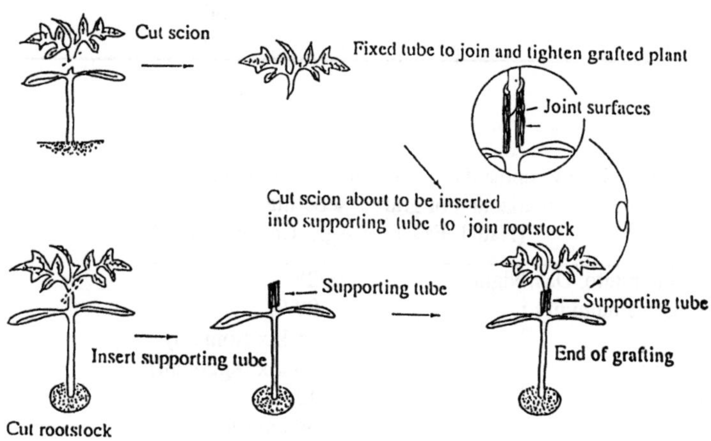

Figure 5. Steps involved in "modernized" whip grafting.
(Natn'l. Feder. of Agri. Co-op. Ass'n.)

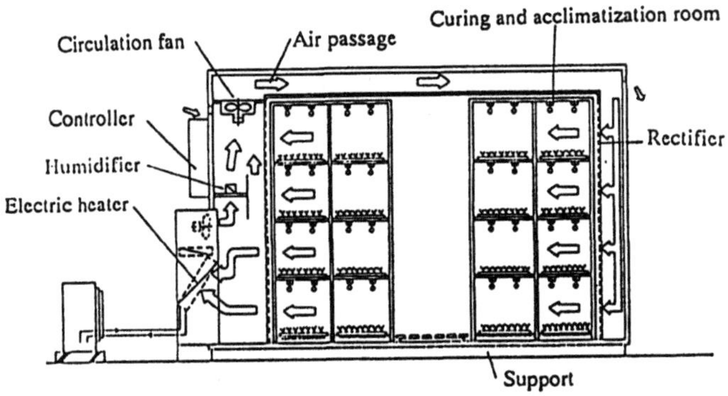

Figure 6. "Nae–pit", growth chamber for curing and acclimatization of grafted plants.
(Natn'l. Feder. of Agri. Co–op. Ass'n.)

TABLE 18. Comparison of labour efficiency between different grafting methods in eggplant. (Natn'l. Feder. of Agri. Co–op. Ass'n.)

Methods of Grafting	No.of leaves for Grafting	Efficiency (plants/person/hr)
Tongue Approach	5–6	65
Wedge	4–5	70
Hole	4–5	80
Whip	2–3	150

TABLE 19. Environmental conditions recommended for curing and acclimatization of grafted seedling.
(Natn'l. Feder. of Agri. Co–op. Ass'n.)

Temperature, Day/Night :	28/25 °C
Relative humidity :	90 %
Light :	5 klx (tomato)
Light :	3 klx (eggplant)
Photoperiod :	12 hrs
Wind velocity :	30 cm/sec
CO_2 concentration :	600 ppm

2.3. MICROPROPAGATED TRANSPLANT

Micropropagation of plant tissue has been considered one of the important propagation methods in the horticultural industry, especially for originating virus–free plants and for multiplying vegetatively propagated plants, such as carnation, gypsophila, strawberry, Chinese yam, asparagus, taro and garlic etc. We have no reliable statistic data for the total number of micropropagated plants produced in this country. In the case of strawberry plants in Tochigi Prefecture, about 8,000 stock plants are produced in experimental stations. These plants are sent to 16 Prefectural netted propagation greenhouses for a total multiplication of 1 million mother plants. The farmers then grow their mother plants and multiply each by 50 to 100 daughter plants in individual netted propagation greenhouses for their own use. A total of 585ha of strawberries are under cultivation in the above mentioned Prefecture hence approximately 1,462,500 mother plants per hectare are needed every year thus indicating 70% of the total number of mother plants are regenerated.

3. Forecasting Development in Transplant Production Industry

The future of the transplant production industry looks rosy or bright with the advantages outlined and it is anticipated that the plug business will grow at a rapid rate in the–not–too distant future.

In forecasting developments in the transplant production industry, our Ministry of Agriculture, Forestry and Fishery, estimated that, the areas planted with plug transplants will be increased by 285% in vegetable crops, and by 131% in flower plants by 1994 as compared with those in 1990.

TABLE 20. Estimated number of transplants originated from micropropagated transplants used for horticultural industry.

Plants and crops	Estimated No.
Ornamental plants[z]	
Carnation	55,000 (10^3 plants)
Orchids	15,000
Gypsophila	9,000
Ornamental plants	5,300
Limonium .	2,500
Gerbera	1,600
Others	2,500
Total	90,900
Vegetables[y]	
Strawberry (mother plants)	17,510

[z]Data obtained from Tsurushima, 1990.
[y]Calculated data obtained from total area of strawberry production in Japan relative to that of Tochigi Prefecture.

References

Ando, T. and T. Ito (1988) 'Plug transplant production in the States and European countries' (in Japanese), in Vegetable Seed Production Ass'n. (ed.), Seed Production of Vegetables in Japan, Seibundo Shinkohsha, Tokyo, pp.175–186.

Berninger, Louis M. (1986) ' Forecasting plug growth', Proc. of the 1986 Natn'l Plug Prod. Conference, 68–87.

Ito, T. (1990) 'The status of plug transplant production' (in Japanese), in T. Ito (ed.), Transplant Production Technique in Vegetables, Ass'n Agric. Improv. Fund, Tokyo, 5–10, pp.55–82.

Kobayashi, G. (1992) 'Cell transplant system and growing medium' (in Japanese), in Present State and Perspective of Plug Transplant Production, Organizing Committee of Intern. Sympos. on Transplant Production Systems, Chiba, pp.12–23.

Koranski, David S. (1986) ' Plug update 1986', Proc. of the 1986 Natn'l Plug Prod. Conference, 88–97.

Larson,Roy A. (1986) 'Nutritional needs of Plug during germination and seedling development', Proc. of the 1986 Natn'l Plug Prod. Conference, 14–25.

Minist. of Agric., Forest., and Fishery (1991) 'Report of Transplant Production and Distribution' (in Japanese), Minist. of Agric., Forest., and Fishery, Tokyo, pp.1–110.

Natn'l Feder. of Agric. Co–op. Ass'n. (1991) (eds.) A Manual for Grafted Seedlings (in Japanese), Natn'l. Feder. of Agri. Co–op. Ass'n., Tokyo, pp. 1–33.

Shiokawa, M. (1991) 'N K plug system' (in Japanese), in Present State and Perspective of Plug Transplant Production, Organizing Committee of Intern. Symp. on Transplant Production Systems, Chiba, pp.44–51.

Tsukada, M. (1990) 'N K plug system' (in Japanese), in T. Ito (ed.), Transplant Production Technique in Vegetables, Ass'n Agric. Improv. Fund, Tokyo, pp.11–47.

AUTOMATION-CULTURE-ENVIRONMENT BASED SYSTEMS ANALYSIS OF TRANSPLANT PRODUCTION

K.C. TING and G.A. GIACOMELLI
Bioresource Engineering Department
Rutgers University-Cook College
New Brunswick, NJ 08903-0231
U.S.A.

ABSTRACT. Regeneration of plant materials under controlled environment has been an important process for commercial production or research purposes. The end products, and therefore the production requirements, vary depending on the primary goal of the process. One form of the end product is a transplant for delivery from its regenerative environment to another maturation environment for further growth and development. Such a transplant production system includes hardware and software components which can be categorized as: automation, culture or environment (ACE). Automation deals with information processing and task execution, culture relates to plant growth and development, and environment encompasses surrounding conditions of plants. Systems analysis is a technique emphasizing the understanding of the objectives of a system as a whole, and facilitating the derivation of solutions to pertinent questions deemed critical to the achievement of these system objectives. While the knowledge about individual components within a system is of great importance, the interactions among components frequently determine the workability of the integrated system. Systems level study ensures the investigation and understanding of compatibility among components. This article presents the mechanism of systems analysis, describes the essence of ACE as applicable to transplant production, and proposes an approach to applying ACE based systems analysis to transplant production planning, design and operation.

1. Introduction

Plant production is an important human activity which is undertaken for economic, environmental, educational/recreational, and scientific purposes. This production activity assumes a variety of forms and paces depending on the available technologies and resources. In other words, a plant production system may consist of components ranging from rather primitive to highly sophisticated. For many years, researchers have been studying various aspects of plant production in an effort to achieve better control of the process. The driving force of this research goal mainly comes from the necessity of producing certain quantity and quality plant materials according a given schedule in commercial situations. Because of this reason, research emphases have been placed upon developing technologies suitable for plant production within controlled environments.

Within a plant production system, plants are nurtured for a period of time when they undergo expected biological growth and development. The extent of this time period and the input and output materials of the system depend on the objectives that the system is organized to achieve. A particular type of production system, one which performs plant regeneration

K. Kurata and T. Kozai (eds.), Transplant Production Systems, 83–102.

processes, has recently attracted a significant amount of attention. The purpose of the regeneration processes is to increase plant quantity and deliver a portion of them from their initial environment for further growth and development in another environment. Like other types of plant production systems, the complexity of this transplant production system may vary over a wide range. However, for the purpose of analysis, its components may be conceptually categorized into three aspects: automation, culture, and environment (and thus the acronym ACE) (Giacomelli (1991), Ting and Giacomelli (1991)). Each aspect may vary in type, appearance or complexity but always present to some degree.

"Automation" deals with information processing and task execution related to a system's operation. By definition, automation describes a state of machine operation which requires no human supervision. However, in our discussion, the term automation represents the "extent of automation" within a system. Therefore, a total manual operation and a fully automated process are the two extreme cases of automation.

"Culture" includes the factors and practices which can directly describe and/or modify the biological growth and development of plants. The cultural factors, such as morphological and physiological conditions, of plants are important in monitoring plant growth. The cultural practices may include operations which directly alter plant morphological and/or physiological conditions such as multiplication, rooting, transplanting, watering, pesticide application, etc.

"Environment" encompasses the surrounding of plants, which consists of climatic, nutritional, as well as, structural/mechanical conditions. Environmental control has been perceived as a major engineering challenge in controlled environment agriculture research in the recent past. The climatic conditions which plants experience have been the major topics of research. The degree of complexity in controlling this climate is heavily dependent on the physical structure which separates the controlled environment from its surrounding natural environment. The interaction of the environment on both sides of the structure and the desired level of maintainability dictate the extensiveness of the control system hardware, software, and actuators. Since the objective of a plant production system is to facilitate plant growth/development and information/materials handling, it is obvious that the concept of environment needs to be expanded to interact with automation and culture components within the system.

When various components are integrated to form a system, their compatibility inevitably influences the workability of the entire system. Frequently, piece-wise forms of knowledge about individual components are abundant but the information about the interrelationships among the pieces is lacking. Occasionally, results of systems integration have been reported on specific plant production systems. While those reports carry valuable information on systems of specific purposes, the information is not exactly transferable nor necessarily applicable to other systems even similar ones. In order to facilitate the use of systems analysis technique, a somewhat "generic" analytical tool is needed. This proposed tool should be able to define/unify the representation of common parameters; identify/establish the relationships among necessary knowledge bases; modularize separate knowledge bases such that compatible building blocks for larger knowledge bases can be formed; and apply optimization techniques to the whole system under consideration. Furthermore, by the implementation of systems analysis techniques, the subject areas where new knowledge is needed will be highlighted so that future research and development efforts may be systematically approached. To carry the above idea one step further, the systems analysis and outcome should be computerized to encourage a broad interest of usage requiring minimal effort. The deliverables of such an endeavor would be computer software packages, of proven capabilities, with expansion possibilities, containing decision support systems.

In this article, the mechanism of systems analysis is presented, the essence of automation, culture and environment as applicable to transplant production is described, and an approach to applying ACE based systems analysis to transplant production planning, design and operation is proposed.

2. Systems Analysis

A system is a set of interrelated components organized to achieve certain goals. Strictly speaking, every system is a component in a larger system and every component is a system of smaller components. The components of a system may be physical objects or knowledge bases. Systems analysis is a technique structured to emphasize systems level thinking (Churchman (1968)). The performance of a system as a whole is of particular interest. This requires the understanding of not only the behavior of its components but also the interactions among the components. Many methods have been developed to facilitate information processing for conducting systems analysis. Two distinctly different types of methods, namely procedural and heuristic analyses, will be presented here because they are highly applicable to engineering and scientific systems such as transplant production.

2.1. PROCEDURAL ANALYSIS

Planning and design of a transplant production system require engineering and horticultural work. An engineering design must be guided by technical and economic principles (Stoecker (1989)). The technical aspect of the design is to follow scientific and engineering laws governing the relationships of the parameters involved. The economic concern is, in most cases, to optimize the return on investment. In addition, there are other intangible managerial, political, social, and aesthetical factors to be considered. The method of procedural analysis provides a step-by-step algorithm most suitable for dealing with quantitative information.

For a design task to be successful certain information about the system, such as processes, equipment, available resources, constraints, etc., is needed. This information can be gathered based on the experience from existing systems or by conducting experiments on "model" systems. A model system can be either a scaled physical model or a mathematical model. The main component of a mathematical model is a set of equations allowing the calculation of dependent variable values from given independent variable values. Two approaches are commonly used in developing these equations: (1) theoretical derivation based on physical laws and (2) equation fitting through experimental data, which involves purely number processing operations. The development of mathematical models is called mathematical modeling, and the conducting of experiments using the models is the process of numerical simulation. Both mathematical modeling and numerical simulation have been recognized as powerful techniques in studying physical and biological systems. The advantages of mathematical models are that they facilitate digital computer simulation and allow the application of mathematical optimization theories on the system's performance.

The major steps of numerical system simulation are as follows:

(1) formulate the problem (i.e. identify goals, resources, constraints, components, etc.)
(2) construct mathematical models of all the components in the system
(3) construct the system mathematical model which incorporates all the component models
(4) derive a solution from the system model (i.e. solve the equations for unknown parameters on the basis of given parameters; normally done by performing sequential and/or simultaneous calculations)
(5) verify and validate the model's workability and accuracy
(6) implement the model for production runs

The result of a design is considered workable if it performs the assigned tasks and satisfies the imposed constraints. For a design project, it is very likely to have more than one workable system. Therefore, a procedure for selecting the "best" design from a number of workable designs is needed. This is called the process of optimization, which can be described as follows:

(1) develop an objective function which depicts the relationship between the criterion of selection (Y) and the design parameters of the system (X_1, X_2, X_3,......, X_n), i.e.

$$Y = \text{function}\ (X_1, X_2, X_3, \ldots\ldots, X_n) \qquad (1)$$

(2) express the constraints to be satisfied by all workable designs in terms of equality or inequality formulas containing design parameters, i.e.

equality:

$$e_1(X_1, X_2, X_3, \ldots\ldots, X_n) = 0 \qquad (2)$$

....

....

$$e_m(X_1, X_2, X_3, \ldots\ldots, X_n) = 0 \qquad (3)$$

inequality:

$$i_1(X_1, X_2, X_3, \ldots\ldots, X_n) \leq L_1 \qquad (4)$$

....

....

$$i_j(X_1, X_2, X_3, \ldots\ldots, X_n) \leq L_j \qquad (5)$$

where, e_1 through e_m are equality constraint functions, i_1 through i_j are inequality constraint functions, and L_1 through L_j are real numbers.

(3) find a set of conditions (i.e. the values of X_1, X_2, X_3,......, X_n) that satisfy all the constraints and produce an optimum Y value in the objective function.

The optimum frequently means the maximum or the minimum depending on the situation being evaluated. Some examples of the optima are minimum cost, minimum time, maximum profit. The optimization algorithms to be used in step (3), after the problem has been properly formulated following steps (1) and (2), may include linear programming, dynamic programming, calculus method, search method, geometric programming, etc.

2.2. HEURISTIC ANALYSIS

Although the numerical treatment of information is an indispensable technique in systems analysis, there is a wealth of qualitative knowledge which is of equal importance. The utility of this type of knowledge becomes obvious when human experts are called to solve problems. Basically, this human expertise consists of two important parts: facts and heuristics capable of applying the facts in problem solving. In most cases, the rules used by human experts during problem solving can not be readily expressed in mathematical forms. Procedural analysis is therefore inadequate in dealing with this type of information. Human experts accumulate valuable expertise over a lengthy period of time. In order to facilitate the utilization of human expertise, there has been a substantial amount of research on the intricacies of heuristic analysis (Hayes-Roth et al. (1983), Buchanan and Shortliffe (1984), Forsyth (1984), Waterman (1986), Barrett and Jones (1989)).

The characteristics of heuristic analysis, which has been set forth by many researchers, include:

(1) expertise contained in a domain knowledge base (for example, the ability of identifying plant nutritional needs)

(2) heuristics in the form of symbolic reasoning exploiting the value of good

guessing

(3) separation of knowledge base and reasoning mechanism allowing the addition and modification of knowledge base without affecting the procedure of reasoning mechanism

(4) capability of reasoning under uncertainty or fuzzy reasoning

(5) self-knowledge (or knowledge about knowledge) facilitating explanation capacity

(6) capability of self-learning for improvement

The development of mechanisms for heuristic analysis is called knowledge engineering. A knowledge engineer's tasks are therefore to extract experts' knowledge, represent and organize the knowledge, and codify the mechanism on a computer for automation.

3. Computerization of Systems Analysis

Two primary advantages of using computers for information processing are their vast memory and high calculating speed. Systems analysis as described above normally requires the processing of a large amount of information. Since computers have already become indispensable tools in engineering design, it is only logical to access the full advantage of computers in performing systems analysis. Computerization inevitably involves the integration of electronic hardware with software algorithms. The advancement of microcomputers in recent years makes hardware selection a relatively easy task when compared to the acquisition of special purpose software packages. Specialized software for systems level investigation of plant production is especially lacking (Ting (1992)).

The procedural and heuristic analyses require different computer programming paradigms. The traditional algorithmic programming paradigm is suitable for procedural analysis which involves straightforward, deterministic, complex mathematical calculations. Statistical analysis of the effects of environmental conditions on plant growth is a good example of an algorithmic programming paradigm. For heuristic analysis, the expert system paradigm is more suitable. Expert systems are expected to solve problems at a performance level normally achieved with human experts. The systems contain expert knowledge and the method of using the knowledge, and are designed to deal with qualitative problems involving uncertainties. A computer program for solving diagnostic problems may be better developed following the heuristic paradigm. Both paradigms are important for developing information processing software for analyzing transplant production systems. Some research work is currently being done on integrating both paradigms to form hybrid systems (Engel et al. (1990), Perry et al. (1990)).

Some research work has been done on computerization of transplant production system scheduling and management. Simonton and Thai (1988) studied the production cycles of commercial micropropagation and developed a computer model to simulate the process. Optimization techniques were also applied to maximize the growth of production material and match production to demand. Simplified multiplication and transfer rates within and between the stock cycle and production cycle were used in the model. Humphries et al. (1989) further improved Simonton and Thai's model by considering the effect of contamination. Walker (1991) used the linear programming method to determine schedules for optimizing the profit of micropropagation operation under space and labor constraints. All of these three models were based on only the multiplication rates of plant materials with no consideration of environment and material handling factors.

Since 1986, the controlled environment agriculture research group in Bioresource Engineering Department (formerly, Biological and Agricultural Engineering Department), Rutgers University has been developing computer software modules for design and decision support purposes (Fang et al. (1992), Ting et al. (1991)). The project was initiated to develop the primary building blocks for analyzing plant production systems involving components of automation, culture and environment. At present, a total of 21 software modules have been developed in five subject areas: (1) structure and environment control, (2) crop production, (3) greenhouse layout and materials handling, (4) labor utilization and resource allocation, and (5) engineering economics. Although many of these modules were not developed specifically for transplant production systems, some of them can be applied with minimum or no modifications. The software modules of particular interest are shown in Table 1.

TABLE 1. Examples of computerized systems analysis modules

Resource allocation	a procedure for studying the profitability of plant production systems subject to resource constraints (Fang et al. (1990)). The constrained conditions and resources are the crop production schedule, market demand, space, labor, and financial budget. There is a database containing the information for determining the required resources and operating costs for growing various crops. The database also provides the estimated revenue from sales of the crops. An algorithm is available to determine first the feasibility of a given production plan under user defined conditions and then determine the quantities of crops to be grown in order to yield an optimum profit. The result of this algorithm may serve to optimize allocation of resources for year-round production.
Engineering economics	a software package for evaluating the life-time economic feasibility of greenhouse systems (Ting et al. (1989)). The software has a generic greenhouse engineering economic analysis algorithm plus a number of supporting subprograms for calculations of costs and revenues associated with specific design/operation conditions. The potential uses of this software are to (1) conduct case studies of an existing or proposed projects, (2) perform sensitivity analysis for alternative design/operations, (3) conduct parametric studies to cover ranges of input variables, (4) facilitate investment risk analyses, (5) incorporate the economic analysis program with other technical models, such as crop production, controlled environment, automation, etc., and (6) provide a means of evaluating quantitative values of an objective function in a systems optimization process.
Robotic plug transplanting	two software packages for simulating the performance and evaluating the economic viability of the robotic workcell developed by Ting et al. (1990a&b) and Yang et al. (1991). Software WORKCELL performs stochastic simulation of the productivity of SCARA-type robot based transplanting workcells (Ting et al. (1990c)). EART calculates the return on investment of the workcell using the traditional manual transplanting system as the basis for comparison (Fang et al. (1991)).

4. Automation, Culture and Environment in Transplant Production

4.1. AUTOMATION

4.1.1. *Fundamental Concepts*. The reasons for automation in various production situations can be readily found in many publications. Commercially available devices and algorithms for implementing automated systems are abundant. There are numerous situations where automation was thought to be a quick solution to a management problem. However, the inherent challenges facing the designers of automated plant production systems are: (1) the less structured working environment as compared to other types of manufacturing factories, (2) the dynamically changing size and shape of the work objects (i.e. plants), (3) the lack of reliable direct sensing devices for plant growth, (4) the diverse management practices of growers, (5) the desired low degree of complexity in system's operation/maintenance, and (6) the high level of difficulty in justifying a major capital investment. Therefore, the allowable error margin for these automation systems is extremely narrow. The concept of value engineering should always be emphasized in implementing automation for controlled environment plant production systems.

As stated above, an automated system consists of machines which are capable of performing activities with minimum human supervision. In order to accomplish this, the system will need to have one or more of the intrinsic capabilities as listed in Table 2.

Automation may be designed to assume various levels of sophistication based on the situation. Two major categories of automation are fixed automation and flexible automation. Fixed automation systems are cost-effective for mass production of standard items. An automatic seeder in a plug production operation is a machine based on the fixed automation concept. Flexible automation takes the economic advantage of fixed automation while responding to make-to-order batch processing. Computers integrated with generic machine tools make automation systems flexible. The key feature of a flexible automation system is the ability to perform various tasks by changing mainly the software and requiring minimum change on the hardware. A robotic plug transplanting workcell is an example of a flexible automation system.

In the past, mechanization has been implemented to substitute human physical power and achieve uniformity of operations (Garrett, 1985). Recently, computer and sensor technologies have been added to develop automation systems. Automation has become increasingly important in order to modernize operations of plant production and maintain market competitiveness. In a broad sense, the activities of automation may include: (1) implementation of computer technology and instrumentation, (2) sensing of biological and environmental properties, (3) modeling biological processes, (4) automating expert knowledge (including informational, procedural, and heuristics), (5) engineering biological systems, (6) developing strategies for environmental control, and (7) flexible automation, robotics and intelligent machines. Unless it is possible to have a fully automated system, the coordination between humans and machines must be taken into consideration in designing automation systems.

4.1.2. *Plant Handling*. The plant handling capabilities available with the production system is a major design consideration which is related directly to the time and monetary expenditures for labor, and the investment costs for the system. The importance of the mechanization system is directly proportional to the amount of handling and maintenance required to successfully nurture a crop from initiation to delivery. Assuming that the labor requirements can be estimated for every stage of the production, then preparations for materials handling of a highly labor intensive crop, or one with less labor demands could be completed. Each would be treated distinctly differently in the design, implementation and operation of its mechanization and automation system.

TABLE 2. Intrinsic capabilities of an automated system

Perception	an understanding of its surroundings. The capabilities of gathering, processing and interpreting information provide the system a basis for response. In a controlled environment plant production system, the physical quantities monitored include environmental parameters, time and spatial coordinates, crop conditions, and labor/machine status.
Reasoning	the capabilities of logical deduction and mathematical analysis for reaching a decision. This decision making process is necessary so that instructions can be properly issued to guide the system's operation. The automated reasoning power normally resides in software packages executed on computers.
Communication	the flow of information among system components. As the complexity of systems continually increases, the design of automation must consider the integration of many components within the system. The flow of information and the hierarchical order of commands must be carefully directed among the components.
Task planning and mechanical work	execution of control functions and physical work. The majority of physical work is materials handling. To automatically direct the operation of mechanical devices, the ability of task planning is essential. A robotic system is a good example for illustrating the importance of task planning. The execution of an algorithm for environmental control of plant production facility can be viewed as another type of task planning at work.

A primary consideration is whether the crop must be fixed at one location for the entire growth period, or whether it can be mobilized. Some transplant production systems, such as micropropagation, have traditionally required many operations for transferring plant materials from one container to another. Most of these systems are the results of direct emulation of research laboratory practices, which tend to hamper the development of automation. For systems like plug production, many plant handling operations have been automated. The automated devices for growth media preparation, container filling, seeding, watering have become very common. Transplanting of plugs is still predominantly manual work; however, a number of automatic transplanters are currently under development by various groups.

A description of the cultural needs of the plant which are directly serviced by hand labor must be initially completed. Information about the type of task, the frequency of the task during the production period, the duration of the task, the timing of overlapping tasks, and the supporting materials or machinery required for the task, must be collected and compared to the number, expertise and cost of the available labor force. Whether the task requires the crop to be transported, such as for input/output during transplant and harvest, or can be completed in the growing area of the greenhouse, is important. Transport and handling of the crop can become arduous and time consuming, which is a good reason why most crop production systems rely on the hand labor working on the plants in the growing area of the greenhouse bay.

4.1.3. *Layout design and labor planning*. Logistical design and labor management have become very complicated in plant production systems, because of the extensive amounts of information, parameters and interrelated processes involved. However, the operational success of each design for any given crop production system has never been assured by preliminary analysis and design. Most are built from experience gained by trial and error. Research work has begun, at Rutgers University, to develop a systematic procedure for evaluating the operation of mechanization systems, which utilizes extensive databases, dynamic modeling and simulation. The future capabilities of this computer model would include the ability to compare capital and operational costs, for a desirable cultural and mechanization system, with a fixed labor force, within a given production layout.

Finally many producers of transplants gain much pleasure in the successful completion of each crop. Systems which can improve working conditions and reduce the difficulty of the tasks, not only help to avoid unpleasant situations and crop loss, but also can improve the outlook and performance of the laborers during crop production.

4.2. CULTURE

4.2.1. *Fundamental Concepts*. The growth and development of plants are not only influenced by their genetic make-up but also by their biological conditions and the external influences to which they are exposed. The correlation between plant growth and its influencing factors is important in many ways. The qualitative understanding of the correlation guides plant growers' personal judgment while nurturing their plants. Presenting this relationship in a quantitative form makes the knowledge readily transferable. The evaluation of plant growth and adjustment of influencing factors can also be objectively conducted. When the relationship is given mathematically, powerful analytical/numerical techniques such as computer simulation and optimization may be used during planning and operation stages.

Plant response to environmental conditions has been the basis for developing environmental control strategies. The underlying principle is to provide appropriate environmental conditions to achieve a desired level of plant growth over a given period of time. In addition to a proper environment, plants need to be physically and chemically manipulated. Plant growing systems are designed to supply nutrition, water, and air, and provide physical support. Cultural practices are needed for plant initiation, maintenance, harvest, and termination purposes. The timing and location of cultural practices, as well as, the layout of production area affect the configuration and degree of automation of materials handling system. In some cases, traditional plant growth practices are modified to facilitate automation and environmental control.

One key component of an automated cultural system is the sensing capability of plant growth and development. Recently, machine vision techniques have been developed to measure the morphological and spectral features of plants (Ruzhitsky and Ling (1991)). The techniques invariably require high quality plant images as the basic data source. In many cases, the image acquisition is done under a very well structured environment. The environment designed for plant growth may not necessarily be suitable for this purpose. Unless robust image processing algorithms can be developed to ease the requirement of image quality, either sample plants need to be collected and delivered to a preassigned image acquisition location, or mobile image acquisition stations must be transported into the crop production area. To automate either of these methods in real-time operations would require further research and development work.

This is an example illustrating that automated cultural systems of the future may include more than just the components for directly sustaining plant life.

4.2.2. *Plant Cultural System*. Both the root environment and the aerial environment of the plant are directly affected by the characteristics and limitations of the cultural system. The cultural system is often described by the type of mechanism employed. As an alternative to classification by name, the general requirements and the general physical components which are common to many of the successful crop production systems are described in Tables 3 and 4.

Many of the biological processes required for the cultural systems are inherently served by the specific procedure in which water and nutrients are delivered to the plants. The root zone is one of the two major connections between the elements of the physical world outside the plant and the biological process within the plant. Processes such as transpiration, nutrient uptake, gas exchange, and organic waste product removal are enhanced or inhibited as a direct result of root zone watering. The characteristics of the water delivery system influence the controlling parameters of each of these processes, and therefore modulate their relative activity, and indirectly the growth and development of the plant as well.

Several important characteristics of the watering system include: (1) the nutrient solution transport/flow pattern within the root zone, (2) the buffering capacity of the root zone, (3) whether the nutrient solution is recirculated or "drained to waste", and (4) the frequency and duration of application. Each of these will be discussed below.

The nutrient solution can be delivered to the root system in a number of ways. The solution may be flowing or static, or require capillary means to reach the root system. It is important to recognize each specific application method has specific control parameters. Within a flowing system the frequency, duration and volume of flow must be considered. A static system may only have to consider the timing for addition of "make-up" water to the reservoir to assure the proper solution depth. A capillary system has the concerns of both flowing and static systems, and potentially is the least controllable.

The capability provided by the system to buffer the environmental fluctuations and provide a reservoir of water, nutrients and oxygen at the root zone is a primary concern. Growing systems employing a root zone medium typically provide water holding capacity, cation exchange capacity and pore space for oxygen. Alternatively, complete water-based systems with no media (except possibly a small seedling cube for transplant) may provide little or no buffering capacity. Additional precautionary measures should therefore be included in the design of non-buffered growing systems. These may include means of automated monitoring and control, which not only measure critical parameters and attempt to maintain their predetermined setpoint values, but also could provide a physical warning of impending or unexpected problems. The lack of buffering in the root zone, in general, requires that careful attention be given to the frequency of watering and the nutrient, pH and dissolved oxygen levels maintained in solution.

Recirculation (closed system) or "drain to waste" (open system) may be operated with either flowing or static delivery systems. Open or closed systems refer to the management of the nutrient water after application to the plants. Open systems dispose of drainage after each application, whereas closed systems provide for reuse of the solution. Generally open systems have large root zone buffering capacity for the plant, such as the soilless media within potted plants, which attempts to reduce watering frequency by increasing water and nutrient holding capacity. An open system with little buffering capacity would be wasteful of resources and impractical. Closed systems can benefit from the savings of not requiring a root zone buffer, but they require continuous concern for the spread of diseases and pests, and the constant monitoring and control of the nutrient content in the solution.

TABLE 3. General requirements of plant cultural systems

Plant support	In a free-standing, containerized plant crop, the container, filled with substrate, is typically designed to provide adequate support. In hydroponic or soilless culture the root zone of the plant may contain little or no substrate, and thus it may be incapable of securely anchoring the plant. The labor time necessary to support the crop can become large, and if possible, a means to simplify or eliminate this cultural task should be considered. Components used for crop support should not restrict access to the plants for maintenance (hand labor) tasks.
Contained root zone	Application of moisture and nutrients for absorption by the roots, generally require that the root zone be enclosed in some manner. The root zone enclosure is an important means to reduce plant stress imposed by relatively low air humidity, direct solar radiation, high temperatures, algae growth (affecting pH, nutrient and dissolved oxygen levels), and potential contaminants (disease, insects) within the root zone. Within most recirculating nutrient systems, enclosure and containment of the root zone is an absolute necessity to eliminate excessive losses of nutrient solution, and possible contamination of the nutrient solution storage tank.
Distribution of nutrients	Both macro- and micro-nutrients must be dissolved within a water solution and delivered to the root system of the plants. The specific types, proportions and sources of the nutrient salts are plentiful. The means of distribution, such as, by spray, drip, flow, drench, fog, flood, etc. has lead to the generic name of many of the cultural systems.
Root aeration	Dissolved oxygen within the water solution provided to the plant root system is critical for the successful production of most all crops. Several means to provide an oxygen rich nutrient solution include: having open channel flowing water, waterfalls and splashing within the channels, pumping air into the flow path or storage tank, or suction of air (by venturi openings) into the distribution pipes. The practice of raising and then lowering the water level around the root zone media provides a cleansing and replenishing mechanism for gasses within the media. Gasses are forced out from the media on water level rise, and atmospheric gasses are drawn into the media as the water level declines.
Environmental conditions	Aerial and root zone environmental conditions, such as radiation, temperature, humidity and carbon dioxide, are modified, to promote plant growth, limit plant stress, and to restrict predators within an environment which is protected from the natural climate conditions.

TABLE 4. General physical components of plant cultural systems

Growing container/fixture	The container to hold the root zone media or the fixture to support the bare-rooted plant can provide the confinement for the roots, the nutrient/oxygen/root interface, the root foundation, and the darkened environment. These may be troughs, tubes, channels or gutters made of plastic film, rigid plastic, aluminum, concrete, or some combination. They may be a series of individual rows, which are grouped on top of containers such as benches, or they may be arrayed on the concrete floor of the greenhouse. They must provide sufficient volume for the growth of the roots, and be capable of conveying the nutrient solution to contact the root system, yet provide good drainage to remove excess water and assure adequate gaseous exchange. If reusable, then they must be easily cleaned and sterilized.
Nutrient solution	Nutrient solution may be previously prepared and stored in bulk, or concentrated nutrients may be combined with fresh water only as needed. A recirculating system would require a tank for holding the nutrient solution returned after each watering, while an "on demand" procedure requires a proportioning/mixing system to incorporate the nutrients with fresh water as needed. In a recirculating system the storage tank holding capacity, its safety from leakage, cleanliness and a secure location must be considered. The "on demand" nutrient solution source must assure accurate proportioning and sufficient volumetric flow, but requires much less storage volume.
Pump and plumbing	A means to move the solution from the source and distribute it uniformly to each plant or group of plants is provided by pumps and piping systems. Plastic components are generally chosen because of the corrosive nature of the nutrient solution to metals. Pumping distances, system pressures, flow volume, duration and frequency of application, all combine to determine the pump size and its energy requirement.
Collection system	A means is provided, primarily in recirculating systems, to collect excess nutrient water drained from the plants. The drainage is returned to storage within covered channels or pipes either by gravity flow or by pumping. In some cases the drain and the distribution system are one in the same. A collection system is not utilized for "on demand" nutrient systems. The typical strategy is to provide the precise amount of nutrient solution with only a small percentage of overwatering, or "drain to waste".

TABLE 4. (cont'd)

Controls and monitors	Control of plant cultural systems may be as simple as a time clock to activate the nutrient solution pump on a fixed frequency. They can become as complex as a computer monitoring and control system which automatically adjusts the nutrient solution concentration, frequency and duration of application to the existing environmental conditions and physiological age of the plant. Although continuous monitoring allows for more automated control of the nutrient solution, manual checking of the solution pH, conductivity and temperature can be sufficient in some circumstances for successful crop growth. Automatic addition of fresh "make-up" water is a valuable safety precaution against system failure due to leakage or evapotranspirational loss of water. Alarms to signal problems of low water level, pH, conductivity, and environmental abnormalities provide additional security.

Frequency and duration of nutrient solution flow is dependent upon the characteristics of the cultural system listed above. Flow pattern and buffering capacity at the root zone, and procedure of open or closed circulation, all must be considered. Most important is the characteristics of the plant within the cultural system. These are dependent upon the physiological age, environmental conditions, and the general water demand of the species under production.

4.3. ENVIRONMENT

4.3.1. *Fundamental Concepts.* The types of plant environment which can be controlled or modified are radiative, thermal, gaseous, liquid, nutritional, physical, chemical, and biological environments. In many cases, plants require different environmental conditions during different stages of growth. The sequential change of environmental requirements can be achieved by either varying the environment of a fixed location over time, or transporting the plants out of and into locations of different environments. The latter will then directly affect the layout of the production area and the pattern of internal transport.

As mentioned above, the purpose of environmental control is to regulate crop growth and development within the environment. Three basic components to be considered in establishing the control strategy for any environmental factor are its set-point, range, and time duration. The selection of these components is heavily dependent on the cultural information available to the strategy builder. For a number of reasons, it is sometimes necessary to design a control system which is allowed to deviate from the desirable states. The available technologies and affordable costs are two of the limitations commonly encountered. Therefore, the information on how plants respond to unfavorable environmental conditions is also valuable in the design considerations.

During crop production, the environmental control is done by first comparing the controlled and the desirable states. If the discrepancy is greater than the predetermined tolerance, some actions will be taken to attempt to force the controlled state towards the desirable state. The controlled state can be either a measured (feed-back control) or a predicted (feed-forward control) value. Many theories have been developed for establishing control actions. Almost all of the theories require the understanding of dynamics of the system being controlled. Numerical simulation has been a proven technique for predicting the behavior of a physical system; especially when computers are used as part of a control system. A valid numerical model of the controlled system is necessary for the simulation task to be reliable.

Recently, attention has been directed towards minimizing plant production systems' impacts on the natural environment. Plant production systems interact with the natural environment by receiving resources, delivering products, and discharging byproducts. Each of these interactions will affect the natural environment to some extent. The adversity and benefit of the plant production system to the natural environment deserve a systematic evaluation. Some research activities have been under way to develop "totally enclosed" controlled environment plant production systems. The development of these enclosed systems, and the findings from their impact evaluations, will inevitably influence future design criteria.

4.3.2. *Plant Environment*. Environmental conditions, primarily air temperature, solar radiation, vapor pressure deficit and carbon dioxide, determine the plant microenvironment. This can subsequently be divided into the top, aerial portion, and the bottom, root zone portion, of the plant. These are discussed in Tables 5 and 6.

Automated monitoring of the plant environment has become readily available for the parameters of temperature, vapor pressure deficit, radiation duration and intensity, atmospheric CO_2 concentration, and nutrient solution characteristics of pH and total salts. The information, to date, has been primarily utilized for maintaining tighter controls on the desired setpoint air temperature and for providing a clear, daily description of the plant microclimate. However with the desire to reduce energy consumption, and more recently the benefit of manipulating plant development without hormonal applications, this flow of information for control purposes is becoming somewhat more indispensable.

The environmental control system as described, must provide the desired setpoint values, and in the process help reduce energy consumption. These should not occur at the expense of maintaining a uniform environment throughout the entire production area. Uniformly distributed environmental conditions will help assure that the plants will respond to the growth enhancement (i.e. temperature and nutrient regimes) and control practices (i.e. hormonal and pesticide applications) in a similar manner. Overall crop quality and uniformity of production can therefore be improved.

TABLE 5. Plant microenvironment - aerial portion

Radiative environment	the quantity of solar or artificial energy, quality or spectral distribution of the radiative energy, and duration of exposure. Radiation may be absorbed, transmitted or reflected as it travels through the atmosphere and the glazing before it reaches the plant. The "useful" portion (PAR waveband) which does arrive at the surface of the leaf is mostly dissipated as heat with only a small percentage utilized for photosynthetic activity. The net photosynthetic capacity of the leaf can further be reduced by photorespiration induced under high radiative intensities (and typically high leaf temperatures). The relative proportion of specific wavebands (red/far red) can affect morphological developments. In general growth and development is (within limits) proportional to the duration of exposure to radiation, assuming no additional environmental limitations. The light environment can be modified by the type of glazing (intensity and quality), the type of artificial lamp(s) (quality), and the mechanical shading procedure (intensity reduction).

TABLE 5. (cont'd)

Gaseous environment	water vapor, oxygen, carbon dioxide, undesirables such as ethylene, air pollutants (sulfur dioxide). The leaf energy balance is primarily modulated by evaporation of moisture at the leaf, which has been transported through the plant. In the evapotranspiration process water is taken up by the plant, and indirectly the rates of CO_2 and oxygen exchange at the leaf are modified. Evapotranspiration and gas exchange are both affected by the rate of ambient air movement. Evaporation is directly affected by surrounding moist air properties. Leaf and air temperature modulate rates of photosynthesis, respiration, assimilation, evapotranspiration, and nutrient uptake.
Thermal environment	temperature and humidity. Temperatures can be modified by heating, convective and evaporative cooling, and/or shading. Moisture removal from the air (dehumidification) is essentially impractical, except by dilution with outside air. The local ambient climate due to geographical location provides the basis for the particular control strategy and systems utilized within the greenhouse.

TABLE 6. Plant microenvironment - root zone portion

Liquid/gaseous environment	proportion of water and gases; availability distribution and duration of water solution. Root zone and nutrient solution temperatures directly affect water uptake and gas exchange. Nutrient solution flow rates and duration of flow affect gas exchange by modifying the boundary flow layer at root. A smaller boundary layer produced by higher flows enhance gas exchange. Increased time of exposure to oxygen rich nutrient solution, resulting from adequate flow duration, will also enhance water and gas exchange. Osmotic pressure (proportional to salt concentration) of nutrient solution modulates water uptake and evapotranspiration capacity of the plant. Bench, floor or nutrient solution storage tank heating systems are readily available to control minimum root zone environment temperature. Cooling of the root zone is more difficult. It is best achieved by preventative measures such as shading from direct radiation, reflecting films to cover the root zone, having large root zone buffers and insulating the troughs and containers, or placing the nutrient storage tank below grade. The nutrient solution can be enriched with dissolved oxygen by agitation or splashing of the flowing solution.

TABLE 6. (cont'd)

Nutritional environment	mineral content, salinity, and pH of the growing medium and nutrient solution. Nutrient uptake is dependent on temperature and proportional to the concentration of the nutrients. The availability of specific elements is affected by pH. The long term increase of nutrients within the growing medium (for example salts from surface water evaporation) can negatively affect plant growth, even though the nutrient solution is maintained at the prescribed levels.
Physical environment	physical properties of root zone medium include cation exchange capacity, water holding capacity, and density. Air temperature can affect root zone temperature quickly or with a lag time depending on volume and the heat transfer properties of the root zone medium. The reactive or inert nature of the medium will determine buffering capacity for nutrients. Physical properties such as density and structure of the medium will influence water holding capacity and the flow paths through the root zone.
Chemical/biological environment	waste products removal, beneficial and harmful organisms. The root zone is alive with biological organisms. The primary concern is to minimize the potential for epidemic levels of disease organisms or pests. Careful sanitation and/or exclusion are the first defenses against such problems. Exudates of organic compounds from the plant root system may be diluted and distributed by the nutrient solution throughout the system.

Examples of components of environmental control systems, listed with some of their attributes, include: heating systems which distribute the heat at lower temperatures (such as bench or floor heating systems), or which can modulate the temperature more quickly (such as overhead hot water or hot air systems); ventilation systems which incorporate controlled shading systems to reduce the radiant load when possible, and which match fan air flow capacity to inlet openings to improve cooling uniformity; evaporative cooling systems with shorter air travel distances from inlet to fan to minimize the air temperature rise; greenhouse structures which minimize the amount of shading of solar radiation; and, supplemental lighting systems which effectively distribute the light energy over the entire plant canopy.

5. ACE Based Systems Analysis

In order to facilitate this ACE based systems analysis, we propose to develop a computerized tool for performing systems analysis on the mass production of transplants under controlled environments, which integrates the components of automation, culture and environment. This tool will be a computer software package containing various information modules and functional modules. The information modules will contain the necessary databases related to components of transplant production systems. The functional modules are the algorithms which access and process the data in the information modules for accomplishing the desired objectives. The desired features of this systems analysis tool are shown in Table 7.

TABLE 7. Desired features of the ACE based systems analysis tool

Completeness	the capability of dealing with a large number of situations for a given production system. Although it is very unlikely that an analytical tool will be able to handle all the imaginable situations of a particular production system, the ultimate goal is to include algorithms and databases for handling all practical cases. This may mean that innovative schemes for categorizing and representing production system characteristics need to be established. It is highly desirable to have ways to unify and simplify the description of any production system, so that there is good probability for a manageable size tool to be able to analyze a large number of situations.
Consistency	the assurance of reaching non-conflicting conclusions for similar conditions. This tool will be mainly used as a decision support system. The result produced by this system should not be contradictory. One way to avoid this is to maintain the consistency of the data and algorithms involved.
Flexibility	the ease of customized configuration of this analytical tool. The modularization of computer programs will allow the software to be packaged in different sizes and forms. Due to the difference in emphases and interests, and possibly the constraints of computer resources, not every user needs to have the access to the entire systems analysis tool. This flexibility in software module integration is definitely valuable in order to encourage a broad application of the tool and not over-burden a user.
Expandability	the readiness for future modification and expansion. The knowledge and technologies of transplant production will continue to develop. The capability of this systems analysis tool must improve when new information becomes available. Therefore, this tool should be designed such that its capability is expandable, even by the users.
Universality	the non-limiting applicability to different production systems. This tool should be designed with an underlying principle that it will be able to analyze all different types of transplant production systems, such as for plugs, cuttings, embryos, micropropagules, etc.

One possible organization of the information modules is in multidimensional matrix form with windows located at each element. The columns of the matrix will be separated into three categories: automation, culture, and environment. The columns within each category will be further divided into sub-groups, and sub-groups within each sub-group, depending on the level of sophistication. Some examples of sub-groups are given in Table 8.

TABLE 8. Examples of column grouping in the organization matrix

Automation	Perception	Sensors	Camera
			Thermocouple
			Pyranometer
	Reasoning	Hardware	Microcomputer
			Data logger/controller
		Software	Generic
			Customized
Culture	Plant support	Substrate	Soil mix
			Rockwool
		Container	Plug tray
			Test tube
	Root aeration	Hydroponics	Open channel flowing water
			Aerated storage tank
Environment	Radiative	Solar	Quantity
			Duration
		Artificial	Quantity
			Quality
	Thermal	Temperature	Set-point desired
			Actual measured

Each row of the matrix will represent a specific transplant production system. These rows may also be grouped in a way similar to the grouping of columns. For example, there may be different types within the categories of plug production system, micropropagation system, cutting system, embryogenesis system, etc.

The intersection of a row and a column is the element which contains as much detail information as practical relating the corresponding ACE component (column) within the particular production system (row). The information contained in each element can be further separated into groups and represented by windows. For example, one window may show the potential physical devices required, and another window may provide their cross reference in terms of where else a particular parameter of these devices appears. This cross reference scheme is important in representing the relationships among parameters.

With the establishment of the information matrix, functional modules may be developed to perform specific tasks of systems analysis for various transplant production systems. These tasks may include: (1) investigation of workability, (2) estimation of productivity, (3) evaluation of economic feasibility, (3) optimization of a particular system, (4) comparison among different production systems, etc. Both procedural and heuristic analysis techniques will need to be implemented in developing these functional modules.

6. Conclusions

Systems analysis is both a way of thinking and a problem solving technique. It considers a system as a whole; therefore, the behavior of individual components, as well as, their interrelationships are investigated simultaneously. Recent social, economic, and technological developments have increased the complexity of controlled environment plant production systems. In the recent past, the production of transplants within controlled environments has become an important commercial and research activity. It is important that a transplant production system be planned, designed, and managed as an integrated system. The components to be included in the transplant production systems can be categorized as automation, culture, and environment. The advancement made to improve any part of this production system will most likely affect the other parts. Systems analysis is a well developed technique which can be applied to evaluate transplant production for drawing conclusions at the systems level. This will improve and promote the application of plant production systems within controlled environments.

This is an information age; but frequently, information needs to be processed into knowledge and presented in an effective format to become useful. Therefore, information alone will not necessary be helpful. As a matter of fact, too much "raw" information can do very little more than become overwhelming. However, information together with the ability to make use of the information effectively will be a great advantage in a decision-making process. The ACE based systems analysis concept and its computerized form are proposed as an approach to accomplish this goal.

7. References

Barrett, J.R. and Jones, D.D. (Editors) (1989) Knowledge Engineering in Agriculture, American Society of Agricultural Engineers, St. Joseph, MI.

Buchanan, B.G. and Shortliffe, E.H. (1984) Rule-Based Expert Systems - The MYCIN Experiments of the Stanford Heuristic Programming Project, Addison-Wesley Publishing Company, Reading, MA.

Churchman, C.W. (1968) The Systems Approach, Dell Publishing Co., Inc., New York, NY.

Engel, B.A., Beasley, D.B. and Barrett, J.R. (1990) 'Integrating multiple knowledge sources', Transactions of the ASAE 33(4):1371-1376.

Fang, W., Ting, K.C. and Giacomelli, G.A. (1990) 'Optimizing resource allocation for greenhouse potted plant production', Transactions of the ASAE 33(4):1377-1382.

Fang, W., Ting, K.C. and Giacomelli, G.A. (1991) 'Engineering economics of SCARA robot-based plug transplanting workcell', ASAE paper no. 917026, American Society of Agricultural Engineers, St. Joseph, MI.

Fang, W., Ting, K.C. and Giacomelli, G.A. (1992) 'Computer software development for greenhouse design and management', The 4th International Conference on Computers in Agricultural Extension Programs, January 28-31, Lake Buena Vista, FL.

Forsyth, R. (Editor) (1984) Expert Systems -Principles and Case Studies, Chapman and Hall Ltd., London.

Garrett, R.E. (1985) 'Social impacts of automation in agriculture', In Proceedings of the Agri-Mation™ 1 Conference & Exposition, February 25-28, Chicago, IL, 15-20.

Giacomelli, G.A. (1991) 'Introduction to the horticultural aspects of hydroponic crop production', In Proceedings of the 12th Annual Conference on Hydroponics, April 12-14, Pheasant Run Resort, St. Charles, IL, 38-49.

Hayes-Roth, F., Waterman, D.A. and Lenat, D.B. (Editors) (1983) Building Expert Systems, Addison-Wesley Publishing Company, Reading, MA.

Humphries, S., Simonton, W. and Thai, C.N. (1989) 'Computer aided management of plant tissue culture production', ASAE paper no. 89-4050, American Society of Agricultural Engineers, St. Joseph, MI.

Perry, C.D., Thomas, D.L., Smith M.C. and McClendon, R.W. (1990) 'Expert system-based coupling of SOYGRO and DRAINMOD', Transactions of the ASAE 33(3):991-997.

Ruzhitsky, V.N. and Ling, P.P. (1991) 'Machine vision based plug quality estimation for robotic transplanting', ASAE paper no. 913505, American Society of Agricultural Engineers, St. Joseph, MI.

Simonton, W. and Thai, C.N. (1988) 'Plant tissue culture production planning using computer simulation/optimization', Transactions of the ASAE 31(5):1616-1622.

Stoecker, W.F. (1989) Design of Thermal Systems, Third Edition, McGraw-Hill Book Company, New York, NY.

Ting, K.C. (1992) 'Mechanization, automation and computerization for greenhouse production', HortTechnology (in press).

Ting, K.C., Dijkstra, J., Fang, W. and Giniger, M. (1989) 'Engineering economy of controlled environment for greenhouse production', Transactions of the ASAE 32(3):1018-1022.

Ting, K.C., Fang, W. and Giacomelli, G.A. (1991) 'Decision support for integrated greenhouse production systems', in J. Prakash and R.L.M. Pierik (eds.), Horticulture-New Technologies and Applications, Kluwer Academic Publishers, Dordrecht, pp 293-298.

Ting, K.C. and Giacomelli, G.A. (1991) 'Systems integration of automation, culture and environment within CEA', In Proceedings of the 1991 Symposium on Automated Agriculture for the 21st Century, December 16-17, Chicago, IL, 518-526.

Ting, K.C., Giacomelli, G.A. and Shen, S.J. (1990a) 'Robot workcell for transplanting of seedlings part I-layout and materials flow', Transactions of the ASAE 33(3):1005-1010.

Ting, K.C., Giacomelli, G.A., Shen, S.J. and Kabala, W.P. (1990b) 'Robot workcell for transplanting of seedlings part II-end-effector development', Transactions of the ASAE 33(3):1013-1017.

Ting, K.C., Yang, Y. and Fang, W. (1990c) 'Stochastic modeling of robotic workcell for seedling plug transplanting', ASAE paper no. 901539, American Society of Agricultural Engineers, St. Joseph, MI.

Walker, P.N. (1991) 'Linear programming for a micropropagation enterprise', ASAE paper no. 91-7533, American Society of Agricultural Engineers, St. Joseph, MI.

Waterman, D.A. (1986) A Guide to Expert Systems, Addison-Wesley Publishing Company, Reading, MA.

Yang, Y., Ting, K.C. and Giacomelli, G.A. (1991) 'Factors affecting performance of sliding-needles gripper during robotic transplanting of seedlings', Applied Engineering in Agriculture 7(4):493-498.

8. Acknowledgments

New Jersey Agricultural Experiment Station Publication No. J-03232-01-92, supported by State Funds.

ISSUES IN ROBOTIC SYSTEM DESIGN FOR TRANSPLANT PRODUCTION SYSTEMS

WARD SIMONTON
The University of Georgia
Georgia Experiment Station
1109 Experiment Street
Griffin, GA 30223-1797 USA

ABSTRACT. Design considerations are presented for the development of agricultural robotic handling/processing equipment with emphasis on nursery and greenhouse applications. Topics include human and resource allocation in automation systems, the impact of nonuniformity, sensing to modify and control robot actions, accessing products for processing, economic issues, and transplant system applications.

1. Introduction

A number of potential automation incentives and benefits for the U.S. commercial nursery and greenhouse industries have been discussed in recent years. Among these are to (1) improve global market competitiveness through increased productivity (Bolusky, 1989); (2) improve quality for a marketing advantage by providing consistent grading and marketing procedures (USDA, 1989); (3) increase standardization throughout the industry and enable faster machine development as the needs of the industry change (Simonton, 1991a); (4) enhance commercial operation stability by reducing labor turnover, reducing layoffs and rehires, and promoting worker retraining (Bolusky, 1989); (5) improve safety of the workplace by reducing chemical exposure to the workforce (Moncaster, 1985); and (6) improve production practices as related to the environment by reducing the amount of chemicals required for production and eliminating run-off (USDA, 1989). Some or all of these apply to various countries which have significant production of nursery and/or greenhouse crops.

Other incentives for automation are based on rapidly changing technology. Faster and less expensive computers, more reliable computer networks, improved sensing techniques including machine vision, and improved motors/drives combine to give us better tools with which to design, test, and produce new automation equipment. Incorporating these tools with innovative software may produce for the nursery and greenhouse industries truly cost-effective and productive automation equipment which can adapt to the changing needs and conditions of commercial operations. Robotic systems are a form of automation with particular promise for the commercial nursery and greenhouse. Robots offer flexibility, reliability, and are proven tools in numerous industries.

The goal of automation in agriculture, including nursery and greenhouse systems, should be to provide the most efficient operation that is practical by optimizing the combination of human resources and machine resources using productivity, economic return, and workforce satisfaction as the performance criteria. The primary challenge is to develop

K. Kurata and T. Kozai (eds.), Transplant Production Systems, 103–115.

automation including robotics which can perform in plant production systems with the typical variations and uncertainties among crops and environment. Research is underway to establish methods for design of machines, e.g. robots, which can autonomously or semiautonomously operate in uncertain environments. At the present time, however, there is no formal methodology for design of such automation systems. This lack of standard design procedures opens opportunities for research with significant technical challenges for agricultural robotic systems development. This paper presents design issues important for the development of agricultural robotic handling/processing equipment with emphasis on nursery and greenhouse applications. Topics include human and resource allocation in automation systems, the impact of nonuniformity, sensing to modify and control robot actions, accessing products for processing, economic issues, and transplant system applications.

2. Agricultural Robotic System Design Issues

2.1. Human and Machine Resource Allocation

2.1.1. *Manufacturing and Agriculture Relations.* Human activity in manufacturing is shifting from skill-based and physical functions such as machining and welding to knowledge-based functions such as machine cell supervision and quality control (Winter, 1986). Robotic systems have become prevalent in the automotive, electronic, general fabrication, and other industries for conducting repetitive, precise, and/or hazardous tasks. However the appropriateness of using manufacturing as the future operational paradigm of agricultural production and processing is subject to question. Significant differences between manufacturing and agriculture exist in the nonuniform and perishable nature of products, workforce utilization, economic system, environmental impact, and level of production technology.

Fortunately regarding automation, issues examined in manufacturing research and development can provide useful insights for agriculture. One of these issues is allocation of human resources and machine resources, including computers, for making decisions and performing tasks. All automation systems regardless of the level of autonomy ultimately function as a combination of human and machine. Agriculture, relatively new to robotics, has applications significantly more challenging to automate than those applications commonly addressed for years in other industries. Therefore the near-optimal combination of human and machine resources may be pivotal for successful agricultural robotics applications.

In the traditional manner of design of automation systems which consist of humans and machines, individual decisions and tasks are divided *a priori* based upon such factors as intelligence, frequency, and supervisory requirements (Brooks (1980); Sheridan (1985)). In this framework human responsibilities may include planning actions, teaching operations, initiating execution, and intervening during operation errors. Machine responsibilities may then include interpreting instructions, assimilating inputs, relaying messages, and performing physical task(s). This design technique has proven successful for automated systems on aircraft, in chemical plants, and in manufacturing. Computer models have been developed for describing decisions and tasks and thus assist in optimizing system design (Brooks, 1980). Various levels of human and robot interaction have been defined and analyzed (Parsons and Kearsley, 1982).

2.1.2. *Artificial Intelligence Considerations.* A different approach of automation system design combines artificial intelligence, operations research, and control theory in a mathematical model for use in formulation and analysis. The structure of one such system as developed by Saridis (1989) is hierarchical and is based upon his intelligent control system theory of "increasing precision with decreasing intelligence." The Saridis framework, while emphasizing artificial intelligence, is appealing because it allows for flexibility concerning human versus machine responsibilities, plus it acknowledges the distinctions between upper level decisions (and perhaps tasks) such as planning and lower level tasks (and perhaps decisions) such as feedback control.

Developments in artificial intelligence will certainly influence the design of automation systems in the coming decades and thus alter man-machine relationships. The degree of this influence is unknown, however, due to the relatively slow progress of the field. One of the particular shortcomings of artificial intelligence software which affects automation systems is the inability to deliver real-time performance in dynamic environments. Typical artificial intelligence systems follow a sense, model, plan, and act framework which is computationally intensive yet unable to handle significant uncertainty and dynamic conditions.

Such shortcomings in mainstream artificial intelligence research has led to development of novel techniques such as "behavior producing modules" which *react* rather than *plan and execute* (Brooks, 1991). Key advancements in the Brooks work include incorporating perception, modelling, and planning facilities within each subsystem and linking input states (e.g. sensors) to output states (e.g. actuators) with short, parallel connections. This design approach has produced small robots with insect-like behavior. It is likely that this framework will supplant neither traditional man-machine system design nor more computationally intensive artificial intelligence approaches, yet will strongly influence each. Regardless of the advancements of artificial intelligence, humans will continue to play a vital role in agricultural production systems for many years. The type and level of human activity will have a significant affect on the three performance criteria listed in the introduction as productivity, economic return, and workforce satisfaction.

2.1.3. *Workforce Management.* For many years the design of manufacturing operations in the United States was based on a concept called Taylorism, the use of technology and human resource allocation to limit manufacturing to a number of simple operations requiring little skill or initiative. Taylor (1907) envisioned systems "which the workman is told in minute detail just what he is to do and how he is to do it, with any improvements which he makes upon the orders given to him being seen as fatal to success." While the de facto philosophy in the past, Taylorism has now been discredited by social scientists and industrial engineers because it dehumanizes employees, ignores employee motivations other than financial, increases labor costs through inefficient management structures, reduces the ability of companies to respond quickly to changes in the marketplace, and fails to fully utilize the resources of a workforce (Kidd, 1988).

An alternative philosophy is based on organizing the workforce in groups that control and are responsible for a set of tasks (Kidd, 1988). Such organization requires consideration of ergonomics, decision support systems, and opportunities for worker training and advancement. In an analysis of 24 U.S. manufacturing facilities, adoption of new automation technology was more successful in terms of productivity, cost reduction, and quality improvement in the companies which had the more flexible workforce organizations (Carlopio et al., 1988). Such flexibility was considered to be represented by the breadth of tasks assigned to individuals along with degree of responsibilities.

The common threads of the above ideas for automation system design are resource development, resource allocation, and resource management. Resources may be humans, computers, simple machines, or six degree-of-freedom robots. Agricultural production systems, such as those in large commercial nurseries and greenhouses, are complex and defy a standard procedure for decision and task allocation. Given the particular needs of an application, several questions must be addressed. What are the general roles of the computer(s), sensor(s), manipulator(s), and human(s)? To which should specific responsibilities be allocated? How should people and equipment be most effectively combined?

2.2. Nonuniformity

Robustness may be defined as the capability of a system to operate within particular performance criteria in the presence of noise, unknown loads, or other external disturbances. In an agricultural robotics application, sufficient robustness is a key concern due to the many potential variable parameters which may impede system effectiveness. Nonuniformity, or variability, refers to random differences in features among relevant entities. Nonuniformity may exist in many different facets including the product to be processed, the layout of the work area, the environment, and the required throughput. Baylou (1987) noted that agricultural robots will frequently need to perform tasks with fragile biological objects in environments of dirt, dust, high moisture, and uncertain spatial parameters. Some researchers believe that robots will never be able to successfully cope with totally unexpected situations including specific difficulties concerning decisions requiring real-time modification as circumstances, environment, etc. change (Parsons, 1988).

In the nursery and greenhouse industries there exists tremendous diversity among commercial operations in production layout, trays/flats/pots, crops, and of course even within crops. This diversity makes cost-effective, functional automation difficult and time consuming to develop (Moncaster, 1985). The great majority of current industrial robotic applications, for example spray painting and spot welding, are performed with known items in a known position/orientation. In the nursery and greenhouse frequently this luxury is not practical. Even while performing the most monotonous duties, workers are making decisions and modifying their tasks to fit the specific needs of the plants. Whether handling different pot spacings on a bench or in a field, or preparing cuttings for propagation which may vary 300% in stem caliper and 600% in stem length, people have the ability to adjust their actions based upon the requirements of the situation. Neither fixed (dedicated equipment) nor flexible (robotic) automation systems are generally equipped to modify their own operations to this degree of robustness.

To minimize the degree of nonuniformity that an automatic system must operate within, several steps should be considered. The task(s) to be automated can be constrained to limit certain variations. The structure of the working environment can be improved to remove certain variations. Task(s) upstream from that to be automated can be modified to reduce variations. Design goals should include minimizing required complexity of sensing, decision making, and mechanical manipulation while maintaining robustness.

2.3. Sensing

2.3.1. *Role of Sensing in Robotic Control.* A common feature of most robots is the ability to repeat a preprogrammed sequence of operations; however, for robots to operate effectively in a changing or uncertain environment, the machines must be equipped with sensors (Koren, 1985). Regardless of success in limiting nonuniformity through system design, sensing will be required in most agricultural robotics applications to measure key parameters and allow the automated system to accommodate variable conditions. Sensors may be necessary for such measurements as proximity, position, force, weight, area, volume, shape, color, or structure. Sensing can be divided into proprioceptive (manipulator position, joint torque) and exteroceptive (product position, product color) information (Baylou, 1987). Measurements may be contact (tactile, force) or noncontact (proximity, vision).

A key activity in robotics research is examining methodologies of coordinating information from various sensors to control robotic actions (Ghallab, 1985). Two of the more technically interesting agricultural robotic applications related to sophisticated sensing are sheep shearing and citrus harvesting. For controlling the height of a cutter assembly above a sheep skin surface, both capacitance and inductance sensing was incorporated into a robotic control system (Trevelyan, 1985). Also resistance contact sensing was used to detect skin penetration and initiate withdrawal of the cutter assembly. For controlling the trajectory of a robotic citrus harvesting manipulator (Harrell, 1990), color vision information for locating oranges within a tree canopy and ultrasonic data for range detection was utilized. Both the sheep shearing robotic system and the tree fruit harvesting system successfully combined exteroceptive data with proprioceptive data to accommodate variability within the robot workspace; however, both systems had to incorporate heuristics to overcome particular sensing and control difficulties.

2.3.2. *Nursery and Greenhouse Applications.* Recent developments in automation systems for nursery and greenhouse plant handling and processing have relied on sensing to attempt to achieve satisfactory performance. Ting et al. (1990) described a robotic system for transplanting bedding plants such as impatiens and marigolds from trays to flats. The robotic workcell used a custom gripper which included a capacitance sensor mounted in a manner to detect the presence or absence of a seedling of sufficient size for transfer into a flat. Ferrand et al. (1990) developed a fairly sophisticated machine dedicated to transplanting bedding plants from trays into flats or pots. This device, which ejected plugs from trays and maintained singulation while handling numerous plugs concurrently, used a visual sensing technique for sorting out poorly developed seedlings. Beam et al. (1991) described a simple but effective robotic system also for tray to flat transplanting. Infrared photoelectric sensors assisted in proper positioning of the containers, and machine vision was used to examine the flat and help ensure a successful transplant had been obtained for each cell. I reported in Simonton (1991b) our development of a robotic workcell for preparing cuttings for propagation. This system used (a) machine vision for identifying position/orientation of plants along with determining the geometric structure of each, (b) position and force sensing in the gripper which enabled compensation for variable stem diameters while maintaining quick, uniform grasp cycles, and (c) a custom light screen sensor for measuring the degree of bend in the stem which assisted in proper plug insertion. Each of the above applications was somewhat successful in using sensing to accommodate nonuniform conditions of plants or media, yet each application also had varying degrees of difficulty providing the goal level of robustness.

2.4. MATERIALS HANDLING AND MOBILITY

2.4.1. *Means of Accessing Products.* Issues related to providing a robotic system with the objects to be manipulated are critical to the success of many robotic applications. In a nursery or greenhouse operation, the issue may be stated as: How to get the plants or plant products to the robot? Or conversely, how to get the robot to the plants or plant products? In the first instance, materials are transported to a stationary robotic workcell. In the second instance, a mobile robot accesses stationary products.

If plants or products are to be transported to a robot, system design is centered on materials handling. Materials handling has been defined as "using the right method to safely provide the right amount of the right material in the right condition at the right cost at the right time in the right position" (White and Apple, 1985). While perhaps overdoing the point, the definition indicates the importance of materials handling in overall system design. In manufacturing facilities, feeding parts to and from robotic workcells can cost half of entire robot assembly systems (Schreiber, 1991). Further complicating materials handling equipment design are those frequent instances where robots are integrated into existing and ongoing operations in which changes to processes and surrounding equipment may be difficult to accomplish (Tanner, 1985). Another key impediment in many agricultural systems is product singulation. Equipment for the automatic singulation of propagation stock (e.g. unrooted cuttings, rooted cuttings, microcuttings) from trays, bags, or bundles is complex to develop.

If a robot is to be transported to various locations for task performance, system design is centered on mobility. Mobile robots may have linear, area, or spatial mobility depending on the number of degrees of freedom. Mobility is achieved through an automated transport device and may be employed to increase workspace utilization, solve materials handling problems, or access otherwise unapproachable work areas (Warnecke and Schuler, 1985). Much of the research in this area has been on the difficulties of autonomous navigation, frequently in uncertain environments (Pugh et al., 1990; Hoffman, 1991). Japanese researchers have been active in development of mobile robotic systems (Ting, 1989). Needs in the nuclear power and space industries have been the impetus for many developments, whereas work in service robots is now identifying many other applications (Englehardt, 1990). Mobile platforms for robots are being developed for greenhouse applications, for example Balerin et al.(1991), and indicate calibration and coordinate transformation between mobile platform, manipulator, and vision system are key challenges.

2.4.2. *Industry Specifics*. Differing production systems for various nursery and greenhouse crops indicate distinct solutions will be required on a case by case basis for materials handling and mobility (van Weel, 1991). For example, transplanting may be best performed in a centralized work area with materials transported to a robot. On the other hand, harvesting a vegetable crop may require a mobile system to access the plants. For each application, certain questions must be addressed. What will control throughput rate of incoming and outgoing material(s)? In what manner will the robot access the plants - will the robot be mobile and move to the plant or will the plant be presented to the robot? How much labor, supervisory or otherwise, will be necessary for all materials handling associated with the robotic system? If required, how will singulation of plants or plant material be performed?

2.5. ECONOMICS

2.5.1. *Robot Justification and Feasibility*. Economics and marketing issues play a significant role in choosing a successful approach for automation. In general automated systems are utilized by industries to produce goods faster, cheaper, and/or with improved quality. Robotics in particular are utilized to provide flexibility in production, eliminate health and safety hazards, and reduce total labor costs. The choice of fixed versus flexible automation is based on such factors as production volume, required cycle time, number of different products involved, and the dynamics of an industry's market (Maus and Allsup, 1986).

When an economic evaluation of a robotic application is performed, it is important to consider both direct costs (e.g. total labor costs including burden and overtime, productivity, maintenance) and indirect costs (e.g. inventory, waste, scheduling, operator training, inspection) (Meyer, 1982). Indirect savings due to robotics in food and agricultural applications can be extremely significant in economic evaluations, with the possible exception of inventory effects being negligible due to the perishable nature of products (Shoup and Macchio, 1985). Intangible benefits such as market leadership and growth opportunities can be used to weight alternatives (Hasegawa, 1985). However as manufacturing studies have shown (Mital and Genaidy, 1988), technical feasibility does not equate with economic feasibility. If cash flow to and from society is included in terms of unemployment benefits and tax effects, robotic installations advantageous to a company may be disadvantageous to society. Also, evidence exists that capital intensity may have an inverse relationship to profitability for some industries (Smith, 1991).

2.5.2. *Challenges and Recent Analyses*. The nursery and greenhouse industries have internal challenges related to economics and automation. The manner in which demand for product varies from season to season and year to year, with the accompanying price fluctuations, makes growers very cautious. This caution leads to resistance to change production methods. Likewise, historically the plant production industry has preferred operating costs over capital spending. In the United States the cost and availability of labor have played an important role. In a survey in Georgia, nursery and greenhouse growers cited capital as the chief limiting factor to expansion (Turner and Mixon, 1990), perhaps indicating the effect of interest rates relative to profit margins within these industries.

Few economic studies have been published recently on nursery and greenhouse system automation. Fang et al. (1991) evaluated the effects of annual production volume, equipment cost, labor cost, and other parameters on the economic feasibility of robotic transplanting of bedding plants. Expected payback varied from 2.6 to 6.4 years and return on investment varied from 8% to 61% depending on the assumptions. Van Waarde (1991) examined the effects of efficiency, robot number, equipment cycle time, labor cost, and other parameters on robotic preparation of geranium cuttings for propagation. The analysis provided design goals for an economically feasible system. Results included predicting (a) a 100% increase in allowable cost per robot resulting from a 33% decrease in equipment cycle time and (b) a 23% decrease in allowable cost for a 2% reduction in efficiency.

2.6. Transplant System Applications

2.6.1. *Nursery and Greenhouse Cultural Practices*. Day-to-day operation within most commercial nurseries and greenhouses does not fit the traditional factory where economic justification for automating may be simpler. For example, many of the labor intensive tasks in a greenhouse are not performed daily for eight or more hours, but rather are periodic. This includes work required 2 hours/day, weekly, or even seasonally. An example for some greenhouses would be shipping. If an automated system performed shipping only once a week, it may be sitting idle 80% of the time. Obviously 20% usage on a machine has a significant effect on break-even costs. A flexible, multipurpose system or a low-cost, single-use alternative may be required in this case.

Other cultural practice challenges to automation include few standards (bench and field layouts, work areas, procedures), multiple crops grown within the same range or field, and the environment (heat, moisture, dust/dirt). Disorder of propagation stock (e.g. geranium, chrysanthemum, and azalea cuttings) is yet another difficulty due to a lack of singulation equipment for cuttings. Also, there is the need for retrofitting. Many companies are understandably reluctant to alter their entire operation to accommodate a new machine. However even in cases where operations embrace new developments, care must be taken to minimize so-called 'islands of automation', the bottlenecks of material flow due to isolated points where automation equipment is in use.

2.6.2. *Automation Equipment Survey*. While there are many challenges to automate in commercial nursery and greenhouse operations today, a number of researchers and equipment developers are busy readying new ideas and machines. The most active area of development is transplanting (Neal, 1991). Equipment for plug or seedling transplanting within a greenhouse is being sold by several companies including Timmer Industries *(Jenison, MI,*

U.S.A.; Plug Transplanter; 10,000 plugs/hour} and Visser *{'s-Gravendeel, The NETHERLANDS; Plant-o-Mat; 10,000 plugs/hour}*. Transplanting systems under development for greenhouse use include those by Bouldin and Lawson *{McMinnville, TN, U.S.A.; Coverplant, a unit described in Ferrand et al., (1990); 7-10,000 plugs/hour}*, Agrobotics *{West Lafayette, IN, U.S.A.; Rotran; handles standard trays or flats, detects missing plugs}*, Precision Measurements *{Raleigh, N.C., U.S.A.; high speed unit with open bottom trays}* and Rutgers University *{New Brunswick, NJ, U.S.A.; experimental robotic workcell; innovative gripper for handling plugs}*.

In-field transplanting machines have been developed by USDA/ARS *{Tifton, GA, U.S.A.}*, University of Florida *{Gainesville, FL, U.S.A.}*, North Carolina State University and Precision Measurements *{Raleigh, NC, U.S.A.}*, Bud of California *{CA, U.S.A}*, Lannen Plant Systems *{FINLAND}*, and Iseki & Company *{JAPAN}*. A review of field transplanters is contained in Brewer (1991). Each of the above listed machines, whether for greenhouse or field use, has different production rates, efficiencies, support personnel requirements, and flexibility. All of the field machines and a majority of the greenhouse machines are fixed automation rather than flexible, with the distinction being allowable crop type, crop size, and container adaptability. Dedicated equipment for pot and flat handling systems developed by Dutch companies such as Visser, Hawe, and Javo and United States companies such as Bouldin & Lawson and Gleason Equipment have been improving productivity in nursery and greenhouse operations for many years. Flexible container handling systems are being researched by groups such as Ohio State University *{Columbus, OH, U.S.A.; robotic system}*, I.M.A.G. *{Wagenigen, The NETHERLANDS}* and E.N.I.T.A *{Bordeux, FRANCE; table handling gantry system}*. Other applications which are under development include spraying by Bouldin & Lawson *{mobile, self-guided unit}* and cutting propagation by The University of Georgia; Griffin, GA; USA *{robotic workcell for geranium propagation}*. Future potential robotic tasks in nursery and greenhouse operations that may not require direct plant contact include unpackaging of incoming pots and trays, packaging of the outgoing product, general materials handling in the headhouse or field, and product grading and sorting. Those robotic tasks that require direct plant contact include harvesting (e.g. tomatoes, cut flowers), pruning (e.g. cacti, ornamental shrubs), and grafting/budding (e.g. fruit trees).

2.6.3. *Micropropagation.* Micropropagation, the use of intensive vegetative propagation under aseptic conditions, is an industry with close ties to the commercial nursery and greenhouse. Many ornamental and floriculture crops grown in nurseries and greenhouses are from micropropagation. Given the potential, the labor intensive nature, and high cost of micropropagation, many researchers in both public institutions and private companies have examined automation alternatives, including robotics, for micropropagation. However, this method of propagating plants has proven very challenging in regards to automation for a number of reasons: common production practices were originally developed for the laboratory and not commercial use; different crops have very different growth habits in culture; tight control of growth habit even within one cultivar can be difficult; the process requires sterile conditions; and manually processing the culture material frequently requires considerable dexterity and skill. The latter difficulty is primarily due to performing the combined tasks of multiplication and harvest simultaneously on nonuniform cultures.

Technically several interesting systems have been developed in attempts to automate some of the more labor intensive tasks of micropropagation. Kozai et al. (1991) contains a review of many of the efforts to date. Toshiba developed a robot for excising nodes from axillary cultures with long internodes; this system effectively used 3-d laser ranging for controlling the robot. Silsoe Research Institute also developed a robot workcell for axillary cultures; machine vision was used to identify stem material and nodes to be excised. Professor Miwa at Waseda

University developed a novel bulb scale processing system which included singulation of the scales and vacuum gripping for handling. Kirin Brewery developed a robot for multiplying bushy adventitious cultures by dicing multiplication material into cubes and transferring the cubes into fresh containers.

Several other major efforts at automating micropropagation have also been made: some for specific types of crops such as ferns (e.g. Plant Biotech Industries in Israel) and some for specific cultivars such as daisy (e.g. Commonwealth Industrial Gases in Australia). Each of these systems has had limited success in terms of performance, economic feasibility, and/or applicability for a wide range of crops. Conversely, relatively simple technology can be applied to commercial micropropagation in selective ways and have considerable impact. Adaptive Equipment (Gainesville, FL, U.S.A.) developed workstations which automated operations such as counting microcuttings, data entry of production and shipping, and monitoring of employee and crop productivity. These workstations which used off-the-shelf hardware combined with custom software increased transfer hood productivity by over 20% in a large commercial laboratory.

Automation of micropropagation will not be solved with the brute force of engineering technology. Production practices will likely have to change dramatically, and this will require strong cooperation between plant scientists and engineers. Different solutions for various types of crops will emerge, with differing levels of automation. Some keys to assisting automation may be photoautotrophic production (Kozai, 1990), and production practices which utilize larger vessels and more uniform crop growth (Aitken-Christie and Davies, 1988) along with separating distinctly the cycles of multiplication and harvest.

3. A Robotic Plant Propagation Case Study

3.1. Preparation of Zonal Geranium Cuttings

3.1.1 Workcell Organization. A robotic workcell for preparing geranium cuttings for propagation was developed as a case study for automatically performing handling and processing operations on live plant material (Simonton, 1990). The robot grasps cuttings from a conveyor and performs tasks of leaf removal, stem trimming, stem bend measurement, and plug insertion. Trays of automatically prepared cuttings are then taken to the greenhouse for the propagation period. The workcell was designed such that fixtures and sensors mounted within the workspace assist in the processing and minimize complexity of operations required by the robot. The workcell consists of three primary subsystems: supervisory computer which coordinates tasks and performs machine vision, robot manipulator and controller, and end effector mechanism and controller.

For workcell operation, a conveyor brings singulated geranium cuttings into the robot work envelope. The supervisory computer locates a cutting and classifies all primary plant parts (Simonton and Pease, 1990). Knowledge of the plant structure is used to grade and determine an appropriate processing strategy. Control information is passed to the robot and end effector controllers. The cutting is grasped by the robot, moved to a pneumatic device for leaf removal if necessary, to a pneumatic cutter for trimming the main stem, to a sensor for measuring the stem bend, and finally to a plug tray for insertion into a propagation medium. Mechanical processing of a cutting by the robot occurs concurrently with the visual analysis of the next cutting on the conveyor by the supervisory computer.

3.1.2. Decision Structure. The following briefly explains the structure of the automatic decisions required by the system. The grade of the cutting, used to specify a desired post-processing main stem length and the appropriate tray in which to perform the plug insertion, is determined from the average main stem diameter 25 mm below the growth tip. There are three categories of grades, each within a particular diameter range. The ideal grasp position is calculated along the main stem a fixed distance from the cut location. To protect against certain machine vision errors, the ideal grasp is checked to ensure the plant part size to be grasped is reasonable for a main stem. A local scene analysis technique (Simonton, 1991b) is performed for near-optimal finger placement. Once the final grasp location and orientation have been determined, the data are transformed into a set of robot coordinates and joint angles.

The final step of the decision structure is to determine the need for leaf removal. Image data is checked for petiole-to-main stem branching locations within the stem region between the grasp location and the cut location. Results of the processing strategy are transmitted from the supervisory computer to the robot controller including: grade, grasp position and orientation, and leaf removal requirement. Also, the required end effector finger width for grasping is transmitted from the supervisory computer to the end effector controller. Software implementing these calculations requires 0.7 to 1.5 s depending on complexity of the machine vision image. Actual robotic processing time varies from 5 to 8 s.

3.2. Workcell Performance

3.2.1. Evaluation Results. Observations during 1991 testing indicated that all cuttings were grasped from the conveyor and handled effectively regardless of plant size. Grasp locations varied from approximately 5 to 15 mm in diameter, and the workcell handled the cuttings with petiole damage occurring on only 4% of all plants. No signs of damage to main stems were evident. However, processing effectiveness was 94% with all losses resulting from improper plug insertion. A large majority of the insertion errors was due to foliage interfering with either the visual processing, the bend sensor, or the sticking operation directly. Propagation efficiencies were 8% less than with manual propagation due to high foliage density causing transpiration and disease problems. Tray uniformity was generally poor due to widely varying prepared cutting main stem lengths. The stem length variations were due to vision errors caused by improper identification of the growth tip region.

Results demonstrate the challenge of accommodating nonuniformity. The robotic system was shown to be capable of preparing a wide size range of plants using sensing and control techniques based on the end effector, vision software, and bend sensor. However relatively simple variations in the morphology of the cuttings (i.e. increased foliage density) prohibited satisfactory or improved performance relative to manual operation.

3.2.2. Performance Summary. Given the capabilities and problems associated with the robotic workcell for propagation, what has to happen before the automatic system operates at human level performance? The answers lie in terms of both technology and cultural practice. The use of machine vision to completely identify the structure of a geranium cutting in a robust manner, including the growth tip, is critical if randomly oriented plants are to be presented to the robot. Color vision appears to be indicated for this application. For uniform trays after the propagation period, a grading method must be developed. Grading based on machine vision data would appear to offer a moderate probability of success. Determining the optimal media (consistent in form, texture, density, compression, and water holding capacity) used for robotic insertion of cuttings is also required.

Limiting the degree of nonuniformity of cuttings presented to the robot for processing is in the realm of cultural practice. Designing an automatic system for preparing geranium cuttings with multiple growing tips, from one to eight leaves, and from 25 to 100 mm in main stem length requires a very high level of sophistication and cost requiring multiple manipulators and three-dimensional machine vision. While less interesting technically, a more pragmatic approach is to reduce the complexity of geranium cutting morphology by limiting the size and leaf number of cuttings that are fed to the workcell for processing.

4. Summary

The commercial nursery and the commercial greenhouse each have relatively structured environments and large numbers of labor intensive tasks as compared to other agricultural production systems. These factors increase the opportunities for automation including robotics to provide economic, safety, and sociological benefits. However, there are many challenges for those in commercial production, for equipment developers, and for researchers. Systems must be developed which combine human, computer, and machine resources in innovative ways which contribute to the individual companies, to the workforce, and to society. Systems must be developed which are robust and can therefore perform tasks dealing with biological materials of variable size, shape, color, position, and strength in a reliable and cost-effective manner. Systems must be developed which can accurately sense biological properties and modify robot system behavior. Systems must be developed which provide efficient access to products for the robot.

Long term solutions will require close cooperation between plant scientists and engineers, as new methods of propagating and growing plants will likely emerge. Successful automation systems in the near term will likely be driven by high volume tasks performed frequently, by labor availability and stability problems, and by global competition. Also in the near term, constraining the problem and the environment to minimize the degree of nonuniformity while incorporating sensory systems to measure parameters and thus accommodate remaining variations is an appropriate design philosophy for agricultural robotics. In practice, this is the same approach used currently for industrial applications. For agriculture these systems will likely consist of a synergistic combination of computers, machines, and humans to satisfy performance criteria of productivity, economic return, and workforce satisfaction.

5. References

Aitken-Christie, J. and Davies, H. (1988) 'Development of a semi-automated micropropagation system', Acta Horticulturae 230:81-87.

Balerin, S., Bourely, A. and Sevila, F. (1991) 'Mobile robotics applied to fruit harvesting', in Proceedings of the 1991 Symposium on Automated Agriculture for the 21st Century, ASAE, St. Joseph, MI, pp. 236-244.

Baylou, P. (1987) 'Agricultural robots', in Proceedings of IFAC 10th Triennial World Congress, pp. 111-119.

Beam, S.M., Miles, G.E., Treece, G.J., Hammer, P.A., Krutz, L.J. and Richey, C.B. (1991) 'Robotic transplanting: Simulation, design, performance tests', ASAE Paper No. 917027.

Bolusky, B. (1989) 'Testimony to the U.S. Congress by the American Association of Nurserymen', April 25, Washington, D.C.

Brewer, H.L. (1991) 'The use of automated field transplanters in plant propagation', in Scaleup and Automation in Plant Tissue Culture, Vol. 8 of the series Cell Culture and Somatic Cell Genetics of Plants, Academic Press, San Diego, CA, pp. 245-259.

Brooks, R.A. (1991) 'New approaches to robotics', Science 253:1227-1232.

Brooks, T.L. (1980) 'Supervisory manipulation based on the concepts of absolute vs. relative and fixed vs. moving tasks, in Seireg (ed.), Advances in Computer Technology, Vol. I, pp. 185-196.

Carlopio, J., Fleisher, M. and Roitman, D. (1988) 'Computerized manufacturing technology and work organization effects of labor relations and worker satisfaction', in Karwowski, Paraei and Wilhelm (eds.), Ergonomics of Hybrid Automated Systems, Elsevier Science Publishers, Amsterdam, pp. 713-717.

Englehardt, K.G. (1990) 'Service robots: An overview', in Proceedings of the 1990 International Robots and Vision Automation Conference, RIA, Detroit, pp. 7-1 through 7-21.

Fang, W., Ting, K.C., and Giacomelli, G.A. (1991) 'Engineering economics of SCARA robot-based plug transplanting workcell', ASAE Paper No. 917026.

Ferrand, C., Baylou, P. and Grenier, G. (1990) 'A non-damaging effector for high speed transplanting robots', ASAE Paper No. 907501.

Ghallab, M. (1985) 'Task execution monitoring by compiled production rules in an advanced multi-sensor robot', in Hanafusa and Inoue (eds.), The 2nd International Symposium on Robotics Research, MIT Press, Cambridge, MA, pp. 393-402.

Harrell, R.C., Adsit, P.D., Poole, T.A., and Hoffman, R. (1990) 'The Florida robotic grove lab', TRANSACTIONS of ASAE 33(2):391-399.

Hasegawa, Y. (1985) 'Evaluation and economic justification', in Nof (ed.), Handbook of Industrial Robotics, John Wiley & Sons, NY, pp. 665-687.

Hoffman, R. (1991) 'Autonomous legged robots in agriculture', in Proceedings of the 1991 Symposium on Automated Agriculture for the 21st Century, ASAE, St. Joseph, MI, pp. 218-225.

Kidd, P.T. (1988) 'Human and computer-aided manufacturing: The end of Taylorism?' in Karwowski, Paraei, and Wilhelm (eds.), Ergonomics of Hybrid Automated Systems, Elsevier Science Publishers, Amsterdam, pp. 145-152.

Koren, Y. (1985) Robotics for Engineers, McGraw-Hill, NY, pp. 229-258.

Kozai, T. (1990) 'Micropropagation under photoautotrophic conditions', in Debergh and Zimmerman (eds.), Micropropagation: Technology and Application, Kluwer Academic Publishers, Dordrecht, pp. 447-469.

Kozai, T., Ting, K.C. and Aitken-Christie, J. (1991) 'Considerations for automation of micropropagation systems', in Proceedings of the 1991 Symposium on Automated Agriculture for the 21st Century, ASAE, St. Joseph, MI, pp. 503-517.

Maus, R. and Allsup, R. (1986) Robotics: A Manager's Guide, John Wiley & Sons, NY, pp. 189-200.

Meyer, R.J. (1982) 'A workbook approach to robotics and automation justification', in Proceedings of the Robots VI Conference, SME, Dearborn, MI, MS82-192, pp. 1-29.

Mital, A. and Genaidy, A.M. (1988) 'Automation, robotization in particular, is always economically desirable - fact or fiction?' in Karwowski, Paraei and Wilhelm (eds.), Ergonomics of Hybrid Automated Systems, Elsevier Science Publishers, Amsterdam, pp. 743-750.

Moncaster, M.E. (1985) 'Applications of robotics to agriculture', Farm Electronics and Computing International Symposium, October 20-25, Stratford-upon-Avon, Great Britain.

Neal, K. (1991) 'Automatic transplanters', Greenhouse Manager 9(11):99-101.

Parsons, H.M. (1988) 'Human nature and robot nature', in Karwowski, Paraei, and Wilhelm (eds.), Ergonomics of Hybrid Automated Systems, Elsevier Science Publishers, Amsterdam, pp. 331-363.

Parsons, H.M. and Kearsley, G.P. (1982) 'Robotics and human factors: Current status and future prospects', Human Factors 24(5):535-552.

Pugh, D.R., Ribble, E.A., Vohnout, V.J., Bihari, T.E., Walliser, T.M., Patterson, M.R. and Waldron, K.J. (1990) 'A technical description of the adaptive suspension vehicle', International J. of Robotics Research 9(2):24-42.

Saridis, G.N. (1989) 'Analytic formulation of the principle of increasing precision with decreasing intelligence for intelligent machines', Automatica 25(3):461-467.

Schreiber, R.R. (1991) 'Assembly robots build quality', Manufacturing Engineering 107(6):47-51.

Sheridan, T.B. (1985) 'Forty-five years of man-machine systems: History and trends', in Mancini (ed.), Design, Analysis, and Evaluation of Man-Machine Systems, IFAC, Varese, pp. 1-9.

Shoup, W.D. and Macchio, V.R. (1985) 'A model for establishing opportunity values of robotics/automation in food processing' in Agri-Mation I, ASAE, St. Joseph, MI, pp. 69-80.

Simonton, W. (1990) 'Automatic geranium stock processing in a robotic workcell', TRANSACTIONS of the ASAE 33(6):2074-2080.

Simonton, W. (1991a) 'Robotics in the greenhouse', Professional Plant Growers Association Newsletter 23(7):2-5, 10-11.

Simonton, W. (1991b) 'Robotic handling and processing for agricultural systems, in Proceedings of the 1991 Symposium on Automated Agriculture for the 21st Century, ASAE, St. Joseph, MI, pp. 226-235.

Simonton, W. and Pease, J. (1990) 'Automatic plant feature identification on geranium cuttings using machine vision', TRANSACTIONS of the ASAE 33(6):2067-2073.

Smith, I.C. (1991) 'Automation or people: Learn the right balance for financial success', Grower Talks 54(12):22-31.

Tanner, W.R. (1985) 'Product design and product planning', in Nof (ed.), Handbook of Industrial Robotics, John Wiley & Sons, NY, pp. 537-548.

Taylor, F.W. (1907) 'On the art of cutting metal', Transactions of American Society of Manufacturing Engineers 28:31-35.

Ting, K.C. (1989) 'Japan Trip Report', U.S. Department of Agriculture, Office of International Cooperation and Development, Scientific and Technical Cooperation Division, 11 pp.

Ting, K.C., Giacomelli, G.A., Shen, S.J. and Kabala, W.P. (1991) 'Robotic workcell for transplanting of seedlings; Part II: End Effector Development', TRANSACTIONS of ASAE 33(3):1013-1018.

Trevelyan, J.P. (1985) 'Skills for a sheep shearing robot: Dexterity and sensing', in Hanafusa and Inoue (eds.), The 2nd International Symposium on Robotics Research, MIT Press, Cambridge, MA, pp. 273-280.

Turner, S.C. and Mixon, B.J. (1990) 'A market examination of Georgia's ornamental plant industry', University of Georgia Agricultural Experiment Station Research Report 593.

United States Department of Agriculture. (1989) 'USDA/ARS Sponsored Report of the Greenhouse and Nursery Mechanization Advisory Committee Workshop', Boynton Beach, FL., November, 1988.

van Waarde, P. (1991) 'Economic feasibility study: Geranium workcell', University of Georgia Agricultural Engineering Research Report.

van Weel, P. (1991) 'Integrated crop production systems', in Proceedings of the 1991 Symposium on Automated Agriculture for the 21st Century, ASAE, St. Joseph, MI, pp. 458-467.

Warnecke, H.J. and Schuler, J. (1985) 'Mobile robot applications', in Nof (ed.), Handbook of Industrial Robotics, John Wiley & Sons, NY, pp. 1002-1010.

White, J.A. and Apple, J.M. (1985) 'Robots in material handling', in Nof (ed.), Handbook of Industrial Robotics, John Wiley & Sons, NY, pp. 955-970.

Winter, H. (1986) 'AI in man-machine systems and automation', in Proceedings of Deutsche Forschungs-und Versuchsanstalt fur Luft-und Raumfahrt, Springer Verlag, Berlin, pp. 1-22.

Measurement and Control in Transplant Production Systems

Y. Hashimoto and H. Nonami
Dept. of Biomechanical Systems
Ehime University
Tarumi, Matsuyama, 790
Japan

ABSTRACT. We have introduced a concept of a growth–induced water potential field which governs water influx to the expanding cells. In growing tissue, the xylem had the highest water potential and cells located near the epidermal surface had the lowest water potential. When plants were subjected to environmental stresses, the growth–induced water potential field collapsed, resulting in growth inhibition. Practically speaking, the magnitude of a growth–induced water potential field can be estimated from differences between water potentials of the elongating region and the water source. Because the growth–induced water potential field recovers prior to growth recovery, if the growth–induced water potential field can be determined appropriately, growth behavior of crops can be predicted under controlled environments. Further, we have introduced modelling and control of long–term plant growth from the aspect of photosynthetic mechanisms. By using computer networks, an artificial intelligence system for plant cultivation has been proposed, and this system can be applied to plant factory automation.

1. Introduction

Transplant production systems deal with processes involved in tissue culture and processes of transplanting tissue–cultured seedlings to greenhouse conditions. In such processes, acclamation of tissue–cultured plants to greenhouse environmental conditions is the most critical process for seedling production systems. Because plant growth is one of the physiological mechanisms of plants most sensitive to environmental stresses, slight changes in environmental conditions in plant factories and greenhouses cause changes in growth of crop seedlings. In order to cultivate high quality seedlings under a controlled environment, methods for adjustment of environmental conditions have been studied. Significant changes in cell volume expansion occur in a matter of hours in most plants, and thus, are the most appropriate parameter for growth regulation. Because water occupies more than 90% of the total molecules in expanding cells of most plants, water movement into cells regulates cell expansion rates in most cases. Thus, we have studied plant–water relations of crop seedlings under controlled environments, and will suggest new types of growth regulation methods for crop seedlings grown under controlled environments. Because physiological principles for plant growth should be applicable to any kind and any size of plants, physiological regulation for growth should be applicable for cells under tissue culture and also for cells of crops grown in greenhouse conditions.

Growth of crops is most commonly measured in terms of dry matter accumulation. However, measurements of dry matter accumulation are not suitable for use as a parameter

K. Kurata and T. Kozai (eds.), Transplant Production Systems, 117–136.

for control in order to regulate growth of crops in plant factories and greenhouses, as such measurements must be evaluated over a long period of time such as days and weeks. Because dry matter accumulation is caused by the results of photosynthetic activities in plants, photosynthesis was analyzed by spectrum analysis with respect to variations of light intensity and CO_2 concentrations, and further, such modelling and control was applied to plant factory automation. Because physiological responses of plants to changes in environmental conditions are complicated, artificial intelligence systems for plant cultivation have been introduced and discussed in the present paper.

2. Water Status Measurement Techniques in Plant Science

Because water constitutes 80–90% of the living plant body and has a large heat capacity, changes in water status affect the growth and metabolism of plants more than any other environmental factors. In particular, when tissue–cultured plantlets and seedlings are transplanted from a high humidity condition in tissue–culture containers to a low humidity atmosphere, adaptation to the resulting water stress is critical for survival of the plantlets and seedlings.

Water must be moved to the leaves each day to replace what is lost by transpiration. This requires water potentials to decrease so that water will move into the plant from the water source, e.g., hydroponic culture solution, tissue culture media and soil. Cell enlargement, which is the primary means by which plants increase in size, is sensitive to these water potentials. Moreover, plant metabolic activities such as photosynthesis, nitrogen metabolism, translocation and protein synthesis are affected by changes in water potentials in plants.

Thermodynamically, solute and pressure forces in plants, tissue–culture media and soil are most exactly described by water potential, defined as the chemical potential divided by the partial molal volume of liquid water. Measurements of water potential are always compared to a reference water potential, which is pure free water at atmospheric pressure, a defined gravitational position, and the same temperature as the system being measured. The reference is defined to have a water potential of zero. The water potential indicates how much the free energy differs from that in the reference state and is the sum of solute forces and local plus external pressure forces acting on the water as shown in the following manner;

$$\psi_w = \psi_s + \psi_p \qquad (1)$$

where ψ_w, ψ_s and ψ_p are water potential, osmotic potential and pressure potential, respectively. Values of both ψ_w and ψ_s are always negative, and that of ψ_p can be positive or negative depending on whether the pressure is above or below atmospheric pressure. In the case of plant cells, ψ_p usually represents turgor inside the cell, which is a positive quantity.

Although Eq. 1 appears in all water–relation textbooks, the validity of Eq. 1 has not been evident until recently (Hashimoto and Nonami, 1991). Nonami et al. (1987) compared turgor measured with a cell pressure probe to that measured with an isopiestic thermocouple psychrometer in mature regions of soybean stems. The probe measured turgor directly in cells of soybean stems whereas the psychrometer measured the water potential and osmotic potential of excised stem segments and turgor was subsequently obtained by calculating the difference between water potential and osmotic potential. Both methods gave similar values of turgor whether the plants were dehydrating or rehydrating (Nonami et al., 1987). Boyer and Potter (1973) compared water potential, osmotic potential and turgor measured with the pressure chamber and the isopiestic psychrometer in mature sunflower leaves, and obtained virtually the same values with both methods. Nonami and Schulze (1989) measured turgor and osmotic potential directly in cells of well–expanded leaves of *Tradescantia virginiana* with a cell pressure probe combined together with a nano–liter osmometer, and the cell water potential, osmotic potential and turgor determined with the probe and osmometer were similar to those obtained with the isopiestic psychrometer. Because the three methods (i.e., the pressure chamber; the isopiestic psychrometer; the cell pressure probe and the nano–liter osmometer) rely on different principles of measurement, and because different plant materials were used for the comparison, it is unlikely that all would be subject to inherent but undetected inaccuracies that would cause water potential, osmotic potential and turgor to deviate by the same amount from the true values of the potentials. Therefore, the agreement between the methods suggests that Eq. 1 should be valid in all plant cells, and it is safe to say that any of the three methods can be used to measure the water status of plants accurately.

3. Plant Growth Regulation under Water Stress

3.1. ORIGIN OF GROWTH–INDUCED WATER POTENTIAL

Cell enlargement involves a combination of biochemical changes in the cell wall coupled with a physical extension of the wall. Sustained enlargement is accompanied by differentiation, a proliferation of membranes and organelles, and increases in protein and cell wall materials. In most cases, however, these biochemical processes influence physical cell enlargement only indirectly. Turgor has long been considered crucial in cell expansion, supplying the necessary pressure from inside. Much of the original evidence was obtained in connection with auxin effects on *Avena* coleoptiles (Cleland, 1971), although the restriction of growth by water stress in general had earlier been considered in terms of turgor reduction (Kramer, 1959; Vaadia et al., 1961). More recently, several reports (Meyer and Boyer, 1972, 1981; Cutler et al., 1980; Matsuda and Riazi, 1981; Cavalieri and Boyer, 1982; Michelena and Boyer, 1982; Mason and Matsuda, 1985; Westgate and Boyer, 1985) showed little relationship between turgor and growth when water stress was imposed. Some physical factors other than turgor may be more critical for cell expansion in certain cases.

In theory, the extension of cell walls by turgor creates a demand that must be satisfied by water from the xylem (Boyer, 1985). As extension occurs, cell turgor decreases until the water potential is low enough to cause water to enter from the xylem at a rate that satisfies this demand. The difference in water potential between the xylem and the enlarging cells is called a growth–induced water potential (Molz and Boyer, 1978). Nonami and Boyer (1987b) have proposed that the growth–induced water potential originates because the walls extend during cell enlargement and prevent turgor from reaching its maximum. This implies that a substantial frictional resistance exists between the enlarging cells and the water supply.

In the enlarging cells, the water–relations for the protoplasts and apoplast can be dealt with separately as follows;

$$\psi_w^{pro} = \psi_s^{pro} + \psi_p^{pro} \qquad (2)$$

$$\psi_w^{apo} = \psi_s^{apo} + \psi_t \qquad (3)$$

where the superscript, pro, indicates protoplasts and the superscript, apo, indicates apoplast. ψ_p^{pro} is the pressure potential in the protoplasts, i.e., the turgor, and ψ_t is the pressure potential in the apoplast, generally negative and thus a tension. In the elongating region, the apoplast solute concentrations were very low (–0.03 to –0.04 MPa) and negligible compared to the water potential of the apoplast (–0.15 to –0.30 MPa) measured directly in intact plants (Nonami and Boyer, 1987b). Most of the apoplast water potential in the elongation zone consisted of a negative pressure that could be measured with a pressure chamber in the similarly grown plants, and ranged from -0.15 to -0.28 MPa (Nonami and Boyer, 1987b). Because water potential of the protoplasts and that of the apoplast are usually in local equilibrium (Molz and Boyer, 1978; Molz and Ikenberry, 1974), relations between Eqs. 2 and 3 can be represented diagrammatically in Fig. 1 (using data from the work of Nonami and Boyer, 1987b).

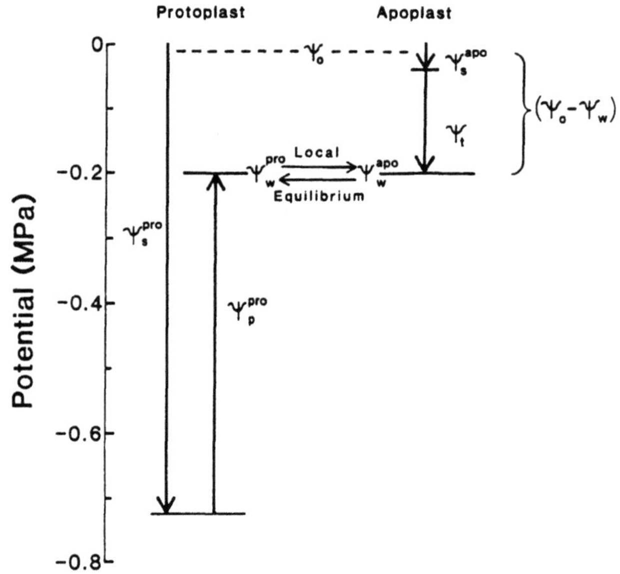

Figure 1. Diagrammatic representation of the average forces affecting water in rapidly elongating stem tissue of dark grown, nontranspiring soybeans (adopted from Nonami and Boyer, 1987b). Downward pointing arrows indicate decreased potential, upward arrows show increased potential. Ψ_w^{pro}, Ψ_s^{pro}, and Ψ_p^{pro} are water potential, osmotic potential, and turgor of the protoplasts, and Ψ_w^{apo}, Ψ_s^{apo}, and Ψ_t are water potential, osmotic potential, and tension of the apoplast, respectively. Ψ_o is the xylem water potential. The growth-induced water potential ($\Psi_o - \Psi_w$) is the difference between the water potential of the xylem and the average water potential of the elongating cells.

Figure 1 shows the lack of full turgidity in the protoplasts and the large apoplast tension in the tissue in the elongation zone. The only hypothesis consistent with the data is that the growth-induced water potential originates primarily from the extension of the cell walls, preventing turgor from reaching its maximum and thus creating a water potential below that in the xylem in response to which water moves into the cells at a rate that satisfies enlargement. The magnitude of the water potential difference would be determined by the frictional resistance between the enlarging cells and the water source. The water potential varies with regard to position in the growing tissue. During rapid growth, the water potential field has a distinctive distribution in the soybean stem; highest potential in the xylem and lowest potential in the outer cortical cells, and the largest gradient near the xylem and smallest gradient in the outer cortical cells. In general, the water potential will be the same at uniform distances from the xylem and will form a set of isopotential cylinders around the xylem. At progressively larger distances from the xylem, each cylinder represents a progressively lower water potential. The water potential gradient is normal to the cylinder surfaces.

The dependence of flow on the potential field implies that cell enlargement will be sensitive to alterations in the field. The xylem water potential which represents the initial part of the potential field would be a particularly important determination of water entry and growth rate. Nonami and Boyer (1989) suggest that a collapse of the water potential gradient necessary for the growth process is responsible for the initial inhibition of growth at low water potentials without turgor loss in the elongation zone. When stem growth was completely inhibited by transplanting plants to low water potentials, turgor of the majority of cells in the elongating tissue was unchanged, but turgor decreased in a small number of cortical cells near the xylem. The reduced turgor in these cells and constant turgor in most of the cells confirmed that no general turgor loss had occurred but also indicated that gradients in water potential extending from the xylem into the enlarging tissue were reduced, thus decreasing the movement of water into the tissue for cell enlargement (Nonami and Boyer, 1989). Therefore, it is possible that plant growth can be predicted if the growth-induced water potential can be determined in plants.

3.2. THE GROWTH-INDUCED WATER POTENTIAL APPLIED TO HYDROPONIC CULTURE

The rate of cell enlargement should increase proportionally with the amount of water flux into the enlarging cell. If the concept of the growth-induced water potential is valid, the water potential gradient between the xylem and the enlarging cells should increase as the rate of cell enlargement increases. Because the growth-induced water potential field is initiated from the xylem water potential and formed by cell water potentials in the growing tissue,

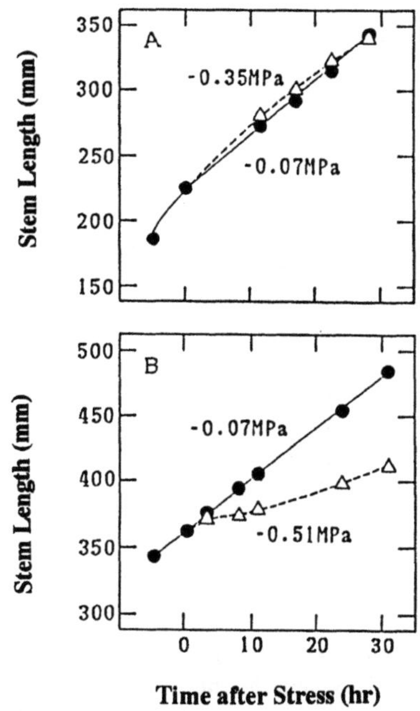

Figure 2. Stem length of kidney bean seedlings when the nutrient solution water potential was increased from −0.07 MPa to −0.35 MPa (A) and from −0.07 MPa to −0.51 MPa (B).

water potential difference between the elongating region and xylem may be a good estimate of the magnitude of the growth–induced water potential field. In order to test this hypothesis, kidney bean seedlings were grown hydroponically, and growth and water potentials of the elongating region and mature region were measured simultaneously. Water stress was applied by increasing the concentration of nutrient solution (Nonami et al., 1991).

When water potential of the culture solution was decreased from −0.07 MPa to −0.35 MPa, the growth of kidney bean seedlings was slightly promoted (Fig. 2A). However, immediately after water potential of the culture solution was decreased from −0.07 MPa to −0.51 MPa, the growth of kidney bean seedlings was inhibited (Fig. 2B). After 10 h of the water stress, gradual recovery of growth of the seedlings was observed (Fig. 2B).

In order to understand the mechanism of growth inhibition by water stress created by increasing the concentration of the nutrient solution, the growth–induced water potential was estimated from the differences between water potentials of the elongation region and the mature region. Because growth is not occurring in the mature region, the water potential of the mature region can be considered to be equilibrated with the xylem water potential. Therefore, the difference in water potential between the xylem and elongating cells is considered to be similar to the difference in water potential between the elongation region and the mature region.

In plants which did not have growth inhibition under −0.35 MPa of the culture solution, the growth–induced water potential was unchanged (Fig. 3B). In plants whose growth was inhibited under −0.51 MPa of the culture solution, the growth–induced water potential decreased immediately after the stress, and began to recover prior to growth recovery (Fig. 3D). It is evident that growth of plants is regulated by the growth–induced water potential field.

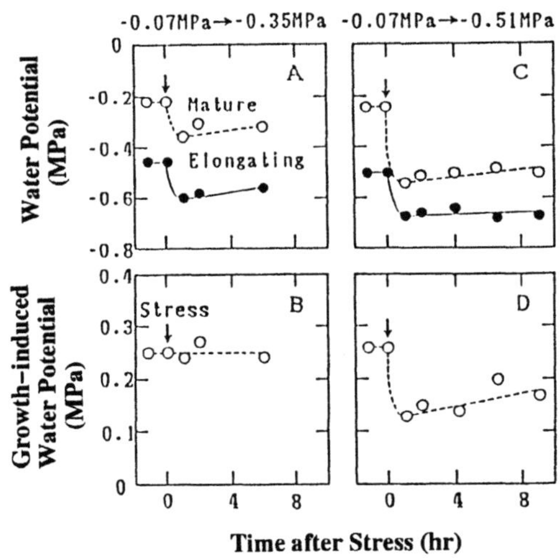

Figure 3. Water potentials of the mature region and the elongation region (A, C) and the growth–induced water potential (B, D) when the nutrient solution water potential was increased from −0.07 MPa to −0.35 MPa (A, B) and from −0.07 MPa to −0.51 MPa (C, D). The growth–induced water potential was estimated by water potential difference between the elongation region and the mature region.

3.3. CONTROL OF CELL ELONGATION UNDER WATER STRESS

Although water entry into cells is the most central cause for cell elongation, properties of cell wall and membrane influence the rate of water flow into cells. The process of cell elongation depends on the extension of the cell walls together with the uptake of water and solute. Turgor (ψ_p) extends the walls when a threshold turgor, the yield threshold (Y), has been exceeded (Cleland, 1959; Probine and Preston, 1962; Green et al., 1971). This extension depends on the extensibility (m) of the walls that is determined by metabolic activity (Probine and Preston, 1962; Green et al., 1971; Green and Cummins, 1974). The parameters affecting cell enlargement can be expressed as averages for the tissue in which case the rate of wall extension for growth can be related to m ($s^{-1} \cdot MPa^{-1}$) and ($\psi_p - Y$) (MPa) by the equation (Green et al., 1971):

$$G = m(\psi_p - Y) \qquad (4)$$

where G is the relative growth rate $(1/V_{te})(dV_t/dt)$. V_{te} is the total volume of the elongating tissue. During growth in localized growing region, V_{te} is constant. The growth rate, (dV_t/dt), is then the rate at which the volume of new mature tissue is produced. Equation 4 implies that demand for water is determined both by the amount of turgor above the yield threshold and by the extensibility of the wall.

The simultaneous uptake of water is necessary because turgor would otherwise decrease as the walls extend, inhibiting cell enlargement. However, water uptake is not sufficient to completely maintain maximum turgor because there are resistances to water movement into the enlarging cells (Nonami and Boyer, 1987). This creates a water potential (ψ_w) that is lower than in the water source (ψ_o), usually the xylem. The ($\psi_o - \psi_w$) is the growth-induced water potential. The rate of water uptake necessary to support cell enlargement can be related to hydraulic conductance (L) ($s^{-1} \cdot MPa^{-1}$) and ($\psi_o - \psi_w$) (MPa) by the equation (Boyer and Wu, 1978; Boyer et al., 1985):

$$G = L(\psi_o - \psi_w) \qquad (5)$$

where G is the relative growth rate $(1/V_{we})(dV_w/dt)$. V_{we} is the volume of water in the elongating tissue. During growth in localized growing regions, V_{we} is essentially constant. The growth rate, dV_w/dt, is then the rate at which water enters the elongating region and is converted to an equivalent volume of water in the mature tissue. Equation 5 implies that supply for water will be determined both by the potential difference and the conductance.

It should be noted that, because tissue volume is proportional to the water volume contained by the cells, $(1/V_{we})(dV_w/dt) = (1/V_{te})(dV_t/dt)$, and thus, G is the same for Eqs. 4 and 5.

Combining Eqs. 4 and 5 by using relationship $\psi_w - \psi_s = \psi_p$ yields:

$$G = (mL/(m+L))(\psi_o - \psi_s - Y) \qquad (6)$$

which is the rate equation governing tissue enlargement. Equation 6 shows both the effects of wall extensibility and tissue conductance which determine water demand and water supply for the enlarging cells.

All parameters in Eqs. 4, 5 and 6 could be measured simultaneously in intact seedlings by using the guillotine thermocouple psychrometry (Boyer et al., 1985; Nonami and Boyer, 1990a). Nonami and Boyer (1990a) studied the kinetics of physiological parameters regulating plant growth under water stress by using stems of soybean seedlings which were exposed to water stress. Figure 4 shows the sequential changes in most of the growth parameters which are expressed in relative values compared with non-stressed seedlings.

The results show that ($\psi_o - \psi_w$) was the first to respond to the water stress. The simultaneous decrease in G and ($\psi_o - \psi_w$) indicates that the change in growth–induced water potential was likely to have caused the decrease in G. A similar conclusion was reached from the hydroponic culture experiment shown in Figs. 2 and 3. The recovery of ($\psi_o - \psi_w$) and G after some time is further evidence that ($\psi_o - \psi_w$) played a regulatory role. However, despite the complete recovery of ($\psi_o - \psi_w$), the recovery of G was incomplete. This indicates that another factor was responsible for limiting for G.

After the coincident decrease in G and ($\psi_o - \psi_w$), m and L decreased (Fig. 4). The recovery of ($\psi_o - \psi_w$) was followed by m, (ψ_p – Y), and L. G recovered as the last of the growth parameters (m and L) recovered. Therefore, the factors limiting G must have been m and L.

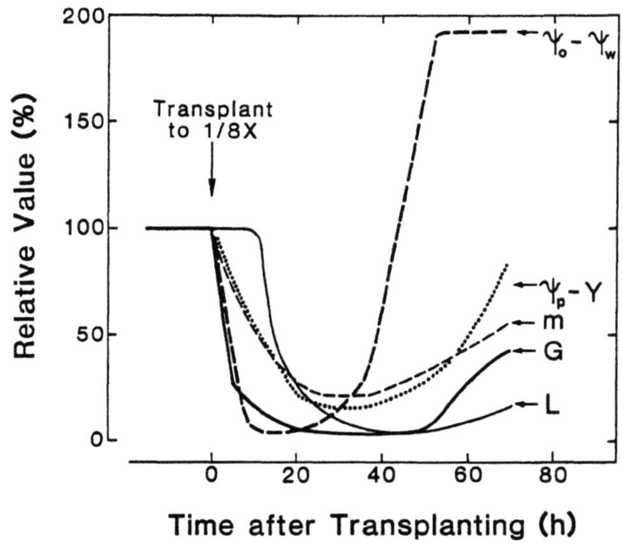

Figure 4. Relative values of the stem elongation rate (G), growth–induced water potential ($\psi_o - \psi_w$), growth–active turgor (ψ_p – Y), wall extensibility (m), and tissue conductance for water (L) at various times in stems of intact soybean seedlings transplanted to water deficient vermiculite (i.e., vermiculite having 1/8x of water content of the control), compared with the control (100 %, before transplanting). (Adopted from Nonami and Boyer, 1990a)

This order indicates that the inhibition of cell elongation was caused first by a physical factor ($\psi_o - \psi_w$) and secondarily by metabolically controlled factors (m and L). Although the physical factor recovered, G continued to be suppressed until the metabolically controlled factors recovered. Therefore, both during inhibition and recovery, the sequence of changes was consistent with the transmission of a physical signal to the cells that resulted in a metabolic response. The transmission required about 15 h during the inhibition and 10 to 20 h during recovery.

In order to check the observations which indicate that the inhibition of stem growth at low water potentials is accompanied by decreases in cell wall extensibility and tissue hydraulic conductance to water, Nonami and Boyer (1990b) developed a theory to relate rheological properties of cell walls measured with an extensiometer and thermodynamical properties of cell walls measured with psychrometry. They found that the plastic properties of the cell walls determined cell wall extensibility (Nonami and Boyer, 1990b).

Both cell wall extensibility and hydraulic conductance are under metabolic control, and their low levels under water stress indicate that substantial metabolic change occur in the enlarging cells (Nonami and Boyer, 1990a). Wall extensibility is probably determined by the action of wall enzymes on the polymerization of wall constituents and the cross-linking of structural proteins with wall constituents (Cassab and Varner, 1988; Fry, 1986; Theologis, 1986). In seedlings similar to those used here, polyribosome and mRNA levels decreased a few hours after water stress (Mason et al., 1988). Increased amounts of proteins were extractable from the walls and a 28 kD protein present in the cytoplasm accumulated in the walls (Bozarth et al., 1987). Translation products of the mRNAs were enhanced in the elongation zone of water-deficient soybean seedlings, and some of them were preferentially associated with membrane-bound polysomes (Mason et al., 1988). These findings suggest that metabolic alterations involving the cell walls and membranes take place after the collapse of the growth-induced water potential associated with the growth inhibition at low water potentials. Such metabolic alterations in cell walls and membranes may be mediated by ABA (Bensen et al., 1988).

The metabolic changes represented by m and L are clearly at the cellular level because m is a property of the cell walls and the changes in L are similar to the changes in hydraulic conductivity observed in individual cells with the miniature pressure probe (Nonami and Boyer, 1990b), which must have involved cell membranes. It is noteworthy in this regard that proton secretion to the cell walls is thought to affect the extensibility of the walls (Cleland, 1971, 1975, 1983; Cleland and Rayle, 1978) and is curtailed by exposure of the plants to water stress (Van Volkenburgh and Boyer, 1985). Thus, it is possible that decreased proton secretion is the manifestation of a metabolic signal that is being transmitted to the walls.

After water stress is applied to plants, growth recovery occurs when the xylem water potential is equilibrated with the medium. This is caused by osmotic adjustment in cells (Nonami and Boyer, 1989, 1990a). Prior to the growth recovery, recovery of the growth-induced water potential is observed. The recovery of the growth-induced water potential induces metabolic changes in cells, resulting in increases in wall extensibility and hydraulic conductance (Nonami and Boyer, 1990a, 1990b). Afterward, growth is resumed. Thus, the growth-induced water potential can be considered to be a primary physiological factor regulating growth under water stress.

Because the expanding cells must absorb water from the water source in order to increase their volume, the growing tissue must form a growth-induced water potential field. The concept of the growth-induced water potential field must be applicable to any growing tissue. Boyer (1988) reduced the rate of cell enlargement by lowering the temperature, and measured the growth rates and the water potential gradient with the pressure chamber and the isopiestic psychrometer. He observed that the water potential gradient decreased as the growth decreased, indicating that the reduction of cell expansion was caused by the reduction of the growth-induced water potential. A similar result was obtained when the growth was reduced by exposure to high pressure (Boyer, 1988). Therefore, a collapse of the growth-induced water potential accounts for the inhibition of cell enlargement caused by environmental stresses such as water deficiency, low temperature and high pressure. It is worth noting that the growth-induced water potential is a physicochemical quantity and changes in the growth-induced water potential can be considered as the first signal received by plants when environmental stresses are imposed on them. After growth inhibition occurs due to the collapse of the growth-induced water potential, subsequent metabolic alteration takes place, and properties of cell walls and membranes will be changed (Nonami and Boyer, 1990a, 1990b).

For practical purposes, the magnitude of a growth-induced water potential can be estimated from water potential differences between the elongating tissue and the water source. When the growth-induced water potential was measured by using hydroponically grown kidney bean seedlings, it was found that growth of the seedlings was correlated with changes in the growth-induced water potential.

To apply the concept of crops grown in plant factories and greenhouses, we grew kidney bean seedlings hydroponically and estimated the magnitude of the growth-induced water potential from water potential differences between the mature region and elongating region. Because growth conditions of these kidney bean seedlings simulate most crops grown under controlled environment and because the estimated growth-induced water potential agreed well with growth of kidney bean seedlings, it is evident that the concept of the growth-induced water potential can be applicable for the regulation of crop growth cultivated in plant factories and greenhouses.

4. Identification, Modelling and Control of the System

Though many papers have been published about identification, modelling and control of non-biological systems, there are few papers about identification for plant growth. Long-term growth of plants or crops is closely related to photosynthesis. For photosynthesis, CO_2 concentration and light intensity could be thought the most important factors among environmental conditions. Therefore, CO_2 uptake in response to environmental stresses is the first target of identification and modelling of the environmental control system. This has been partly identified based on both spectral analysis and the method of least squares.

4.1. SPECTRAL ANALYSIS

Spectral analysis was used to determine the effect of variations in light intensity upon photosysnthesis in sunflowers (Hashimoto et al., 1984). The method can be described in brief as follows:
The light intensity is described by the signal x(t) and the net CO_2 uptake by y(t) in the time domain. In the case, we suppose that the Fourier transform of the light intensity is X(f) and the Fourier transform of the net CO_2 uptake is Y(f) in the frequency domain. Further, we assume that the relation between input and output is described by linear filtering based on g(t) or G(f) as in Fig. 5.

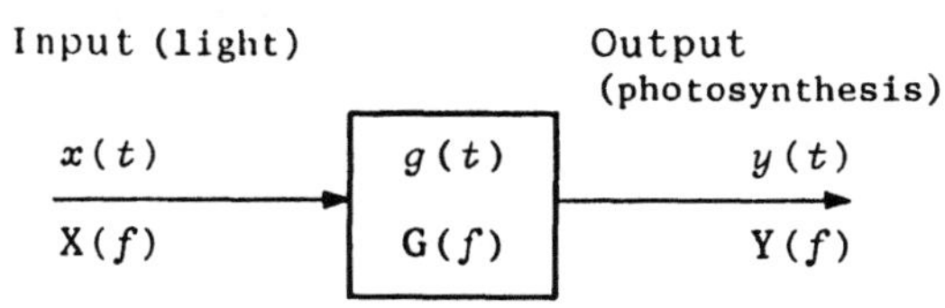

Figure 5. Input–output relation of the system. In the time domain, input is x(t), output is y(t) and g(t) of black box is impulse response. In the frequency domain, input is X(f), output is Y(f) and G(f) of black box is frequency response function.

In general, the time series of x(t) may be random, or stochastic under natural condition. Spectral analysis may be considered one of the most effective identification methods in a system where the input and the output are random variables. Actually, we may compute any

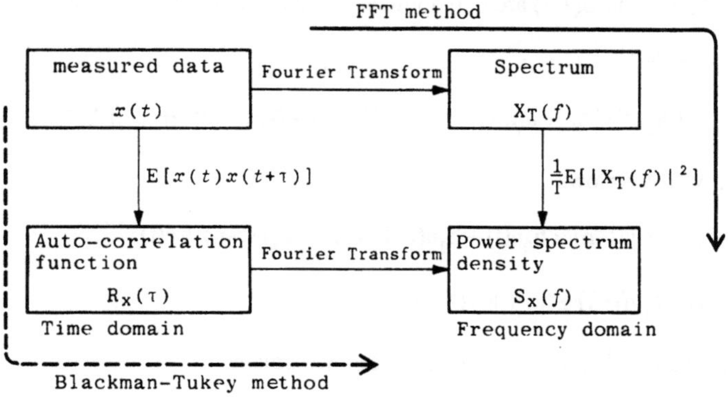

Figure 6. Flow chart for the estimation of power spectrum density. There are two methods. One is FFT (Fast Fourier transform) method and another is Blackman–Tukey's method. FFT method is indicated by the right arrow of solid line and Blackman–Tukey's method, by the left arrow of dotted line.

kind of power spectrum by using FFT (Fast Fourier Transform) algorithm as in Fig. 6.

Finally, we can obtain G(f) from the equation given in

$$Sxy(f) = G(f) \cdot Sx(f) .$$

Thus, g(t) can be calculated from G(f) by operating inverse Fourier transform, and the system is identified.

4.2. METHOD OF LEAST SQUARES BASED ON ARMA

The method of least squares may be explained as follows (Hashimoto and Morimoto, 1985; Torii et al., 1991). This method is based on ARMA (Auto Regressive Moving Average). Let us suppose that the input signal is described by u(t), the output signal by y(t) and the noise by v(t) respectively. We assume that v(t) and u(t) are statistically independent.

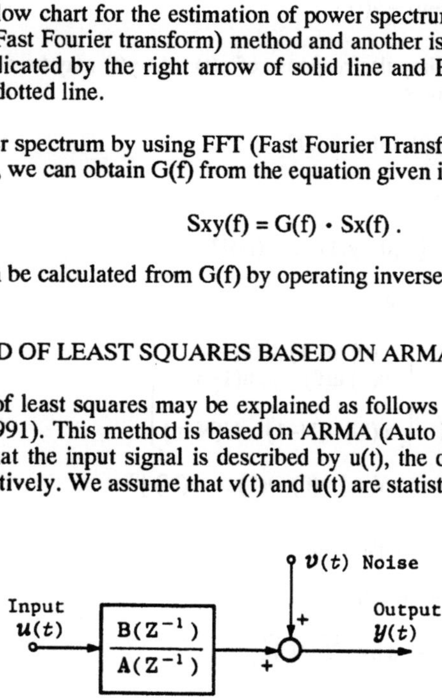

Figure 7. System configuration required for the identification.

We obtain the following input–output relation from Fig. 7.

$$y(t) = B(z^{-1}) \cdot u(t)/A(z^{-1}) + v(t),$$

We assume that $A(z^{-1})$ and $B(z^{-1})$ are polynomials in the backward shift operator z^{-1}.

That is:

$$A(z^{-1}) = 1 + a_1 z^{-1} + ... + a_n z^{-n}, \qquad B(z^{-1}) = b_0 + b_1 z^{-1} + ... + b_n z^{-n}$$

Then, we obtain y(t) as follows.

$$y(t) + a_1 y(t-1) + ... + a_n y(t-n) = b_0 u(t) + b_1 u(t-1) + ... + b_n u(t-n) + e(t)$$

where $e(t) = v(t) + a_1 v(t-1) + ... + a_n v(t-n)$

Suppose that each value of $a_1, a_2, ..., a_n$ is zero, thus we obtain

$$y(t) = b_0 u(t) + b_1 u(t-1) + ... + b_n u(t-n) + v(t)$$

This equation shows y(t) given by impulse response of u(t). As y(t) may be obtained based on time series, then y(t) is described as

$$\{y(t), t=0,1,2,...N\}$$

Or, we can describe Y(t) in vector as

$$Y = [\, y(0), y(1), ..., y(N)]^T$$

Similarly, U(t) in matrix, b and V(t) in vector as

$$U = \begin{bmatrix} u(0)\ u(-1) ...... u(-n) \\ u(1)\ u(0) ...... u(1-n) \\ .... \\ u(N)\ u(N-1) ... U(N-n) \end{bmatrix}$$

$$b = [\, b_0, b_1, ..., b_n]^T, \qquad V = [\, v(0), v(1), ..., v(N)]^T$$

Therefore, we can obtain the following equation.

$$Y = Ub + V$$

We may denote the optimal least squares by $\hat{Y} = U\hat{b}$.
Then,

$$\partial J / \partial \hat{b} = 0 \ , \ \text{where } J = \| Y - \hat{Y} \|^2$$

Thus, $\hat{b} = (U^TU)^{-1}U^TY$

Therefore we can predict any output by using $\hat{b}$ and the system is also identified.

4.3. GMDH (GROUP METHOD OF DATA HANDLING)

Dynamic model of CO_2 uptake is identified based on GMDH (Hashimoto and Yi, 1988b). It is reported that GMDH is developed by Ivakhnenko based on the pioneering work of Wiener and Gabor for nonlinear system representations and the work of Rosenblatt for the perception concept (Toldalagi, 1982). The method is effective for identifying the nonlinear systems of unknown structure. A schematic diagram of GMDH is shown in Fig. 8.

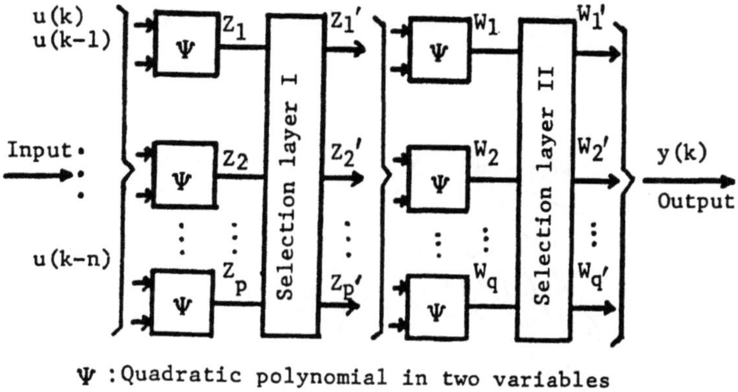

Figure 8. The GMDH model structure.

The algorithm can be described in brief as follows: Let y(k) denote the system output, u(k) denote the system input. Choose u(k−i) and u(k−j) arbitrarily from u(k), u(k−1),..., u(k−n), where, n is the input variable number, and combine the two variables with a quadratic polynomial Ψ, then $n!/(2(n-2)!)$ equations, which are called partial description, can be obtained. Denote those partial descriptions with Z_1, Z_2, ..., Z_P, and select them using a selection criterion in selection layer I. Suppose that p' equations are passed through selection layer I, and let Z_1', Z_2',..., Z_p' denote these equations. Combine Z_i and Z_j with the same quadratic polynomial Ψ, then, $p'!/(2(p'-2)!)$ partial descriptions are obtained. Denote them as W_1, W_2, ..., W_q, and make a selection with the criterion in the selection layer II. Combine W_i, W_j with the same quadratic polynomial Ψ from W_1, W_2,..., W_q passed through selection layer II and select the partial descriptions again as described above. Repeat those steps until they are passed through a satisfactory criterion, then, the optimal representation of the input–output relation can be obtained.

4.4 APPLICATIONS OF NEURAL NETWORK TO IDENTIFICATION

Applications of neural network to the identification are examined (Morimoto et al., 1991a ; 1991b). The method is explained as follows: Suppose that the input–output relation are described by ARMA. Thus,

$$y(t) = - a_1 y(t-1) - .... - a_n y(t-n) + b_0 u(t) + b_1 u(t-1) + ... + b_n u(t-n)$$

From this equation, it seems evident that the output y(t) is obtained from the weighted sums of the n th historical output data { y(t−1),..., y(t−n)} and (n+1) th historical input data { u(t), U(t−1),..., U(t−n) }. The structure of the neural network is built based on the equation. That is composed of three layers as shown in Fig. 9.

These are input layer, hidden layer and output layer. The input layer is composed of (2n+1) data given as { y(t−1),..., y(t−n)} and { u(t), U(t−1),..., U(t−n) }. Therefore the input layer is thought to have neurons composed of (2n+1) data. It may be noted that the neural network can acquire the complex dynamics such as non-linearity and time-varying parameters by its own high learning capability. The learning procedure used here is based on the error back propagation, where the optimal weights (W_{ij}) and biases (H_i) are obtained by computing the squared error between the observed and the calculated values to its minimum value.

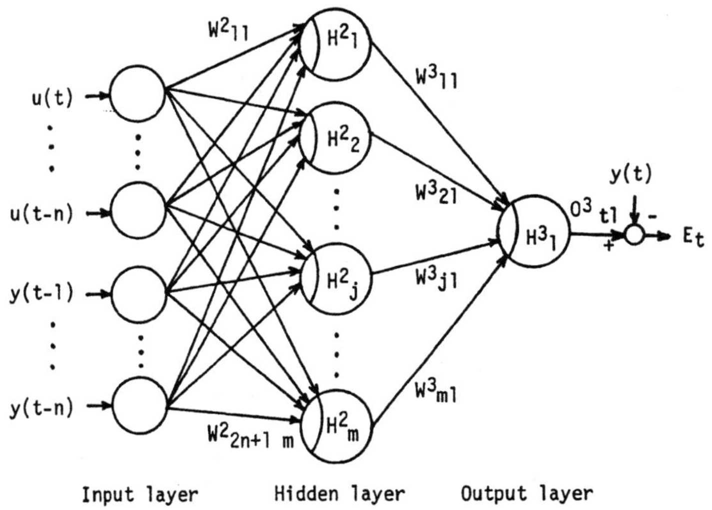

Figure 9. Structure of neural network.

4.5. ENVIRONMENTAL CONTROL

Environmental control in the agricultural field has been developed by greenhouse engineers (Takakura, 1991). Now a lot of controlling technologies are introduced to the advanced system of greenhouse and plant factory from process industries. From the aspect of computer control of the system, the routine of the process control is established based on micro-computer (micro-processor). Control algorithm is also ready in PID algorithm based on feedback control system. Computer calculates the deviation between the set-point for the environment and the data monitored from the change of the environment in each sampling interval, and decides the manipulating signal by operating PID action on the deviation. Environmental factors including the light, temperature, relative humidity and CO_2 concentration can be divided into several sub-systems when those factors are controlled respectively. Then, several minutes of sampling interval may be sufficient enough to control the environmental factors. Furthermore, only very short time is needed for processing sampled data in PID control algorithm. Therefore, it is not difficult to combine all the sub-systems by one micro-computer based on a time sharing method. Nutrient control, however, has a big dead time. Therefore, it might be necessary that the system should be examined based on feedforward control algorithm as well as feedback one. Furthermore, EC and pH could not identify the state of nutrient solution. Therefore the ion sensors are introduced to the system (Hashimoto et al., 1988). New method for the nutrient supply is also proposed (Kurata, 1991). As for the control of light, the intensity and spectrum of the light have strong effect on plant growth. It seems interesting that the effect of frequency involved in the on–off interval of fluorescent lamp is not meaningless (Hashimoto and Yi, 1988a). These technology of environmental control is now introduced to the process of micropropagation of tissue culture and nursery system (Kozai, 1991).

4.6. ADAPTIVE, OPTIMAL, AND FUZZY CONTROL

To control the environment more flexibly, it is necessary to change the set point of the environment and the system parameters disturbed by the sudden changes in the outside. This belongs to one of the adaptive control problems. To reduce the cost to the minimum in non–biological system, the optimal strategy is needed. This can be solved with the optimal control theory. The optimal control is much more effective for biological system (Hashimoto, 1989a). Though the advanced control algorithm is rather complicated , it could be improved partly from the fundamental PID algorithm. Further, nutrient control is also examined as the example of fuzzy control (Morimoto et al., 1991a ; 1991b).

5. Plant Factory Automation

5.1. GREENHOUSE AUTOMATION

Human labor has been necessary in seeding, transplanting, harvesting and spreading farm chemicals, even in the process control system. Some robots and automated mechanizations begin to take place of these human labour in the agricultural and horticultural field. In England, we can find an interesting example in commercial grower, where several types of automated mechanization substitute farmers as a pioneer of greenhouse automation.

5.2. ROBOT FOR MICROPROPAGATION

Clean and micro–robot are used in the LSI (Large Scale Integrated circuit of semi–conductor) factory, which could get over the unable barrier by human labour. Circumstance of micro–propagation is closely similar to that of LSI industry. The robots for micropropagation and transplanting in tissue culture are developed (Fujiwara, 1991; Kinase, 1991; Simonton, 1991), which will greatly improve nursery industry in agriculture and horticulture.

5.3. PLANT FACTORY AUTOMATION

In the near future, robots, containing intelligent robot and intelligent tele–robot, will substitute the human labor in plant factory including tissue culture. With these robots and mechanization, the plant factory may be called "Plant Factory Automation" just like FA (Factory Automation) in the industrial manufacture (Hashimoto, 1989b).

6. Computer Integrated Plant Factory

In industrial manufacture, the use of robots has elevated the status of the automation system from process automation (PA) to factory automation (FA). Furthermore, the computer has been introduced into the factory management and is referred to as computer integrated manufacture (CIM). Referring the circumstance in industriy, a computer integrated system in agriculture and horticulture has been proposed (Hashimoto, 1991). Figure 10 shows the recent computer integrated system.

The system has the computer network composed of host computer, usual micro-computer for process control and the special computer for knowledge processing based on the artificial intelligence. The necessary menu is chosen through the conversation between operator and computer as shown in Fig. 11.

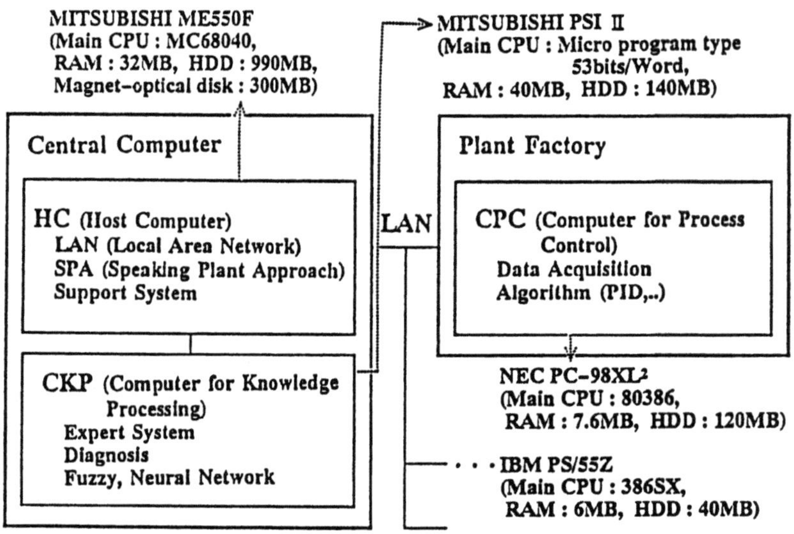

Figure 10. The computer integrated system.

Five main menus are prepared , these are "LAN", "Process Control", "SPA(speaking plant approach)", "Support System", and "Knowledge Processing" as shown in the left part of the figure. Each menu is used for the purpose described in the right part of the figure. The languages used in the main menu are "BASIC", "C" and "PROLOG".

The menu for "LAN" deals with computer communication with the proper protocol. The system is programmed with about several hundred steps. The menu for "Process Control" is always used for monitoring and real time control, depending on mainly mathematical and physical method. The menu for "SPA" is more complicated for the purpose of identification and optimization of total and sub-system based on physiological plant ecology. The menu for "Support System" is prepared for the decision support of horticultural operation and management. Finally, the menu for "Knowledge Processing" is developed for expert system and diagnosis system based on the artificial intelligence. Thus, plant factory is elevated to the level of "Intelligent System".

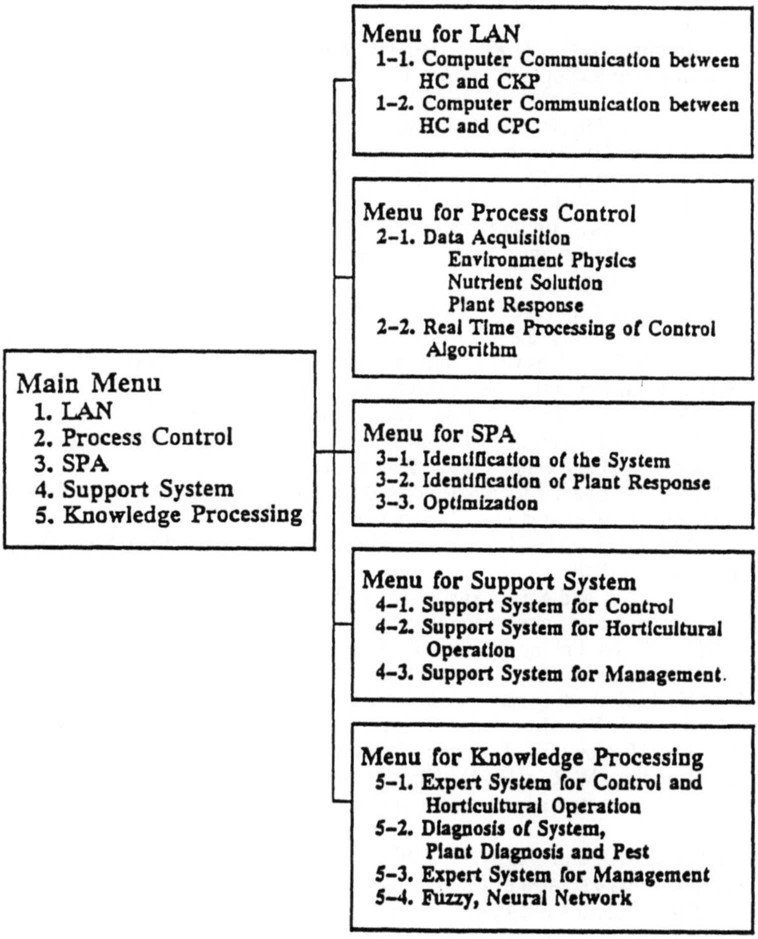

Figure 11. The menu of software for the computer integrated system.

REFERENCES

Bensen, R.J., Boyer, J.S. and Mullet, J.E. (1988) 'Water deficit-induced changes in abscisic acid, growth, polysomes, and translatable RNA in soybean hypocotyls', Plant Physiol., 88, 289-294.

Boyer, J.S. (1967) 'Leaf water potentials measured with a pressure chamber', Plant Physiol., 42, 133-137.

Boyer, J.S. (1988) 'Origin of growth-induced water potentials: Pressure and temperature effects', Plant Physiol. (supplement), 86, 202.

Boyer, J.S., Cavalieri, A.J. and Schulze, E.-D. (1985) 'Control of the rate of cell

enlargement: Excision, wall relaxation, and growth-induced water potentials', Planta, 163, 527–543.
Boyer, J.S., and Knipling, E.B. (1965) 'Isopiestic technique for measuring leaf water potentials with a thermocouple psychrometer', Proc. Natl Acad. Sci., 54, 1044–1051.
Boyer, J.S. and Potter, J.R. (1973) 'Chloroplast response to low leaf water potentials. I. Role of turgor', Plant Physiol., 51, 989–992.
Boyer, J.S. and Wu, G. (1978) 'Auxin increases the hydraulic conductivity of auxin-sensitive hypocotyl tissue', Planta, 139, 227–237.
Bozarth, C.S., Mullet, J.E. and Boyer, J.S. (1987) 'Cell wall proteins at low water potentials', Plant Physiol., 85, 261–267.
Cassab, G.I., and Varner, J.E. (1988) 'Cell wall proteins. Annu. Rev. Plant Physiol', Plant Mol. Biol., 39, 321–353.
Cavalieri, A.J. and Boyer, J.S. (1982) 'Water potentials induced by growth in soybean hypocotyls', Plant Physiol., 69, 492–496.
Cleland, R.E. (1959) 'Effect of osmotic concentration on auxin action and on irreversible and reversible expansion of the *Avena* coleoptile', Physiol. Plant., 12, 809–825.
Cleland, R. (1971) 'Cell wall extension. Annu. Rev.', Plant Physiol., 22, 197–222.
Cleland, R.E. (1975) 'Auxin-induced hydrogen ion excretion: Correlation with growth, and control by external pH and water stress', Planta, 127, 233–242
Cleland, R.E. (1983) 'The capacity for acid-induced wall loosening as a factor in the control of *Avena* coleoptile cell elongation', J. Exp. Bot., 34, 676–680.
Cleland, R.E. and Rayle, D.L. (1978) 'Auxin, H^+ excretion and cell elongation', Bot. Mag. (Spec. issue), 1, 125–139
Cutler, J.M., Shahan, K.W. and Steponkus, P.L. (1980) 'Influence of water deficits and osmotic adjustment on leaf elongation in rice', Crop Sci., 20, 314–318.
Fry, S.C. (1986) 'Cross-linking of matrix polymers in the growing cell walls of angiosperms', Annu. Rev. Plant Physiol., 37, 165–186.
Fujiwara, H. (1991) 'Discriminating robot system for carnation seedlings with fuzzy logic', in Y. Hashimoto and W. Day (eds.), Mathematical and Control Applications in Agriculture and Horticulture, Pergamon Press, Oxford, pp.231–235.
Green, P.B. and Cummins, W.R. (1974) 'Growth rate and turgor pressure: Auxin effect studied with an automated apparatus for single coleoptiles', Plant Physiol., 54, 863–869
Green , P.B., Erickson, R.O. and Buggy, J. (1971) 'Metabolic and physical control of cell elongation rate. *In vivo* studies in *Nitella*', Plant Physiol., 47, 423–430.
Hashimoto, Y. (1989a) 'Recent strategies of optimal growth regulation by the speaking plant concept', Acta Horticulturae 260, 115–121.
Hashimoto, Y. (1989b) 'Dynamical approach to plant factory', Proc. AGROTIQUE 89, 111–118.
Hashimoto, Y. (1991) 'Computer integrated plant growth factory for agriculture and horticulture', in Y. Hashimoto and W. Day (eds.), Mathematical and Control Applications in Agriculture and Horticulture, Pergamon Press, Oxford, pp.105–110.
Hashimoto Y., and Morimoto, T. (1985) 'Identification of water relation and CO_2 uptake in physiological ecological processes in a controlled environment', Proc. IFAC 7th Symposium on Identification and System Parameter Estimation, 1677–1681.
Hashimoto Y., Morimoto T., Fukuyama T., Watake H., Yamaguchi S., and Kikuchi, H. (1988) 'Identification and control of hydroponic system using ion sensors. Acta Horticulturae 245, 490–497.
Hashimoto Y., and Nonami, H. (1990) 'Overview of current measurement techniques from aspects of plant science', in Y. Hashimoto, P.J. Kramer, H. Nonami and B.R. Strain(eds.), Measurement Techniques in Plant Science, Academic Press, San Diego, pp.7–24.

Hashimoto Y., Strain B.R., and Ino T. (1984) 'Dynamic behaviour of CO_2 uptake as affected by light', Oecologia 63, 159–165.

Hashimoto Y., and Yi, Y. (1988a) 'Control model of plant as affected by the pulsed light illumination', Acta Horticulturae 245, 448–454.

Hashimoto Y., and Yi, Y. (1988b) 'Dynamic model of CO_2 uptake based on system identification', Acta Horticulturae 248, 295–300.

Kinase A., and Watake, H. (1991) 'Robot for masspropagation', in Y. Hashimoto and W. Day (eds.), Mathematical and Control Applications in Agriculture and Horticulture, Pergamon Press, Oxford, pp.225–230.

Kozai, T. (1991) 'Environmental control in plant tissue culture and its application for micropropagation', in Y. Hashimoto and W. Day (eds.), Mathematical and Control Applications in Agriculture and Horticulture, Pergamon Press, Oxford, pp.99–104.

Kramer, P.J. (1959) 'The role of water in the physiology of plants', Adv. Agron., 11, 51–70.

Kurata, K. (1991) 'Roots dividing method for nutrient supply in soilless culture', in Y. Hashimoto and W. Day (eds.), Mathematical and Control Applications in Agriculture and Horticulture, Pergamon Press, Oxford, pp.137–140.

Mason, H.S. and Matsuda, K. (1985) 'Polyribosome metabolism, growth and water status in the growing tissues of osmotically stressed plant seedlings', Physiol. Plant., 64, 95–104.

Mason, H.S., Mullet, J.E. and Boyer, J.S. (1988) 'Polysomes, messenger RNA and growth in soybean stems during development and water deficit', Plant Physiol., 86, 725–733.

Matsuda, K. and Riazi, A. (1981) 'Stress-induced osmotic adjustment in growing regions of barley leaves', Plant Physiol., 68, 571–576.

Meyer, R.F. and Boyer, J.S. (1972) 'Sensitivity of cell division and cell elongation to low water potentials in soybean hypocotyls', Planta, 108, 77–87.

Meyer, R.F. and Boyer, J.S. (1981) 'Osmoregulation, solute distribution, and growth in soybean seedlings having low water potentials', Planta, 151, 482–489.

Michelena, V.A. and Boyer, J.S. (1982) 'Complete turgor maintenance at low water potentials in the elongating region of maize leaves', Plant Physiol., 69, 1145–1149.

Molz, F.J. (1976a) 'Water transport through plant tissue: the apoplasm and symplasm pathways', J. Theor. Biol., 59, 277–292.

Molz, F.J. (1976b) 'Water transport in the soil-root system: transient analysis', Water Resource Res., 12, 805–808.

Molz, F.J., and Boyer, J.S. (1978) 'Growth-induced water potentials in plant cells and tissues', Plant Physiol., 62, 423–429.

Molz, F.J., and Ikenberry, E. (1974) 'Water transport through plant cells and cell walls: theoretical development', Soil Sci. Soc. Am. J., 38, 699–704.

Morimoto T., Cho I., and Hashimoto, Y. (1991a) 'Identification of hydroponics in an advanced control system of the greenhouse', Proc. IFAC 9th Symposium on Identification and System Parameter Estimation, pp.610–615.

Morimoto T., and Hashimoto, H. (1991b) 'Application of fuzzy logic and neural network to the process control of solution pH in deep hydroponic culture', in Y. Hashimoto and W. Day (eds.), Mathematical and Control Applications in Agriculture and Horticulture, Pergamon Press, Oxford, pp.147–152.

Nonami, H. and Boyer, J.S. (1987) 'Origin of growth-induced water potential: Solute concentration is low in apoplast of enlarging tissues', Plant Physiol., 83, 596–601.

Nonami, H., and Boyer, J.B. (1989) 'Turgor and growth at low water potentials', Plant Physiol., 89, 798–804.

Nonami, H., and Boyer, J.B. (1990a) 'Primary events regulating stem growth at low water potentials', Plant Physiol., 93, 1601–1609.

Nonami, H., and Boyer, J.B. (1990b) 'Wall extensibility and cell hydraulic conductivity decrease in enlarging stem tissues at low water potentials', Plant Physiol., 93, 1610–

1619.
Nonami, H., J.S. Boyer, and Steudle, E. (1987) 'Pressure probe and isopiestic psychrometer measure similar turgor', Plant Physiol., 83, 592–595.
Nonami, H., Hashimoto, Y. and Boyer, J.S. (1991) 'Environmental control for plant growth in plant factory operation and greenhouse management from physiological viewpoint', in Y. Hashimoto and W. Day (eds.), Mathematical and Control Applications in Agriculture and Horticulture, Pergamon Press, Oxford, pp. 125–130.
Nonami, H., and Schulze, E.-D. (1989) 'Cell water potential, osmotic potential, and turgor in the epidermis and mesophyll of transpiring leaves: Combined measurements with the cell pressure probe and nanoliter osmometer', Planta, 177, 35–46.
Probine, M.C. and Preston, R.D. (1962) 'Cell growth and the structure and mechanical properties of the wall in internodal cells of *Nitella opaca*. II. Mechanical properties of the walls', J. Exp. Bot. 13, 111–127.
Simonton, W. (1991) 'Automated plant handling and processing in a robotic workcell', in Y. Hashimoto and W. Day (eds.), Mathematical and Control Applications in Agriculture and Horticulture, Pergamon Press, Oxford, pp.213–218.
Takakura, T. (1991) 'Greenhouse Production in Engineering Aspects', in Y. Hashimoto and W. Day (eds.), Mathematical and Control Applications in Agriculture and Horticulture, Pergamon Press, Oxford, pp.19–21.
Theologis, A. (1986) 'Rapid gene regulation by auxin', Annu. Rev. Plant Physiol., 37, 407–438.
Toldalagi P.M., and Mehra R.K. (1982) 'Application of nonlinear identification to long range weather prediction', Proc. IFAC 6th Symposium on Identification and System Parameter Estimation Vol.1, 512–520.
Torii T., Ochiai T., Okamoto T., and Kitani, O. (1991) 'Identification of the water status of the plant-identification of AR model', in Y. Hashimoto and W. Day (eds.), Mathematical and Control Applications in Agriculture and Horticulture, Pergamon Press, Oxford, pp.389–393.
Vaadia, Y., Raney, F.C. and Hagan, M.R. (1961) 'Plant water deficits and physiological processes', Annu. Rev. Plant Physiol. 12:265–292.
Van Volkenburgh, E and Boyer, J.S. (1985) 'Inhibitory effects of water deficit on maize leaf elongation', Plant Physiol. 77, 190–194
Westgate, M.E. and Boyer, J.S. (1985) 'Osmotic adjustment and the inhibition of leaf, root, stem and silk growth at low water potentials in maize', Planta 164, 540–549.

PERSPECTIVE OF MICROPROPAGATION INDUSTRY

IRWIN Y. E. CHU
Twyford International Inc.
15245 Telegraph Road
Santa Paula, California 93060 U.S.A.

ABSTRACT. Recently, tissue culture micropropagation has become one of the most important ways to reproduce crops that are difficult to propagate by conventional methods, such as by seed or cuttings. The speed of tissue culture technology development has been accelerated by its practical commercialization. Tissue culture micropropagation offers many unique advantages over other propagation methods. At the present time, the greatest application of tissue culture micropropagation is to tropical ornamentals. However, systems for successfully producing food crops such as potato, asparagus, and strawberry have been defined and are currently replacing traditional methods of vegetative propagation.

For many crops, the tissue culture process is currently competitive as a method of propagation. However, the widespread application of tissue culture propagation to many other crops is limited, since it is currently not cost-competitive with conventional propagation methods, e.g., seeds, cuttings, division. The major factor limiting the cost-competitiveness of micropropagation for many crops is the quantity of labor and handling currently required. Tremendous development activities have been conducted in these cost reduction and market expansion areas and the major breakthrough will be made in the near future.

1. Advantages and Limitations of Micropropagation

1.1 ADVANTAGES

1.1.1. Faster product introduction. Micropropagation can introduce the product into the market place faster than conventional methods based on the following facts:

a) Single elite plant selection. Theoretically, a uniform large population can be reproduced from any single plant. Therefore, it is possible to select one desirable plant and build up to a large number for commercial purposes, even if it is genetically heterozygous. This application could be adopted to any crop in agricultural, horticultural and forest trees.

b) Fast build up and propagation. In most cases, the current production system offers a fast build up speed, such as from one plant to a million plants, in 12 - 24 months. This advantage provides a great benefit to male sterile parent propagation and also

K. Kurata and T. Kozai (eds.), Transplant Production Systems, 137–150.

single elite clone selections. It is not unusual for a seed and plant producer to save a few years of build up by use of micropropagation.

c) Greater uniformity of products. Micropropagation is basically a cloning technology. It produces greater uniform population either from homozygous or heterozygous plant genotypes. Furthermore, a totally controlled growing laboratory environment also contributes tremendously in the uniformity of the products.

d) Greater volume per production space. Since the system is total control in the laboratory, without any seasonal limitation, and starting plants are smaller in size, the number of plants which could be produced in a certain space is much greater than conventional greenhouse or field production. This creates a great advantage for developed countries where the land and buildings are expensive.

1.1.2. Greater product enhancement. Product value could be increased tremendously based on the following points:

a) Phenotypic improvement. A phenotype of the plant can be manipulated much easier in tissue culture environment than in conventional greenhouse or field conditions. Various growing parameters and different growth regulators could change plant type as desired. For example, well-branch, dwarf, early maturity are used commercially today for increasing the value of the product.

b) Disease tree plants. Bacteria, fungi and virus elimination are very commonly used in micropropagation. Some of the viruses are also indexed in certain crops. For example, lily, geranium, potato and asparagus have benefitted greatly from this application.

1.1.3. Better product marketability. Marketability of micropropagated produce can be enhanced by the following facts:

a) Flexible form of product. A much greater range of product formats can be created from micropropagation system. For example, shoots can be marketed as production stock in agar, unrooted microcuttings, rooted microcutting and greenhouse liner plugs. Bulb crops can be marketed at different sizes and stages, even at a format of a propagated shoot.

b) Shipment advantages. Plants are produced in sterile conditions. Also, they can be certified as free of disease in many cases. Therefore, international phytosanitary regulations are more easily met by this way of plant production than by any other.

c) Full year production. Production can be operated any time of a day and any season of a year. Furthermore, many of the plants could be stored in a certain condition to avoid a peak production season, even though the product has to be shipped in a certain short period of time.

1.2. LIMITATIONS

1.2.1. Restriction of product line. Not all of the crops can be commercially propagated by micropropagation at current technology levels. In spite of the fact that numerous academic reports have been published in different crops, they are often not suited to commercial application yet.

1.2.2. Relatively high production cost. Micropropagation still requires a lot of high skill technical labor. Therefore, the product cost is still higher than conventional methods, such as cutting and seed propagation in many crops.

1.2.3. Happening of off type product. Even though micropropagation is a cloning technology, mutation could happen during the stages of production. Stable production and early detection systems are applied as much as possible in commercial industry, but mutation is always a risk in this method.

2. Production System and Cost

2.1. PRODUCTION STAGES IN PROCESS.

The commercial production of micropropagation has a relatively complicated process that requires many steps or stages. There are five distinct stages, each with specific requirements and considerations.

2.1.1. Stage 0. Stage 0 is the initial part of the production cycle. This stage could take 3 to 6 months under certain conditions. This stock plant conditioning stage includes reducing level of disease contamination and health of plant growing conditions.

2.1.2. Stage I. Stage I is the establishment of an aseptic culture. Procedures of surface sterilizing the explants and media for inducing a healthy growth culture must be established for each species. Plant materials identified as free from contamination and vigorous plants would then be used for Stage II. Stage I usually requires from 3 to 24 months, and at least four subcultures or passages.

2.1.3. Stage II. Once a clean culture is obtained and grown from the explant, the culture enters production stage, or Stage II. A maximum rate of multiplication is expected and culture condition and media should be formulated to achieve that goal. Passage and culture cycle usually ranges from 4 to 8 weeks depending on the crop. Multiplication rates range from 2.0 x to 8.0 times. Shoots are harvested from the Stage II cultures to either be sold as unrooted microcuttings or carried on to Stage III for rooting. Stage II generally lasts 10 - 36 months.

2.1.4. Stage III. At present, a large number of species cannot be rooted directly in the greenhouse. Therefore, Stage III or in vitro rooting is necessary. Media and growing conditions were designed for rooting, but not aggressive multiplication. Stage III generally requires 2 - 8 weeks.

2.1.5. Stage IV. Stage IV is the first step for growing plants outside of the sterile laboratory condition. Plantlets are re-established in soil in the greenhouse. Different methods are used for acclimation process to reduce the shock of environmental change. That is, from sterile and moistured condition to greenhouse and a tougher condition. Plastic covered tents and fine particle fog systems are the most commonly used acclimations. Liner plug plants can be established in artificial growing medias, such as soilless mixes, rockwool and sponges. Stage IV usually requires 4 - 16 weeks.

2.2. PRODUCTION METHOD IN CROPS

2.2.1. Axillary Branching and Cutting. At present, axillary branching method is probably the most common production method used in commercial industry. Both terminal and axillary shoot meristems tissue can be used to initiate cultures. By using particular growth regulators and growing conditions, the apical dominance is suppressed and axillary branching is induced. Plants could be cut as single shoots or divided as shoot clumps. The starting multiplication rate is usually slow, but rate will rapidly increase after several subcultures.

2.2.2. Nodal Growth and Cutting. Individual nodal sections are excised by each subculture. The axillary bud grows and elongates, producing additional nodes for excisim. This method provides the least possible genetic variation of the product.

2.2.3. Adventitious Shoots Separation. Adventitious shoots can be isolated from stem, leaf or any part of plant tissue. Shoot producing cultures can be repeated until the desired number of plants has been obtained. Mutation is more often induced in this system. Therefore, it is not recommended to apply in a critical production such as parent lines for seed production.

2.2.4. Somatic Embryogenesis Multiplication. By passing a callus stage, somatic embryo can be induced and continuously reproduce secondary embryos in a well designed culture system. These embryos may be grown into complete plants in soil. Somatic embryogenesis is considered as the most powerful, ideal system of the future since it would be totally automated and require a minimum of space. Furthermore, both liquid and solid systems could be applied.

2.3. PRODUCTION COST COMPONENTS

2.3.1. Laboratory. Production costs from Stage 0 to Stage III are presented on laboratory cost. As indicated in Table 1, more than 60% of production costs are based on labor.

TABLE 1. RELATIVE LABORATORY PRODUCTION COST COMPONENTS

Cost Components	Percentage
Direct Labor	51
Supervision	13
Materials	10
Utilities	9
Depreciation	12
Misc. Cost	5
TOTAL	100%

2.3.2. Greenhouse - Production costs from Stage IV are presented in greenhouse costs.

TABLE 2. RELATIVE GREENHOUSE PRODUCTION COST COMPONENTS

Cost Components	Percentage
Direct Labor	40
Supervision	10
Materials	23
Utilities	10
Depreciation	10
Misc. Cost	7
TOTAL	100%

Both lab and greenhouse cost calculations are based on Twyford International, Inc., California, Operation Study in 1990. These numbers are based on a total annual production volume of 10 million units. A total of 10 different plant genera and 80 different clones were produced.

3. Operating Facility

3.1. GEOGRAPHIC LOCATION.

When a new operation is to be built, the location should be chosen carefully. Important factors to be considered include: climate for plant production, water quality, labor resource, close to market and convenient transportation. If air transportation is necessary for shipping, then distance to airport needs to be considered. Inexpensive labor should not be the only factor for consideration.

3.2. OPERATION FLOOR PLAN.

Efficiency and work flow need to be considered for the operations floor plan. The building needs to be easy to clean and easy to maintain a sterile condition. Building should be solid so that an accurate temperature control is possible. Positive air pressure should cover the entire building and be initiated from growth room to transfer room and then to other areas.

3.3. LABORATORY

3.3.1. Media Preparation and Cleaning Room. The media preparation room should include a source of purified, reverse osmosis water, balance, pH meter, refrigerator, mix tank, stirrer, adequate counter and storage space. In a large scale commercial laboratory, a culture media dispensing apparatus for multiple culture vessels is necessary. Usually, laboratories are equipped with double door autoclaves. Cleaning room includes a commercial dishwashing machine in an easy to clean condition since this is a common source of contamination.

3.3.2. Cutting, Planting and Harvesting Room. This is the center of the production activity and 80% of employees will be in this room at any given time. Therefore, the room needs to be designed with particular attention to minimizing contamination. Generally, airflow is maintained under positive pressure and sticky floor mats are placed on the floor at entry. Cleaning of filters needs to be done on a regular basis and airflow velocity to maintain aseptic working conditions needs to be checked. Sterilizers for forceps and scalpels should be included and electrically heated ceramic tubes, an ultrasonic cleaner and a heated vessel filled with small glass beads are commonly used.

3.3.3. Growth and Inventory Control Room. Plants are grown and maintained in this artificially controlled room with accurately regulated temperature, light quality and photoperiod. Lighted shelving should be open in structure to permit airflow for better temperature control. Mobile or rolling shelving can be installed to maximize space utilization.

3.4. GREENHOUSE

3.4.1. Planting, Acclimation and Growing Area. These areas are best located together, but away from the laboratory to minimize possible contamination. Microcuttings are transplanted into a soilless mix or other artificial substrate. Then, these plants are generally acclimated in a fog house equipped to produce a very fine particle mist of high humidity condition. Once the shoots are rooted or the plants are acclimated, they are transferred to a conventional production greenhouse.

3.4.2. Packing and Shipping. Usually, plants are grown in plug trays. Special cardboard shipping boxes are designed to fit the tray size and provide adequate aeration . A combination of air and ground transportation is generally utilized. For shipping unrooted or rooted microcuttings, a shipping box with styrofoam insulation may be necessary for additional protection.

4. Plant Tissue Culture Industry

Since the beginning of the micropropagation industry in the 1970's, there has been a steady increase in the number of commercial micropropagation laboratories. There are currently more than 50 such companies in the United States. The U.S. laboratories are distributed as follows, in descending order of rank; Florida, California, Texas and Hawaii. There are 3 to 5 relatively large size companies and others are growing specialized crops or are small size operations. The larger labs employ over 100 employees each and tend to produce large volume crops, particularly foliage crops and ornamental species.

The globalization of the micropropagation industry has been particularly active during the 1980's and commercial laboratories are primarily located in the following countries other than the United States: Australia, Belgium, Canada, France, Israel, Italy, Japan, the Netherlands, the United Kingdom, India and Malaysia. The expansion of the micropropagation industry worldwide has not only increased potential markets but also helped to create new product opportunities.

4.1. CROPS

4.1.1. Foliage plants. Definitely, the largest crop in tissue culture is foliage and it is primarily in America. Total units have been estimated at 60 million units in the U.S. and 100 million units worldwide.

4.1.2. Flowering species. This includes cut flowers and pot plants. Total units have been estimated at 30 million in Europe and 50 million worldwide.

4.1.3. Orchids. Commercial plants tissue culture business was started from the orchid business. Due to the fact of the slow growing nature of orchids, the business units are small and the industry has not grown so much. Total units have been estimated as 15 million in Asia and 25 million worldwide.

4.1.4. Bulbs. Bulbs include cut flower, pot plants and garden products. Total units has been estimated at 15 million in Europe and 20 million worldwide.

4.1.5. Parent lines for seed production. This includes flower and vegetables. Production is especially concentrated on male sterile parent, polyploid parent and heterozygous lines. Total units have been estimated as 5 million in America and 8 million worldwide.

4.1.6. Vegetable clones. Only a few species are produced commercially. These include: asparagus, garlic, potato and melons. Total units have been estimated as 5 million in American, 5 million in Europe and 12 million worldwide.

4.1.7. Fruit. This includes banana, pineapple, vine fruit and large woody fruit. Total units have been estimated at 10 million in America and 23 million worldwide.

4.1.8. Root stocks. This includes root stocks for fruit and vegetables. Total units have been estimated at 2 million units in America and 3 million worldwide.

4.1.9. Plantation crops. This includes all plantation crops ranging from palms and other industry crops. Total units have been estimated at 5 million in Asia and 10 million worldwide.

4.1.10. Forest and industrial trees. This includes tress for forest, paper, landscape and other special purposes. Total units have been estimated at 3 million in America and 6 million worldwide.

4.2 MARKET SIZE WORLDWIDE

As indicated in Table 3, the total tissue culture units worldwide is approximately 250 million units. It consists of 100 million units in both America and Europe and another 50 million units in Asia. This number includes many of internal and self propagator, therefore a real commercial available market size will be much smaller.

TABLE 3. WORLDWIDE MARKET IN TISSUE CULTURE

Crops	America	Europe	Asia	Total
Foliage Plants	60	30	10	100
Flowering species	15	30	5	50
Orchids	5	5	15	25
Bulbs	3	15	2	20
Parent lines	5	2	1	8
Vegetable clones	5	5	2	12
Fruits	10	8	5	23
Root stocks	2	1	1	4
Plantation crops	3	2	5	10
Forest trees	3	2	1	6
TOTAL	111	100	47	258

5. Future Trends of the Tissue Culture Industry

For estimating the trends of business development in the future, various factors have to be considered concurrently. Automation for reducing production cost, extend application into non-ornamental crops, expand market into new geographic areas, develop new products by using tissue culture advantages and replace conventional cutting and sexual propagation by cost and quality all have to be considered as one piece of strategy.

5.1. FUTURE PRODUCTION SYSTEMS

The future expansion of the micropropagation industry to include a greater variety of plant species, such as plantation, forest and vegetable crops, depends upon the reduction of current production costs and ultimate selling prices. More cost-effective production systems will support competitive pricing compared with the more conventional propagation methods of cuttings or seed. Of course, the greater the associated benefits of producing a particular species via micropropagation, the greater the relative selling cost accepted by the marketplace. Furthermore, due to the highly competitive nature of the micropropagation industry, as more tissue culture products are accepted, it will also become increasingly important to protect proprietary technologies and products. This will undoubtedly affect production strategies.

5.1.1. Conventional In-house Production. Even though various automation systems will be developed, offshore arrangements will provide inexpensive direct labor cost, but there will be a certain number of crops that cannot be automated and a certain number of proprietary products and technology would not like to be transferred to offshore for security reasons. Therefore, the crops which are too difficult to automate, too small in number to automate, and high price, high technology items will always be produced in-house. Furthermore, some of the contract production from breeding companies may also request the materials should not leave the country.

5.1.2. Automation and Robotic Production. In the last few years, tremendous effort has been made in development of automation and robotic application in micropropagation. Automation could be applied in almost all phases of commercial production in the form of computerization, materials handling, robotic cutting and planting. Significant cost saving can be realized in both the laboratory and greenhouse based production phases. However, no feasible commercial robotic application in the laboratory has been implemented to date.

a) Laboratory phase. The automation and mechanization in laboratory include; culture media preparation, growth room control, material handling, dishwashing, cutting and planting. Some of them are already in operation by large micropropagation facilities. Much of this technology has been developed for food related and production related industries. Apparently, the most complex area of automation is in cutting and planting. It usually requires computerized sensor judgement, accurate cutting and soft handling.

In many of the development cases, computer vision systems and a CCD camera have been incorporated to locate cultures and identify desired cuts, then a modified industry available robotic arm was attached. The key to overcome the robotic application is not in technology, but in economy. Most of the current systems are too complex, too expensive and, sometimes, even too slow compared with experienced technicians. After all, for such a high investment cost to apply, a complete automation system would need a much larger volume of single plant materials on a whole year basis.

b) Greenhouse phases. Many of the mechanization systems are already available in greenhouses today. Tremendous developments have been made in the last ten years by the seedling plug industry and many of them can be directly utilized in the finishing of tissue cultured liners without any major modifications. These include: soil filling, material transport, irrigation, fertilization and general greenhouse environmental systems. Specific areas, however, which require further development are sorting, grading and planting of unrooted and rooted microcuttings.

5.1.3. Somatic embryogenesis bioreactor. Somatic embryogenesis production could be carried out by a closed bioreactor in a liquid culture media. This character provides a great opportunity for becoming a totally automated production system. However, there are two major limitations to extensive utilization of somatic embryogenesis on a large commercial scale production basis. The first is the possible degree of somaticlonar variation that could be produced. The second limitation is that the system is still limited to only a few crop species.

5.1.4. Low labor cost offshore production. For any labor intensive industry, it is a shrewd business practice to take advantage of inexpensive labor. Since over 50% of tissue culture production costs can be directly attributed to labor, a tremendous cost reduction could be achieved through successful offshore production. Reduced labor costs may also provide an opportunity to produce those crops which presently require too much handling to be commercially viable inhouse. Therefore, in the future, an increase in offshore business is expected. In fact, over the past ten years, many tissue culture laboratories have already been built in Southeast Asia, Latin America and Eastern Europe. These laboratories are strongly supported by local governments anxious to promote biotechnological products and export business.

Offshore laboratories have been greatly affecting the tissue culture industry. Since they can offer microcuttings at lower prices, they have been increasing their share of the market supply. However, to date, the quality of their microcuttings has been quite variable, thereby diminishing product trust. As a result, the majority of offshore laboratories have been forced to shut down. The companies that are successful have been concentrating their efforts on large volume, contract production of foliage products.

While offshore production will provide inexpensive direct labor costs, there will still be a certain number of proprietary products and technologies which, for security purposes, cannot be transferred to offshore facilities. Import/export regulations may also restrict the distribution of some crops across international borders. Therefore, it will always be necessary to produce some crops by hand and in-house. A large size tissue culture company will distribute their production among in-house, offshore and wautomation, according to their product mix and market segment, paying utmost attention to maximizing the company's gross profit. Creating the proper balance between these different production strategies will present the most important challenge to management over the next ten years.

5.2. CROP DIVERSIFICATION

Micropropagation has become a stable and routine propagation system for many of the orchid and foliage species. Recent progress has enabled an extension of micropropagation to other crops as well. Tremendous research efforts in public and private institutions on flowering crops, vegetable crops, plantation crops and tree crops are now being commercially applied and are resulting in adequate gross profits. These research efforts will undoubtedly continue to offer new opportunities to the tissue culture industry.

5.2.1. Flowering species. Many of the cut flower species and potted plants, such as gerbera, statice, gypsophila, carnations, marigolds and african violets have been produced by micropropagation. In the next ten years, the high price items, such as peonies and alstromerias, will be effectively produced by micropropagation. Bulk-up of hybrid roses is also likely to be done through micropropagation as an aid to breeders. In the more distant future, large volume crops such as poinsettias may also be micropropagated.

5.2.2. Bulbs. Currently, lilies, zantedeschia and caladiums are the major bulb crops produced through tissue culture. However, as liquid bioreactor tissue culture systems become more applicable to the production of bulb crops, there may be an increase in micropropagation of popular bulb crops such as tulips, gladiolas, amaryllis, freesias, verines and dahlias. This could be commercially applied in perhaps the next five to ten years.

5.2.3. Vegetable crops. Potatoes and sweet potatoes are already commercially produced through tissue culture. Recently, many seed parents of crops such as asparagus, broccoli and cauliflower have also been produced through tissue culture. More tissue culture systems are being developed and tested with garlic, melon and carrot. However, the production of single clone, hybrid selections offers the greatest potential to the micropropagation industry. This could include asparagus, triploid watermelon and hybrid celery. A plant with heterozygous hybrid vigor would be selected as the commercial variety, micropropagated, and then introduced into the market place several years earlier than conventional plant breeding would allow. This will be a powerful advantage for the entire vegetable and seed industry.

5.2.4. Fruit and root stock. Strawberries, raspberries and blackberries are already produced by tissue culture and recently, some of the root stocks of plants such as peach and orange have joined the list. Micropropagation of root stock plants will definitely increase in the near term and expand this application to crops such as cherry, pear, apple, plum and other stone fruits. Papayas and pineapples also offer big opportunities to plant tissue culture businesses, for there will be high demand for disease resistant hybrids in these two crops.

5.2.5. Plantation crops. Banana and date palms are the only plantation crops currently produced by tissue culture commercially. Since plantation crops have high market potentials, many studies have focused on them, in both public and private institutions.

Research of plantation crops is often slowed down by the fact that most of the crops take several years to produce fruit, thereby substantially delaying system performance evaluations. Plantation crops cannot tolerate off-types, plant quality and stability are crucial. In the future, commercial micropropagation of coconut palms, coffee and cocoa are likely to be achieved. However, the system development of these crops is likely to be slow.

5.2.6. Forest and landscape trees. Woody species used to be classified as a group which is relatively difficult to be produced by tissue culture. However, Ficus, Nandina and Rhododendron have been very successfully commercially produced over the last few years. These trees are used as foliage plants, potted plants and landscape ornamentals.

Since many trees are not used for their flowers or fruit, the requirements for their system developments are less stringent, and can be less timely, thereby enabling more prompt commercial production. For example, if Eucalyptus is desired for reforestation, firewood or paper, uniformity and quality enhancement are not required from the production system. In this crop, specific selections can be made for drought, salt and/or cold resistance, and these selections can then be micropropagated. Some studies have even demonstrated that micropropagated Eucalyptus grow faster and are more uniform than seedlings or cuttings. This crop has gone from system development to commercial production relatively quickly.

Many of the industrial wood plants can be planted as plantation crops. In the future, the interests in greening the environment and replanting stripped forests should accelerate the wood micropropagation businesses. This could bring tissue culture into a new, large scale agricultural and industrial business.

5.3. FUTURE NEW PRODUCT DEVELOPMENT

Micropropagation is a powerful tool for creating and introducing new products into the market place. It can speed up the release of new or improved products.

5.3.1. Elite selection. Hybrid vigor is the fundamental basis behind conventional breeding processes in all horticultural, agricultural and forestry crops. Hybrid vigor is created and maintained by either crossing two inbred lines or asexually propagating heterozygous materials. Development of inbred lines not only takes time from backcrossing over several generations, but also, in many cases, creates inbred weakness due to homozygous breakdown. Conventional asexual vegetative cutting is not only slow in speed, but also tends to accumulate different virus strains and other systemic diseases. Tissue culture can properly overcome these two problems. Theoretically, a single superior plant selection, which could be one of the segregants from an early generation breeding program, or even a single progeny from a natural population, could be cloned and propagated and become a commercially available product. During this process, viruses and systemic diseases could also be eliminated and confirmed by an indexing system. All asparagus, triploid (seedless) watermelons and superior trees would fit into this category. Many horticultural companies are starting to adopt this concept in their breeding programs and it is clear that many new products will be developed from the selection and cloning of single, superior plants.

5.3.2. Improved phenotype. Micropropagation is a controlled production system. The speed of growth and phenotype or appearance of the final product can be manipulated by adjusting various chemical and physical conditions. Characteristics which can be altered include: growth forms (such as dwarf or bush plant types), amount of branching, growth rates, flowering times and sex ratios. Despite substantial research efforts at public institutions, commercial applications of results has been limited. Further studies of plant behavior under various chemical and physical conditions and exploration of more plant growth regulators will generate more opportunities in this area. At least micropropagation lends itself readily to such manipulations.

5.3.3. Transgenic bioproducts. Genetic engineering of crop species has been one of the highlights in the research community within the last twenty years. Even though the progress has not been as fast as expected, many of the promising results are starting to materialize. It is clear that many of these superior plants are sterile, or may have a low seed production capability, thereby being slow to propagate following conventional methods. Micropropagation will definitely play an important role in bringing transgenic, superior varieties into the market place. In fact, micropropagation may be the only way to rapidly and effectively bring high value transgenic products to the market in large volume.

5.4. FUTURE GEOGRAPHIC DEVELOPMENT

At present, the greatest demand for tissue culture products are in California, Florida, Holland and Japan. Wider acceptance of tissue culture products has been limited by the high cost of tissue cultured started plants, as well as the narrow selection of plants currently available. However, this situation may be gradually changing.

5.4.1. America. Recently, agricultural production of cut flowers, vegetables and industrial crops has been growing in Latin America, with only the finished products being shipped back to the North American consumer market. Some of the production is done by local growers and some is managed and owned by Americans. Since many restrictions with free trade have been lifted, and transportation costs have been reduced, these trends are expected to broaden and continue. This may open up further opportunities for tissue culture to expand in Latin America.

5.4.2. Europe. Holland has been Europe's center for production of ornamentals for many years. However, many economic arrangements within the EEC in 1992 may trigger some decentralization. Spain, Italy and other southern countries which have better growing climates and relatively inexpensive labor may pick up some of Holland's business. Also, changes which have been occurring in Eastern Europe may also open up some opportunities there. However, at present, fierce competition among private tissue culture laboratories and excess offshore production have resulted in a serious oversupply situation. The depressed economies are also limiting the purchase of high price consumer products. In the future, it is likely that substantial economic cooperation will create a new business pattern in the European countries.

5.4.3. Asia. Japan is by far the largest consumption market in the entire Asian territory. However, even though the per capita consumption of Japan is very high, total consumption is still very limited. Also, Japan is an isolated island which must contend with high shipping costs and inevitable shipping mishaps. Furthermore, the Japanese government considers agriculture to be their most important industry, so they strongly protect this national interest. This will prohibit the development of agricultural businesses by outsiders. Taiwan and Korea have been rapidly increasing their tissue culture efforts, thereby creating new internal markets. Overall, however, Asia still offers the least business opportunities when compared to America or Europe.

5.5. GROWTH OF REPLACEMENT MARKET

Basically, micropropagation is a method used to reproduce plants. Based on system availability and production cost reduction, it is going to replace a major portion of conventional propagation systems. Furthermore, the tendency of growers using more starter plants than seeds is going to accelerate the replacement speed. Vegetative cuttings could be replaced by microcuttings due to being free of disease, timely supply, product quality and product uniformity. Seed production crops could be replaced due to seed born diseases, new product format and economy of hybrid plant breeding. Therefore, the replacement market of micropropagation will be developed rapidly in the future and tissue culture is going to be one of the important and major business segments in the plant and crop industry worldwide.

MASS PROPAGATION OF STRAWBERRY AND NEW ALTERNATIVES FOR SOME HORTICULTURAL CROPS

Ph. BOXUS
Centre de Recherches Agronomiques
STATION DES CULTURES FRUITIERES ET MARAICHERES
B-5030 GEMBLOUX (Belgium)

ABSTRACT. Strawberry axillary branching is an interesting *in vitro* mass propagation model. Photoautotrophic and heterotrophic conditions are discussed. Strawberry clones with a low number of subcultures have a good field behaviour. Except for chimerical varieties the micropropagated clones are genetically stable, although the adventitious bud origin is also possible.

New alternatives for mass propagation are discussed. Simple models, as nodules or meristematic aggregates, offer the same mechanization possibilities than the embryogenesis techniques. Observations on *Eucalyptus sp.*, *Asparagus* and *Cichorium intybus* are presented.

1. Introduction

Strawberry micropropagation has been applied on large scale commercial production since the mid 1970s. It was the first fruit crop to be micropropagated. The technique was mainly spread in European countries as Belgium, France, Germany, Italy and Spain.

As early as 1975, Hummel, a German strawberry breeder, started a large commercial laboratory, shaped to produce yearly 3 millions of strawberry plants. In the culture room each square metre shelve was producing 20.000 plantlets, i.e. approximately 1.000 m2 in the open (Merkle 1981).

More recently, as underlined by Zimmerman (1991), the production of strawberry microplantlets has increased considerably in United States. At the same time, the production had severely dropped in Italy. Therefore Zimmerman (1991) considers the future of this technique as uncertain.

However, the recent important development of dayneutral cultivars, as Selva, and the high sensitivity to the soil diseases of the very important Dutch cultivar Elsanta, are

K. Kurata and T. Kozai (eds.), Transplant Production Systems, 151–162.

responsible for the new interest concerning strawberry micropropagation. The technology is developping now in The Netherlands. The extension of soil fungi and other new diseases, like BLO (Bacteria Like Organisms) in France and in Spain (Nourrisseau *et al.* 1991), confirms the necessity for *in vitro* propagation techniques.

2. Strawberry micropropagation technique

The techniques used for meristem culture and for mass propagation in Europe, as in United States, are very well known. They were described by Boxus in 1974. Several recent reviews give a lot of details concerning these techniques and their improvements (Boxus 1989, Boxus *et al.* 1984, Jungnickel 1988). Therefore, we are discussing here only some important or controversial points.

The first explants are normally very small meristem tips. So it is possible to eradicate all the virus diseases with a very high percentage of success.

The propagation phase is based only on the axillary branching technique. The benzyladenine (BAP) is used as cytokinin, at the concentrations 0,5 - 1,0 $mg.l^{-1}$. The rooting phase is frequently realized *in vitro*, in absence of cytokinin.

Several teams of research have tried to improve the classical techniques of *in vitro* propagation by use of a very high light intensity, or photosynthetic photon flux (PPF). They demonstrated the need of a very high CO2 concentration in the vessel correlated to this high PPF (Desjardins *et al.* 1987, 1989, Kozai 1989, 1991, Kozai and Sekimoto 1988, Oda 1989).

In presence of 3.000 ppm CO_2 and 250 $\mu m.m^{-2}.s^{-1}$ PPF, the dry weight is increasing by 230%. Moreover, the acclimatization phase of microplants is improved, owing to some anatomical modifications and thicker epidermal cells (Desjardins *et al.* 1989). For this Canadian group, the photosynthetic activity related to the ribulose-biphosphate carboxylase (Rubisco) is weak. In contrast, the strong CO_2-fixation is due to the phosphoenolpyruvate carboxylase (Pep-Case). This enzyme allows an increasing of the reserves which will serve to produce leaves adapted to the *ex vitro* conditions.

Kozai (1989, 1991) try to get autotrophic cultures, by increasing CO_2-concentration at 3.000 ppm and under high PPF, 150-220 $\mu m.m^{-2}.s^{-1}$ (10-20 Klux). The advantage of this autotrophic micropropagation system is a considerable decrease of the total production cost, in spite of the increasing of the lighting, cooling and CO_2 enrichment-costs. Indeed, the sugar elimination decreases the contamination risks, authorizes the use of much larger vessel, robots, and makes the soil transfer easier. Moreover in this propagation system, the use of the exogenous growth regulators is minimized and thus the mutation rate of *in vitro* plantlets can be reduced (Kozai 1989, 1991).

However, as many other searchers, we are not ourself completely sure of the Kozai's assertions. In spite of the heterotrophic *in vitro* growth, the adaptation of the plantlets to the *in vivo* environment can be recorded very early on (De Proft *et al.* 1990). The plantlets transferred in NFT system showed a very strong growth and this acclimatization system could be interesting for early crop production on hydroponics. (Huygens 1989).

3. Performances of micropropagated strawberry plants

3.1. PROPAGATION FIELD

For the nurserymen, the micropropagated plants have a lot of advantages. First, concerning the sanitary problems, this material is free of viruses, BLO, soil-fungi, nematodes. On the other hand, the microplantlets are producing more runners that the conventional ones. This last point, is very important for the dayneutral cultivars which produce very few runners. For this reason, the Dutch authorities accept the micropropagation of cv. Rapella.

3.2. FRUIT PRODUCTION

All comparative experiments carried out up to now on tissue cultured and conventionally obtained strawberry plants showed no difference in yield (Boxus, 1985, Dijkstra 1988, Faedi *et al.* 1989, Navatel *et al.* 1986, Swartz *et al.* 1987), and that independently of the first or second runner progeny (Dijkstra 1988, Navatel *et al.* 1988).

Recent experiment trials conducted by Lieten and Baets (1989, 1990) in the Northern Belgium, show a yield increasing for the first runner generation of cvs Elsanta and Elvira, originated by micropropagation (table 1).
Before the winter, this origin in comparison with the control plants gave very homogenous stronger plants, with larger number of reduced size leaves. At spring time, the micropropagated plants gave more flowers and larger production. Nevertheless the fruit size was partially reduced. For instance, in open field micropropagated Elsanta produced 5% more, but size losses were estimated at 4% for small and malformed fruits. For cv. Elsanta under

large tunnels the production was increased by 15% and only 10% were lost as small fruits.

TABLE 1. Fruit production (first runner generation for micropropagated plants), *following Lieten and Baets 1990.*

	Kg/pl.	Kg/m^2	large size %	middle size %	small size %
ELSANTA					
standard	0.767	4.220	7.34	31.89	57.69
micro.	0.885	4.867	4.26	23.65	68.63
ELVIRA					
standard	0.484	2.662	15.08	13.84	68.32
micro.	0.588	3.232	3.46	19.68	74.25
GORELLA					
standard	0.570	3.137	14.96	17.48	62.59
micro.	0.534	2.939	5.55	21.96	66.00

At economic point of view, the micropropagation origin was judged more interesting (Lieten and Baets, 1990)

However, special cultivars as Gorella, subcultured many times *in vitro*, may have an abnormally high fruit set entailing fruit size reductions (Boxus 1985, Navatel *et al.* 1986) (table 2).

TABLE 2. Fruit production and fruit size of cv. Gorella, comparison between different number of subcultures *in vitro* in France. Mean of 3 years (1984-1986), *following Navatel et al.,1986*

	South Eastern		South Western	
	Commercial grade (g/pl.)	Mean weight (g/fr.)	Commercial grade (g/pl.)	Mean weight (g/fr)
Standard	837	18.2	510	17.5
New clones	892	18.0	602	17.6
Old clones	700	16.1	418	14.8

Standard = never propagated *in vitro*
New clones = < 10 subcultures
Old clones = > 30 subcultures

The physiological aspects of this disorder are studied at Liege University in Belgium (Kinet and Parmentier 1989, 1990). It occured more frequently when the number of *in vitro* subcultures on the multiplication medium was high. It persisted in the successive progenies of runner plants issued from these high fruit set plants. A cold treatment at 4°C. during a few weeks can induce a partial or a total recovery. Although, the induction mechanism was unknown, it seems to be related to habituated tissues.

In the practice, this trouble is now controlled by the generalization of new clones where the total number of subcultures is limited to ten (table 2).

4. Genetic stability and strawberry field behaviour

Except for somatic chlorophyllian mutations, which also occur in conventional plantation, strawberry mutations were not frequently recorded.

According to Swartz *et al.* (1987) the absence of variants in the production fields results from the inability of the off-type plants to survive soil transfer.

But for several breeders the octoploid strawberry plants are notable for having a very stable genetic make-up. Moreover, following D'Amato (1984) the genetic stability can be preserved *in vitro*, if meristems are cultured without callus formation.

However variability could be induced by adventitious shoots which must be discarded during multiplication process. But, it is not always very easy to identify the adventitious origin of buds developped in clusters. Zimmerman (1991) underlined that frequent subculturing of strawberries, i.e. every two weeks, stimulated the formation of adventitious buds on leaf margins.

Jemmali, in our laboratory, observed the presence of photocorm-like structures, giving buds within the proliferation clusters. Moreover, he found very numerous stipular buds at the basis of the petioles of different cultivars. These very small buds, arising between or on the two stipules, can be easily cloned by axillary proliferation. The field behaviour of these buds is now in progress, and looks normal until now (Jemmali, pers.comm.).

Old Japanese papers described as mass propagation technique, the adventitious shoots arising from meristem or anther callus (Nishi and Ohsawa 1973, Ohsawa *et al.* 1974). But there is no information about the genetic uniformity of these plants.

For Zimmerman (1991), large scale long term field evaluation of plants grown from adventitious shoots have to made. Data must be collected and evaluated before use of adventitious shoots can be recommanded for industrial propagation.

An other possibility to increase the variability is the dissociation of chimera. Also Marcotrigiano *et al.* (1987) determined that the leaf variegation of *Fragaria vesca* "albo marginate" was of chimerical origin. They observed than 90% of the plantlets propagated *in vitro* varied from the original explants. Histological evidence indicated that many shoots were adventitious, arising from basal callus tissue or petioles. In this specific case, the micropropagation must be avoided.

Regeneration through callus cultures, or cell cultures, issued from petioles, flower stalks, petals, ovaries, ... is possible but depends on the variety, the explant origin, the hormonal balance,... (Foucault 1990, Navatel pers.comm., Predieri *et al.* 1989).

In general, the regenerated plantlets were true to type and very uniform. The progeny of a few plants showing hyperflowering during the first crop gave normal flowering one year later. Nevertheless, in the different trials, a very few number of plants originated from different explant organs, and from different cultivars, showed similar abnormalities, female sterility and very compact flower trusses.

For G. Risser, breeder at Avignon Research Station, these female sterile plants were polyploid plants. They are issued from polyploid cells, already present in the tissue of the mother plant (Navatel pers.comm.).

But if permanent variants are rarely observed in the production fields, transient variants are more commun. So, unifoliated leaves, high runner production, susceptibility to soil diseases of resistant cultivar Raritan (Shoemaker and Swartz, 1985), don't reflect genetic modifications, but are expressions of the juvenile state of strawberry plant. These phenotypes could be partially recovered by application of abscissic acid *in vitro* (Mohamed *et al.*, 1991). There is probably for the same reason that Rancillac and Nourrisseau (1989) observe a better resistance of micropropagated plants to cold hardiness.

Careful attention should be paid to the "white streaking" symptoms observed on micropropagated strawberry by numerous authors (Boxus, 1989). This trouble is related to the "June yellows" frequently observed on strawberry seedlings. We believe that specific physiological conditions, occuring during the *in vitro* growth of strawberry, induces these symptoms. This disorder can involve later whether the disparition or the recovery of the contaminated variety.

5. New alternatives for mass propagation

To reduce the relatively high production cost of the conventional micropropagation, many researches are made on somatic embryogenesis applied to several horticultural plants. However, this exciting technology is strong dependant of the genetic stability of cell lines in liquid culture or in callus culture.

To avoid the risks of variant, Kozai (1991) proposes a system of micropropagation under photoautotrophic conditions. The methodology was demonstrated with different horticultural plants (see above for strawberry). The use of sugar-free medium, CO_2 enrichment, high light intensity, to propagate nodal cuttings is closely related to nutrient film technique and well adapted to robotization (Kozai 1991).

But, we believe that Kozai loses, by this way, the high proliferation rate partly due to the container confinement. We are more confident in other alternative techniques, as nodule cultures (Mc Cown *et al.* 1988) or meristem aggregates (Aitken *et al.* 1988, Boxus *et al.* 1991, Ziv 1990).

Indeed in the case of meristem aggregates, it is possible to get very large number of "axillary buds", without callus and internode tissues. The passage from one subculture to an other one is made by single dilution for liquid culture medium, or by squashing on solid culture medium.

The nodule model is also very easy to handle. The nodules are divided without great care and spread on a new culture medium as for orchid protocorms.

In our laboratory, *Eucalyptus camaldulensis, E. globulus, E. gunnii* and *E. robusta* are propagated in Petry dishes on MS medium enriched with 30 $g.l^{-1}$ sucrose, IBA 1-10 $mg.l^{-1}$, activated charcoal 1 $g.l^{-1}$, pH 5.7, complete darkness. On this culture medium, and without cytokinin, the proliferation rate is very high, more than 10 times each 3 weeks.

The nodal explants developped 3 different morphologies, (1) long white shoots with sometimes a few short internodes, (2) shoots with a very important fasciation, (3) plus or less important bud clusters, with a light pink coloration, but without callus. After cloning and subculturing, these three polymorphed shoots gave homogenous plantlets in large lighted vessels containing an hormone-free culture medium. These results were got with juvenile and 20 years-old *E. Camaldulensis* (Ossor, pers.comm.).

The *Eucalyptus* bud clusters are rather similar to the *Gladiolus* bud aggregates induced by ancymidol or paclobutrazol (Ziv 1990), and could be related also at the meristem aggregates observed on *Pinus radiata* in presence of 5 $mg.l^{-1}$ BAP (Aitken *et al.* 1988).

Using 8 $mg.l^{-1}$ ancymidol on *Asparagus* in solid proliferation medium, we were able to induce bud aggregates in very short times. In liquid culture medium, our first results showed vitrified bud aggregates (De Ryck 1991).

Up to now, we didn't observe protocorms like structures during *Asparagus* or *Eucalyptus* propagation.

At the contrary, on Brussels cichory leaf explants, numerous meristematic nodules or protocorms were observed, isolated and propagated. These nodules were externally as the poplar nodules observed by Mc Cown *et al.* (1988). Cross-sections of nodules showed meristem organizations near some vascularized centers, budding off, or nodule division by invaginations, as for poplar nodules.

The nodules appear always at the upper lip of wounded leaves, near the main vein, and without callus. The Lepoivre culture medium, without hormonal substance regenerated a little less nodules than in presence of $1mg.l^{-1}$ BAP, 0,01 $mg.l^{-1}$ IBA. The formed nodules can be isolated and divided. They regenerated buds within 4-6 weeks and later adventitious roots. The germinated nodules were sowed like seeds directly on compost substrate. They regenerated plantlets with perfect tap root restitution (Pieron *et al.* 1992).

6. Conclusions

These few examples show that very simplified models could be developped. They offer a very high propagation rate and an easy handling. They are able to compete economically with embryogenesis systems. Of course, before they were used definitively in commercial propagation process, it will be necessary to have more informations on their field behaviour.

7. References

Aitken-Christie, J., Singh, A.P. and Davies, H. (1988) "Multiplication of meristematic tissue : a new tissue culture system for radiata pine", in Hanover, J.W. and Keathley, D.E. (eds), Genetic manipulation of woody plants, Plenum, New-York, pp. 413-432.

Boxus, Ph. (1974) "The production of strawberry plants by *in vitro* micropropagation" J. Hort. Science 49, 209-210.
Boxus, Ph. (1985) "Comportement du fraisier issu de micropropagation *in vitro*" Fruit Belge 410, 106-110.
Boxus, Ph. (1989) "Review on strawberry mass propagation", Acta Horticulturae 265, 309-320.
Boxus, Ph., Damiano, C. and Brasseur, E. (1984) "Strawberry", *in* Ammirato, Evans, Sharp and Yamada (eds), Handbook of plant cell culture, Macmillan, N.Y., pp. 453-486.
Boxus, Ph. Terzi, J.M., Lievens, Ch., Pylyser, M., Ngaboyamahina, P. and Duhem, K. (1991) "Improvement and perspectives of micropropagation techniques applied to some hot climate plants", Acta Horticulturae 289, 55-64.
D'Amato, F. (1984) "Nuclear cytology of tissue cultures", *in* Proceedings Intern. Symp. Plant tissue and cell culture - Applications to crop improvement, Czech. Acad. Sci., Prague, pp. 295-303.
De Proft, M., Boxus, Ph. and Huygens, H. (1991) "Greenhouse transplant of *in vitro* cultured strawberry plants using hydroponics", XXIII Intern. Hortic. Congress, Firenze (Italy), Abstract of contributed papers, Poster n° 4339.
De Ryck, G. (1991) "Vegetatieve vermenigvuldiging van asperges *in vitro*", Memory, Fakulteit der Landbouw wetenschappen, Katholieke Universiteit Leuven (Belgium), 88 p.
Desjardins, Y., Gosselin, A. and Yelle, S., (1987) "Acclimatization of *in vitro* strawberry plantlets under CO_2 enriched, environnement and supplemental lighting" J. Amer. Soc. Hort. Sci. 112, 846-851.
Desjardins, Y., Laforge, F., Lussier, C. and Gosselin, A. (1989) "Echanges gazeux en culture *in vitro*", *in* La Culture *in vitro*, une technologie au service de la biologie végétale, C.R. Colloque sur la culture *in vitro*, Université Laval, Québec, pp. 4-21.
Dijkstra, J. (1988) "Samenvatting onderzoekresultaten vollegrond-aarbeien 1987", Intern Verslag, (unpublished report).
Faedi, W., D'Ercole, N., Turci, P., Bazzocchi, C. and Siroli, M. (1989) "Effect of different propagation systems on strawberry plant performance", Acta Horticulturae 265, 321-326.
Foucault, C. (1990) "*In vitro* néoformation sur cals dans le cas du fraisier (*Fragaria x ananassa*, Duchesne), étude du comportement au champ de plants régénérés", Thèse U.F.R. des sciences de l'environnement, Université d'Angers (France), 212 p.
Jungnickel, F. (1988) "Strawberries (*Fragaria spp.* and hybrids)", *in* Bajaj, Y.P.S. (ed.), Biotechnology in Agriculture and Forestry, vol.6, Crops II, Springer-Verlag, Berlin, Heidelberg, pp. 38-103.

Huygens, H. (1989) "De konditionering by de rechtsreekse overgang van *in vitro* geteelde aardbeiplantjes (*Fragaria X ananassa* Duch.) naar produktieplanten, benaderd via vergelijkend groei- en gasuitwisselingsonderzoek", Verslag navorsingsbeurs IWONL, Katholieke Universiteit Leuven (Belgium), 105 p.
Kinet, J.M. and Parmentier, A. (1989) "The flowering behaviour of micropropagated strawberry plants cv. "Gorella" : the influence of the number of subcultures on the multiplication medium", Acta Horticulturae 265, 327-334.
Kinet, J.M. and Parmentier, A. (1990) "Characterization of an abnormal flowering behaviour in micropropagated strawberry plants", XXIII Intern. Hortic. Congress, Firenze (Italy), Abstract of contributed papers, Poster n° 3207.
Kozai, T. (1989) "Photoautotrophic (sugar-free) micro-propagation under high CO_2 and high photosynthetic photon flux conditions" *in* La culture *in vitro*, une technologie au service de la biologie végétale, C.R. Colloque sur la culture *in vitro*, Université Laval, Québec, pp. 59-72.
Kozai, T. (1991) "Micropropagation under photoautotrophic conditions" *in* Debergh, P.C. and Zimmerman, R.H. (eds), Micropropagation, technology and applications, Kluwer Academic Publishers, Dordrecht, pp. 447-469.
Kozai, T. and Sekimoto, K. (1988) "Effects of the number of air changes per hour of the stoppered vessel dioxide concentration inside the vessel and the growth of strawberry plantlets *in vitro*", Environmental Control in Biology 26, 21-29.
Lieten, F. and Baets, W. (1989) "Aardbei", Jaarverslag 1989.
Lieten, F. and Baets, W. (1990) "Eerste nabouw meris-temenplanten geschikt voor produktiedoeleinden?", Tuinbouw Visie 59, 20-21.
Marcotrigiano, M., Morgan, P.A., Swartz, H.J. and Ruth, J. (1987) "Histogenic instability in tissue culture-proliferated strawberry plants", J. Amer. Soc. Hort. Sci. 112, 383-387.
Mc Cown, B.H., Zeldin, E.L., Pinkalla, H.A. and Dedolph, R. (1988) "Nodule culture : a developmental pathway with high potential for regeneration, automated micropropagation, and plant metabolite production from woody plants", *in* Hanover, J.W. , and Keathley, D.E. (eds), Genetic manipulation of woody plants, Plenum, New-York, pp. 149-166.
Merkle, S. (1981) "Erdbeervermehrung durch Meristem-kultur, Erfahrungen aus der Praxis, Obstbau 6, 95-96.
Mohamed, F., Swartz, H.J. and Buta, G. (1991) "The role of abscissic acid and plant growth regulators in tissue

culture induced rejuvenation of strawberry *ex vitro*", Plant Cell, Tissue and Organ Culture 25, 75-84.
Navatel, J.C., Verachaud, G., Roudeillac, P. and Bardet, A. (1986) "Multiplication *in vitro* : comportement agronomique de plants de fraisiers issus de pieds mères produits par micropropagation par rapport au matériel classique. Synthèse de 3 années d'expérimentation", Infos-CTIFL 21, 2-8.
Navatel, J.C., Germain, P., Roudeillac, P. and Bardet, A. (1988) "Micropropagation du fraisier. Incidence du nombre de multiplication au champ après la phase *in vitro* sur le comportement du plant", Compte-rendu d'essais 1987, CTIFL-CIREF, Mars 1988, 6 p.
Nishi, S. and Ohsawa, K. (1973) "Mass production method of virus free strawberry plants through meristem callus", Jpn. Agric. Res. Q.7, 189-194.
Nourrisseau, J.G., Lansac, M. and Garnier, M. (1991) "The marginal chlorosis of young leaves, a new disease of strawberry in France", *in* International workshop, Strawberry diseases and breeding for varietal resistance, Société Française de Phytopathologie, Bordeaux, October 1991 (abstract).
Oda, Y. (1989) "Effect of light intensity, temperature and CO_2 concentration on photosynthesis and growth of strawberry plantlets cultivated *in vitro*", Acta Horticulturae 265, 399-404.
Ohsawa, K., Toka, M. and Nishi, S. (1974) "Studies on the anther culture of vegetable crops. II. Breeding of a great quantity of virus free plants by means of strawberry anther culture method", Bull. Veg. Ornam. Res. Station, Japan, A 1, 41-57.
Pieron, S., Belaizi, M. and Boxus, Ph. (1992) "Nodule culture, a possible morphogenetic pathway in *Cichorium intybus* propagation", Scientia Horticulturae, (submitted).
Predieri, S., Fasolo, Fabbri, Malavasi and Ancherani, M., (1989) "Regeneration of plants from strawberry (*Fragaria x ananassa* Dutch.) unpollinated ovaries and petals", Acta Horticulturae 265, 335-342.
Rancillac, M. and Nourrisseau, J.G. (1989) "Micro-propagation and strawberry plants quality", Acta Horticulturae 265, 343-347.
Shoemaker, N.P. and Swartz, H.J. (1985) "Cultivar dependent variation in pathogen resitance due to tissue propagation of strawberries", HortScience 20, 253-254.
Swartz, H.J., Lindstrom, J.T., Fiola, J.A. and Zimmerman, R.H. (1987) "The use of tissue culture propagation of strawberry in the United States", *in* Boxus and Larvor (eds), *In vitro* culture of strawberry plants, C.E.C. Luxembourg, pp. 79-100.

Zimmerman, R.H. (1991) "Micropropagation of temperate zone fruit and nut crops", *in* Debergh, P.C. and Zimmerman, R.H. (eds), Micropropagation, technology and applications, Kluwer Academic Publishers, Dordrecht, pp. 231-246.

Ziv, M. (1990) "Morphogenesis of *gladiolus* buds in bioreactors. Implication for scaled-up propagation of geophytes", *in* Nijkamp, H.J.J., Van Der Plas, L.H.W. and Van Aartrijk, J. (eds), Progress in plant cellular and molecular biology, Proceedings of a Symposium, 27-29 June 1990, Amsterdam, pp. 119-124.

MICROPROPAGATION OF FOREST TREES

J. AITKEN-CHRISTIE
Forest Research Institute
Private Bag 3020
Rotorua
New Zealand

and

M. CONNETT
Department of Biological Sciences
Waikato University
Private Bag
Hamilton
New Zealand

ABSTRACT. Micropropagation as a delivery system for producing plants for the clonal propagation of genetically superior and genetically transformed forest trees is reviewed. Recent advances in the micropropagation of material from juvenile and mature trees produced via either the organogenesis or embryogenesis pathways, field evaluation of micropropagules, genetic transformation, and automation of micropropagation systems for forest trees are discussed. Micropropagation of juvenile trees from immature embryos, mature embryos and young seedling explants has been successful, with major advancements recently in embryogenesis and in large-scale clonal production via organogenesis. Field trials of trees grown from micropropagules of juvenile trees planted in the early 1980s showed advanced maturation characteristics in some cases; some of the factors involved in this have now become apparent. Micropropagation of material from mature trees has also had some major breakthroughs: reliable techniques have been developed for plantlet formation from mature buds, and embryogenesis from male reproductive tissues has been demonstrated for a few hardwood species. Some advances have been made in the genetic transformation of forest trees for transient expression, particularly with the use of DNA-coated microprojectiles ("biolistic" methods). The simultaneous development of better regeneration systems is essential for stable transformation. There is still much to be researched on the long-term stability of the transforming DNA and identification of genes of major interest to forest managers and researchers for transformation. Automated systems to reduce the cost of micropropagated trees are being developed for both organogenesis and embryogenesis systems.

1. Introduction

The study of micropropagation is a sub-field of forest biotechnology which is attracting the attention of scientists, forest managers and conservationists alike. The possible application of

K. Kurata and T. Kozai (eds.), Transplant Production Systems, 163–194.

micropropagation techniques in forestry is seen as a great challenge around the world, particularly for producing large numbers of conifers and hardwoods for reforestation. Micropropagation in this article is reviewed as a delivery system for the clonal propagation of genetically superior and possibly genetically transformed trees; the more fundamental aspects of tissue culture in forestry are not covered. Micropropagation is also recognized as being only one form of vegetative propagation aiming for faster forest growth and improved productivity and genetic gains; other forms of forest-tree propagation are reviewed by Menzies and Arnott (1992, this volume).

Micropropagation can be simply described as an *in vitro* method of producing clonal offspring identical to a superior stock plant. Micropropagation is in fact a complex multistep process involving numerous different starting tissues and cell types, ranging from cotyledons and vegetative meristems to reproductive tissues, which can be induced to differentiate into new plantlets via either the organogenesis pathway or the embryogenesis pathway (Figure 1). The starting tissues, or explants, may be taken from either juvenile or mature trees or trees of any age in between. Tissues from juvenile seedlings or seeds are frequently used in a new programme because they are usually easier to propagate and form the basis of development of suitable nutrient media and methods for a new species. When a breeding programme is in place, micropropagation can be carried out from genetically superior control-pollinated seed (Aitken-Christie and Thorpe 1984). However, the plantlets produced from juvenile tissues of seedlings or seeds must be considered untested clones, unless the clonal tissue has been stored in a juvenile state by low temperature or cryogenic storage and previously tested in a field clonal trial. There are no examples of plantlets produced from tested juvenile clones to date, although several programmes around the world have adopted this approach (Aitken-Christie and Singh 1987, Klimaszewska *et al.* 1992, Gupta and Durzan 1991). Plantlets produced from tissues selected from mature elite trees have the advantage that the clones are likely to have been tested in a tree breeding programme and thus may be genetically superior. Alternatively, they may be produced from selected untested mature trees (Figure 1). The starting tissues can be either more juvenile (e.g. stump sprouts on a redwood tree) or more mature (e.g. buds from the top of the tree). The micropropagation of mature trees is carried out to provide plants for clonal forestry if the micropropagules are rejuvenated, or for seed orchard establishment if micropropagules are more mature.

Genetic transformation of forest trees is a new area of forest biotechnology offering the incorporation of new traits into forest trees without the time constraints of a conventional breeding programme, or incorporation of traits such as pesticide resistance that may not be known in these species. A basic molecular biology programme comprising gene mapping, characterization, transformation systems and gene manipulation is necessary to realize the full potential of this new biotechnology, which can use either the organogenesis and embryogenesis routes for delivery of plants (Figure 1).

This review paper highlights particularly the micropropagation of juvenile and mature trees via the organogenesis and embryogenesis pathways to bring to the reader's attention the key factors for success, and the recent advancements in embryogenesis and applications in genetic transformation. No attempt has been made to repeat the excellent review by Thorpe *et al.* (1991). There have been at least 60 new references since then, many of which are in the areas of embryogenesis and micropropagation of mature trees and the significant findings are covered in this paper. The field evaluation of micropropagules, viewed as one of the most important components in any forest tree micropropagation programme, is also yielding results as trees grow older. New technologies such as genetic transformation and automation, which will form an integral part of any forest biotechnology programme, are also reviewed. For more complete discussions on how and why micropropagation can be used in forestry and genetics programmes, readers should consult recent reviews and articles by Dunstan (1988), Mascarenhas *et al.* (1988), Cheliak and Rogers (1990), Thorpe *et al.* (1991) and Paranjothy *et al.* (1991). For further details of micropropagation protocols for numerous hardwood and coniferous species, readers are advised to consult the books on tissue culture and biotechnology in forestry by Bonga and Durzan, Vols. 2 and 3 (1987) and Bajaj (1989, 1991).

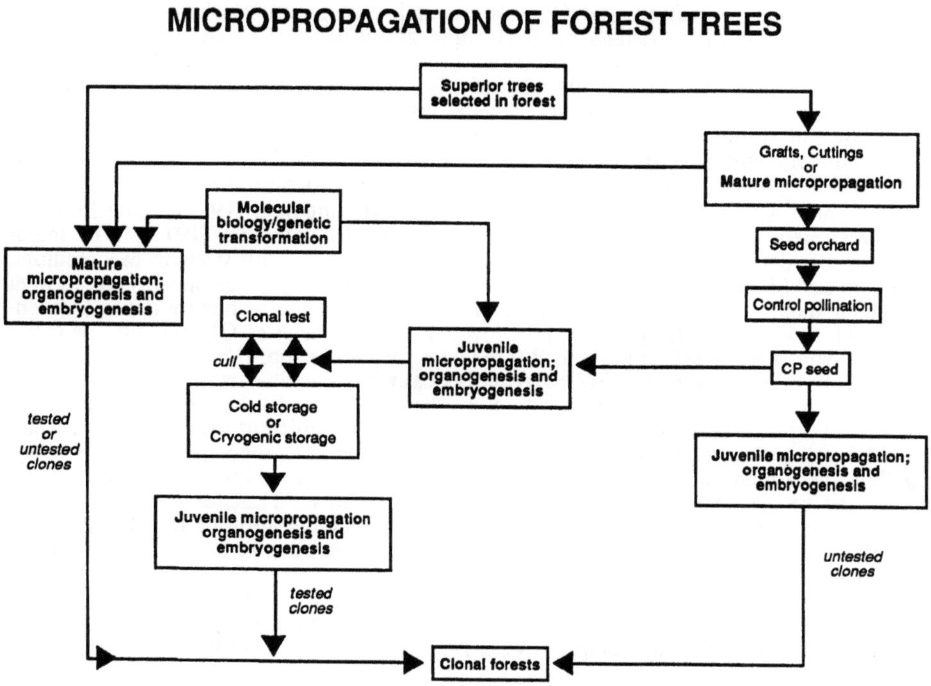

Figure 1. General scheme for micropropagation of juvenile and mature forest trees. This scheme is being used at the Forest Research Institute (FRI), Rotorua, New Zealand for *Pinus radiata.*

2. Micropropagation of Juvenile Trees

2.1. GENERAL

"Juvenile" in this section refers to all the tissues and developmental stages found from the immature embryo stage up to an approximately 1-year-old seedling. The micropropagation of material from juvenile trees via either organogenesis or embryogenesis has been demonstrated for a wide range of hardwoods and softwoods. The major route presently used for reliable regeneration of plantlets and trees for clonal forest plantings and research trials is via organogenesis. Organogenesis differs from embryogenesis in that axillary or adventitious shoots are induced first, and roots are induced later to form complete plantlets. In embryogenesis, somatic embryos (complete with cotyledons and root axes) are formed and these "germinate" to form plantlets. Some species have been easier than others to micropropagate by embryogenesis, e.g. spruces (*Picea* spp.).

2.2. ORGANOGENESIS

2.2.1. *Stages and Nutrient Media.* Micropropagation of juvenile trees via organogenesis can be divided into four stages, namely (1) initiation of shoots on explants, (2) shoot elongation and multiplication, (3) rooting, and (4) acclimatization of plantlets. The basic methods outlined for *Pinus radiata* (radiata pine) (Horgan and Aitken 1981, Aitken-Christie and Thorpe 1984) and for *Eucalyptus* spp. (Mehra-Palta 1982) have been extensively used elsewhere for conifers and hardwoods respectively. However, optimum requirements for each stage have had to be determined for each species (Thorpe *et al.* 1991).

The choice of nutrient media for all *in vitro* stages is usually either a "low salt" or a "high salt" medium depending on the species. Low salt media include Gresshoff and Doy (GD) medium (1972), Lloyd and McCown's Woody Plant (WP) medium (1980) and various dilutions ($\frac{1}{4}$ and $\frac{1}{2}$ strength) of Schenk and Hildebrandt (SH) medium (1972), Murashige and Skoog (MS) medium (1962) and Quoirin and Le Poivre (LP) medium (1977). The major high salt media employed in forest tree micropropagation include full strength MS medium, LP medium and SH medium. Examples of more optimal media for some species are: a high salt medium such as LP or SH medium for *Pinus radiata* juvenile or mature tissue (Aitken-Christie and Thorpe 1984, Horgan 1987); a low salt medium such as GD medium for *Pinus taeda* (Amerson *et al.* 1988); a low salt WP medium for *Betula* spp. (McCown 1989); and a low salt White's (W) medium for *Tectona grandis* (teak) (Mascarenhas *et al.* 1987).

Many researchers have made modifications to these basic media for various species and for different stages of development. For shoot initiation, the choice of explant and cytokinin type and level are more important than the nutrient medium composition, but for shoot elongation, shoot multiplication and root initiation, the nutrient medium can have a profound effect on the rate of shoot growth and the quality of shoots. Transfer from a high salt medium to low salt medium for preconditioning shoots, root initiation or root formation *in vitro* is often beneficial and helps to "wean" the shoots before transfer to *ex vitro* conditions (Horgan and Aitken 1981).

Other physical, chemical and environmental factors (vessel and lid type, supports, carbon balance and metabolism, light, CO_2 concentration etc.) are also gradually being shown to be more important than previously thought (Kozai 1991, Kozai *et al.* 1992 this volume).

2.2.2. *Initiation of Shoots on Explants.* The choice of an explant in the optimum physiological or developmental state is the most important parameter governing the success of shoot formation. Healthy, well-nourished explants with some meristematic tissue present are a prerequisite. Loss of competency has been demonstrated with radiata pine and Douglas fir (*Pseudotsuga menziesii* Mirb.) explants when cotyledon explants from the youngest seedlings formed more adventitious buds than cotyledons from older seedlings (Aitken-Christie *et al.* 1985, Goldfarb *et al.* 1991) and if exposure to BAP was delayed (Biondi and Thorpe 1982, Ellis and Bilderback 1991). Mature zygotic embryos, cotyledons, epicotyls and axillary buds at the base of primary or secondary leaves, all of which contain meristematic cells, have proved to be reliable explants for the initiation of adventitious and axillary shoots from juvenile tree tissues (Thorpe *et al.* 1991, Table 1). The physiology of adventitious shoot and axillary shoot formation has been researched extensively and reviewed by Dunstan and Thorpe (1986), Boulay (1987) and Thorpe (1988) and will not be covered here. Axillary shoots are almost always formed from an explant containing an existing or pre-existing axillary bud meristem, and adventitious shoots can form on all the explant types. The induction of shoots from a callus is less reliable, but has been recorded for *Pinus eldarica* and Engelmann spruce (*Picea engelmannii* Parry) (Phillips and Gladfelter 1991, Lu and Thorpe 1988, Harry and Thorpe 1991) and a number of hardwood species.

Shoot initiation on explants occurs most frequently when the cytokinin, benzylaminopurine (BAP), is incorporated into the gelled nutrient medium at concentrations between 0.5-5.0 mg/l (2.2-22 μM BAP) At higher levels of BAP (e.g. 5 mg/l), meristematic nodules and large numbers of slow-growing shoot primordia are produced, which can take 4–6 months to grow out into fully elongated shoots. At lower BAP levels (0.5 mg/l), fewer shoots are produced which require only 2–3 months to grow out. In some cases, a dual cytokinin treatment is better than BAP alone, e.g. in *Thuja occidentalis* L. (Eastern white cedar), 2-isopentenyl adenine (2ip), kinetin (K) or zeatin (Z) are used in combination with BAP (Harry *et al.* 1987). Liquid phytohormone-free medium has also been used for shoot initiation as an alternative to gelled medium containing cytokinin (John and Pearson 1986).

Other phytohormones, e.g. gibberellins and auxins, are not necessary to elicit a good shoot induction response, but auxins are sometimes added in low concentrations of 0.05-0.5 mg/l NAA (0.27-2.7 μM NAA). Recently, abscisic acid (ABA) has been shown to be beneficial for enhancing the BAP response in *Pinus taeda* (loblolly pine) and *Pinus virginiana* (Virginia pine) cotyledons (Sen *et al.* 1989, Chang *et al.* 1991). The role of ABA is not clear, but the authors suggest that it is not a species-specific phenomenon and may have a primary role in shoot initiation associated with peroxidase induction and changes in turgor pressure of the tissue, as in other tissue culture systems.

It is generally recognized that there are large family and clonal differences in the initiation of shoots from embryos or seedling parts (Abo El Nil and Wochok 1986). Good performers can be identified at the initiation stage since poor performers often do not react, or die, and can be rejected. This response can be useful for early *in vitro* screening of embryos from full-sib or half-sib families.

2.2.3. *Shoot Elongation and Multiplication.* Shoot elongation media are generally the same as initiation media, but with cytokinin absent. The addition of 0.5-1.0% activated charcoal to the media is often beneficial, as is use of gelrite or a gelrite/agar blend during shoot elongation (Nairn 1987). Sucrose concentration is reduced from 3% to 2% prior to rooting in some cases (Berlyn *et al.* 1991), but use of 3% throughout the shoot elongation and multiplication stages is more widespread. Gelled media are most commonly used, but liquid nutrient media have been used to enhance shoot elongation as a double layer treatment, liquid on agar, for *Pinus radiata* (Aitken-Christie and Jones 1987). Until problems of contamination and vitrification are better understood and control of the microenvironment is more precise, liquid media will not be used on a large scale.

The multiplication of elongated adventitious or axillary shoots to produce a new crop of shoots can be done by at least two methods, namely (1) nodal section methods — cutting of shoots into nodal segments containing one or more meristems or by topping the shoot to stimulate side shoots (cytokinin may or may not be necessary); (2) shoot cluster methods - using cytokinin in the medium to enhance axillary branching of single shoots or shoot clumps and suppression of apical dominance. The latter are more common with hardwoods than with conifer cultures, but have recently been demonstrated for Norway spruce cultures (Kunze, unpublished results). A third type of method is the multiplication of meristematic tissues or nodules on media containing cytokinin; nodules retain the capacity to regenerate large numbers of shoots when placed on a shoot elongation medium. Meristematic nodule-type multiplication systems have now been reported for *Pinus radiata* (Aitken-Christie *et al.* 1988), *Picea glauca* and *Populus* spp. (McCown *et al.* 1988), *Pinus eldarica* (Phillips and Gladfelter 1991) and *Eucalyptus grandis* hybrids (Warrag *et al.* 1991).

2.2.4. *Rooting and Acclimatization.* Rooting and acclimatization procedures for juvenile conifers and hardwoods have been reviewed by Mohammed and Vidaver (1988) and Thorpe *et al.* (1991). For further details, readers should refer to these articles. The authors state that the major influences on root formation include auxin concentration and mode of application, shoot quality, donor age, clone and temperature. Rooting of shoots can be done either *in vitro* or *ex vitro*, or in some cases root initials are formed *in vitro* followed by root elongation in a potting mix.

For rooting, shoots are given an auxin treatment in gelled media with or without nutrients. Auxins used are indolebutyric acid (IBA) or naphthalene acetic acid (NAA), or both, at concentrations in the range of 0.1–5.0 mg/l IBA (0.49–24.6 μM) and 0.1–5.0 mg/l NAA (0.53–26.8 μM). The concentration and length of time on auxin media varies depending on species. The addition of rutin to auxin media at a selective stage during root initiation has also been beneficial (Berthon *et al.* 1990). Liquid media containing auxin, as a dip for a short period, are sometimes used with high auxin concentrations (50–1000 mg/l) (Harry and Thorpe 1991).

An *in vitro* mycorrhizal inoculation treatment incorporated into auxin media can also be beneficial (Malajczuk and Hartney 1986, Rancillac 1991). The mycorrhizal fungi, *Pisolithus tinctorius* and *Hebeloma cylindrosporum* improved the number of roots per shoot and root length on micropropagated maritime pine (*Pinus pinaster*) shoots when inoculated into the medium (Rancillac 1991). This technique could be useful for other species and could aid acclimatization and improve early growth of plantlets. Root formation and elongation generally occur in a peat, pumice, perlite or sand mixture under a high humidity. Typically, for radiata pine a 40:30:30 peat, pumice, perlite medium is used.

In some cases rooting and acclimatization are carried out together when rooting is carried out *ex vitro*, saving an extra step. A gradual reduction in humidity from 95%+ to 30–50% RH, conversion from heterotrophic to autotrophic growth, formation of "hardier" leaves, and root growth occur during the acclimatization phase.

Rooting and acclimatization of shoots formed from juvenile tissue is generally not a problem and 70–80% survival and rooting has been achieved in many cases (Mohammed and Vidaver 1988, Table 1).

2.3. EMBRYOGENESIS

2.3.1. *Stages and Nutrient Media.* A major breakthrough in the micropropagation of forest trees occurred when somatic embryogenesis was reported from hypocotyls of *Liquidambar styraciflua* (sweetgum) (Sommer and Brown, 1980). Following this, in 1985 the first conifer plantlets were produced from somatic embryos of *Picea abies* (Norway spruce) by Hakman and von Arnold (1985) and Chalupa (1985). Similar techniques have now been developed for numerous other hardwoods and conifers, with spruces (*Picea* spp.) being the most amenable to date. The formation of plantlets via embryogenesis is also a multistaged process, like organogenesis, and consists of at least five distinct stages, namely (1) initiation of embryogenic tissue, (2) multiplication of embryogenic tissue, (3) development and maturation of somatic embryos, (4) germination of the embryos, and (5) acclimatization of plantlets (Figure 2). The process going via the multiplication of embryogenic tissue stage is termed "indirect embryogenesis" and is common for conifers. Direct embryogenesis, without an intervening embryogenic tissue stage, frequently occurs in hardwoods and broad leaved forest trees. For a detailed and critical review of somatic embryogenesis in conifers, readers are referred to Tautorus *et al.* (1991).

The choice of nutrient media for embryogenesis is similar to that for organogenesis as far as species and high or low salt media are concerned. However, several additional media have been used for conifer embryogenesis, for example, Gupta and Durzan's medium (1985), von Arnold and Eriksson's LP medium (Hakman and von Arnold 1985) and modified Blayde's medium for hardwoods (Merkle *et al.* 1990). The addition of abscisic acid (ABA) to the medium for development of "mature" somatic embryos has been essential.

2.3.2. *Initiation and Multiplication of Embryogenic Tissue.* There is a considerable body of data which describe the initiation and multiplication of embryogenic callus, and the physiology and biochemistry of somatic embryogenesis (Gupta and Durzan 1991, Thorpe 1991). Embryogenic tissue grows and is subcultured like a callus, but anatomically is composed of suspensor cells with or without embryo heads and is often referred to as embryonal suspensor masses (ESM's) or embryogenic tissue.

For initiation of embryogenic callus, immature zygotic embryos, 3-5 weeks after fertilization has occurred, are the most frequently used and reliable explants for conifers and hardwoods, but a variety of explants can be used (Thorpe *et al.* 1991). It has also been possible to obtain somatic embryos from mature zygotic Norway spruce (*Picea abies*) embryos (von Arnold and Hakman 1986), from mature zygotic *Picea mariana* embryos stored for 13 years (Tautorus *et al.* 1990) and from haploid gametophytes of larch (*Larix decidua*) (von Aderkas *et al.* 1987). Recently, cotyledons and other juvenile seedling explants of conifers have yielded somatic embryos after a shoot initiation treatment with cytokinin (Lelu *et al.* 1990, Attree *et al.* 1990). When immature pine embryos are used as explants, multiple genotypes of embryogenic tissue are formed initially and subsequently one genotype predominates (Becwar *et al.* 1991). As with organogenesis methods, the developmental stage of the explant critically affects embryogenic potential (Sotak *et al.* 1991, Ruaud 1991).

Protoplasts prepared from embryogenic callus have also yielded somatic embryos in sandalwood (*Santalum album* Linn) (Rao and Ozais-Akins 1985), Caribbaean pine (*Pinus caribaea* Morelet var. hondurensis) (Laine and David 1990), *Pinus taeda* (Gupta and Durzan 1987), *Picea glauca* (Attree *et al.* 1987) and hybrid Dunkeld larch (*Larix* x *eurolepsis*) (Klimaszewska 1989). These regeneration systems from protoplasts now pave the way for conifer protoplasts to be used in direct gene transfer and somatic hybridization experiments.

BAP and 2,4-dichlorophenoxyacetic acid (2,4-D) are usually added to the nutrient medium in the range of 0.5–5.0 mg/l BAP (2.22–22.2 µM) and 1.0–10.0 mg/l 2,4-D (4.5–45 µM) (Hakman and von Arnold 1985, Gupta and Durzan 1987, Merkle and Sommer 1986). There are numerous exceptions to this, e.g. no auxin is used with BAP for *Abies* spp. (Norgaard and Krogstrup 1991, Schuller *et al.* 1989), other auxins such as picloram are better than 2,4-D for white spruce (*Picea glauca* [Moench] Voss) (Lu and Thorpe 1987), and a phytohormone-free medium is used for oak (Gingas and Lineberger 1989). The use of phytohormone-free media for somatic embryogenesis has been recommended to allow zygotic embryos to be released from positional or chemical constraints and thus express innate embryogenic potential (Smith and Krikorian 1989). However, more synchrony in development, regulation of embryogenesis and competence of embryogenic tissue to form somatic embryos can be achieved with the addition of auxins (Komamine *et al.* 1990, Carman 1990). Multiplication of embryogenic callus occurs by subculturing onto the same medium every 2–6 weeks, depending on species and protocol. Both gelled and liquid media have been used for this stage. Gel strength and pH of the medium also influence embryogenic induction (von Arnold 1987).

2.3.3. *Development and Maturation of Somatic Embryos.* Development and maturation of somatic embryos in conifers is dependent on a relatively high level of abscisic acid (ABA) added to the medium, for example 10–20 mg/l ABA (37.8–75.6 µM). ABA induces the deposition of the storage lipids, proteins and carbohydrates during embryo development (Hakman and von Arnold 1988, Roberts *et al.* 1990a, Joy *et al.* 1991) which are essential prerequisites for the next stage of germination. The use of a medium with a high osmoticum and modification of the gaseous environment with increased levels of O_2 and CO_2 have also improved the quality and number of somatic embryos produced (Kvaalen and von Arnold 1991). A medium containing 12% sucrose to provide an osmotic environment that duplicates that found in nature was suggested by Gates and Greenwood (1991). Attree *et al.* (1991) used polyethylene glycol (PEG) to provide that osmotic environment for enhanced maturation and dessication tolerance of white spruce somatic embryos. More synchronous development of somatic embryos and a much greater percentage of responding genotypes are necessary to optimize this stage further. With hardwoods, somatic embryos are generally formed on media devoid of phytohormones.

2.3.4. *Germination of Somatic Embryos.* The recovery of plantlets from somatic embryos has improved dramatically since the review by Dunstan (1988), but in general, germination and conversion to plantlets is low (Thorpe *et al.* 1991, Tautorus *et al.* 1991). Germination of somatic embryos has been successfully carried out on a gelled nutrient medium to obtain good development of cotyledons and radicles. Some morphological (fusion of cotyledons) and genetic changes have been observed in Norway spruce somatic embryos (Lelu 1988).

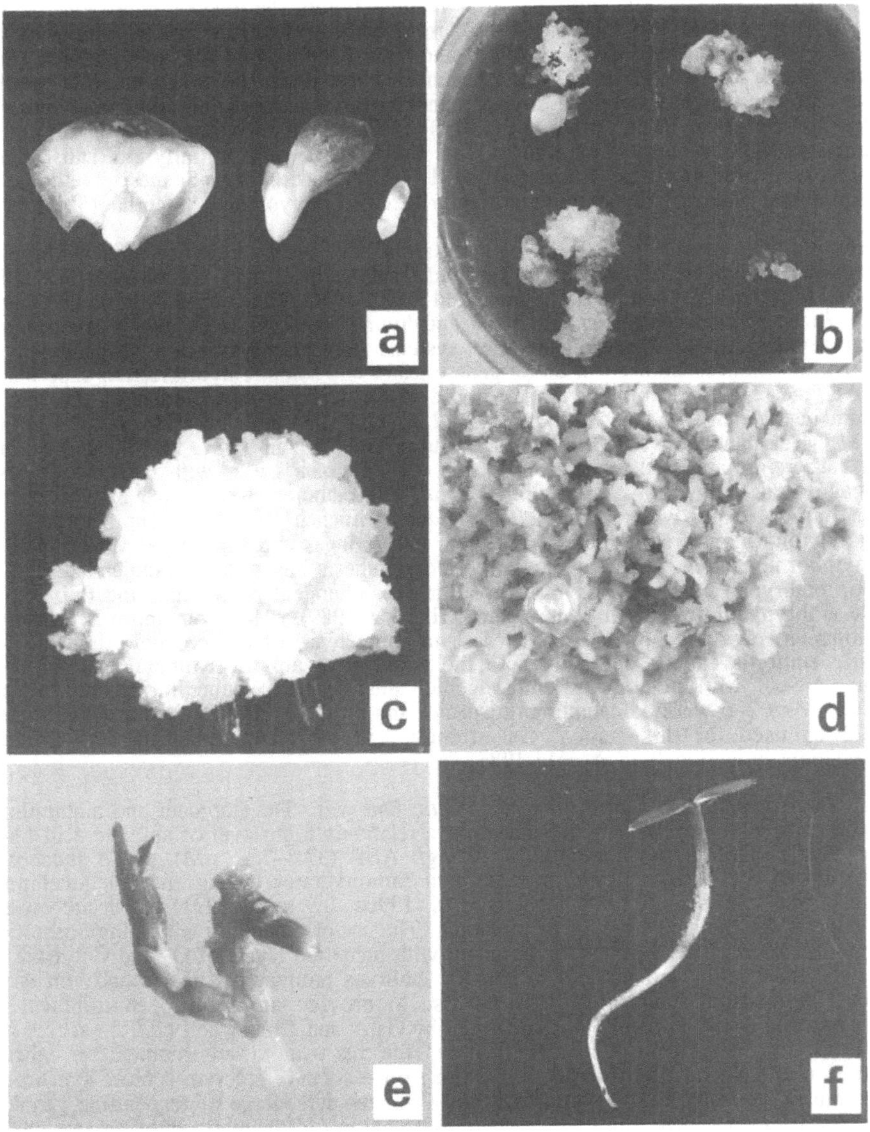

Figure 2. Various stages in plantlet formation via embryogenesis in *Agathis australis* (kauri) (Aitken-Christie, Gough and Davies unpublished). a. - cone scale with immature seed (LHS), immature seed (middle), excised immature gametophyte containing fertilized embryo (RHS). b. - initiation of embryogenic tissue from micropylar end of gametophyte. c. - embryogenic tissue for multiplication. d. - development and maturation of somatic embryos. e. - germination of somatic embryos. f. - cotyledon and radicle development on somatic embryo.

The use of a high relative humidity treatment by placement of mature somatic embryos in sterile wells with moistened filter paper for 1–2 weeks has improved subsequent survival, germination and growth of embryos (Roberts *et al.* 1990b, 1991). The use of an artificial female gametophyte encapsulating somatic embryos of *Pinus taeda* to produce a synthetic seed is being investigated by Gupta and Durzan (1991).

Further information on the complex physiological and biochemical requirements of the developing somatic embryo is necessary before significant progress can be made in this area. It was demonstrated by Joy *et al.* (1991) that somatic embryos of *Picea glauca* (white spruce) behaved metabolically differently from zygotic embryos. The authors suggest that the requirements of somatic embryos induced *in vitro* may be different from zygotic embryos. Gene expression and response to transformation also differ between zygotic and somatic embryos (Ellis *et al.* 1991). Application of information from studies on natural zygotic embryo development to somatic embryogenesis should therefore be treated carefully and may not prove useful.

2.4. FIELD EVALUATION OF MICROPROPAGULES FROM JUVENILE TREES

2.4.1. *General.* Micropropagules regenerated from shoots/plantlets and somatic embryos *in vitro* must perform in the field in a manner comparable to that of seedlings or cuttings. The risk is that the growth of micropropagules may be different due to physiological and/or genetic problems occurring throughout the micropropagation process. Consequently, it is very important to evaluate carefully the performance of micropropagules before the method is used for large-scale propagation of forest trees.

2.4.2. *Micropropagules Produced Via Organogenesis.* The majority of micropropagules produced from juvenile trees for field trials have been formed via the organogenesis method from either axillary or adventitious buds. Micropropagated plants produced via organogenesis methods from hardwoods (birch, oak, elm, eucalypts, teak, willow and poplar) and conifers (radiata pine, loblolly pine, Douglas fir, larch, redwood and maritime pine) have been established in the field in large numbers (1,000–50,000) (Gupta *et al.* 1992, Mehra-Palta and Thomson 1988). In general, micropropagules of hardwood species have grown as well as seedlings and in some cases better (Thorpe 1991, McCown and McCown 1987, Khuspe *et al.* 1987), and are phenotypically very uniform (Chalupa 1987). There are a few exceptions. In teak (*Tectona grandis*), micropropagules raised from seedlings of plus trees flowered, producing normal fruits and seeds, within 2 years of planting in the field (Mascarenhas *et al.* 1987). This is indicative of an advanced maturation state. Eight percent of the birch micropropagules produced via adventitious buds were of a variant type, "snake birch", and this trait remained stable (Jokinen and Tormala 1991). Micropropagules regenerated from juvenile conifer (e.g. Douglas fir, loblolly pine and radiata pine) cotyledons and planted in field trials also displayed an advanced maturation state (forming secondary needles earlier than seedlings) and a tendency to form more closed buds (characteristic of mature trees or 4–5 year old cuttings) (Horgan 1987, Ritchie and Long 1986, McKeand 1985, Frampton and Isik 1987). In more recent trials, radiata pine micropropagules measured at ages 5 and 6 in the field had a similar height growth, a smaller diameter, a higher proportion of trees suitable as final crop trees, less stem and leader malformation and a higher percentage of defect-free trees compared with seedlings of similar age and genetic background (Menzies *et al.* 1991). Trees derived from cuttings were better than those from micropropagules in diameter growth, but similar in all other respects. For trees derived from larch micropropagules, height and diameter growth was comparable with the growth of trees originated from seeds (Chalupa 1991). In loblolly pine, trees derived from micropropagules were also less susceptible to fusiform rust, a major fungal disease of loblolly pine, and formed more female cones compared with seedlings (Frampton and Isik 1987).

2.4.3. *Reasons for Advanced Maturation.* Some of the factors related to the appearance of characteristics of advanced maturation states in micropropagules derived from juvenile conifer tissues via organogenesis are becoming apparent. Formation of plantlets from adventitious buds rather than axillary buds was initially cited as one possible factor in the advanced maturation state of loblolly pine micropropagules (Frampton and Isik 1987), but confirmation of this was never published. Shoot quality, root quality and stress during acclimatization have all been found to be key factors affecting plantlet morphology and maturation state. Shoot quality rating was consistently correlated with high performance in the nursery and field (Wisniewski *et al.* 1986) and it was recommended by the authors that to obtain growth rates similar to seedlings only high quality plantlets with well-branched fibrous root systems should be planted out. A solid connection between the root and shoot with minimal callus development should allow the production of plantlets with normally functioning root-shoot axes (Patel *et al.* 1986). In an effort to improve the root system, loblolly pine plantlets were partially root-pruned and subsequently formed more "seedling-like" characteristics and also equivalent growth rates to seedlings (Wisniewski *et al.* 1991). Root pruning significantly increased the number of root apices. A link between premature maturation in plantlets and deficient root systems has thus been established as had been suggested earlier (Franclet *et al.* 1980, Mohammed and Vidaver 1988). The degree of stress during acclimatization of Douglas fir plantlets was also found to affect plantlet morphology (Mohammed and Vidaver 1990). The role of nutrient uptake and translocation of minerals is less clear, as Eastern white pine plantlets behaved similarly to seedlings (Bender *et al.* 1987), whereas loblolly pine plantlets behaved differently from seedlings (McKeand and Allen 1984). This is probably affected by the quality of the root system. The early flowering of one clone of juvenile birch (*Betula pubescens* Ehrh.) micropropagules after 1 year in the field was also not necessarily attributable to tissue culture conditions (Meier-Dinkel 1991), but to age of the micropropagules since initiation in culture and stress at planting out. Stress had previously been found to be one of the factors causing early flowering in birch.

In many of the field trials described, plantlets were produced in small numbers and were often the result of numerous different *in vitro* and *ex vitro* experiments to optimize conditions, which may have stressed the plantlets. Recently at Tasman Forestry Ltd. in New Zealand, more "seedling-like" micropropagules have been produced after conditions for rooting and acclimatization had been optimized, with the production of 1 million radiata pine plantlets in the year to June 1991 and for a further 2.5 million plantlets to be produced by June 1992 (B. Nairn personal communication). The main factors responsible have been a less stressful environment for the plantlets and roots present when plantlets are transferred *ex vitro* so that plantlets do not stop growing.

It is clear that more research is necessary to determine the effects of *in vitro* conditions on the subsequent *ex vitro* quality and performance of forest-tree micropropagules, similar to that done for other woody plants (Smith and McClelland 1991), and that more effective environmental control systems during rooting and acclimatization are required to avoid plantlet stress.

2.4.4. *Micropropagules Produced Via Embryogenesis.* The performance of micropropagules regenerated from somatic embryos has not yet been thoroughly tested and evaluated. Recently, micropropagules produced by embryogenesis have been established in the field in large numbers (400–4,000 per species) for black and white spruce (Attree *et al.* 1990), interior spruce (Webster *et al.* 1990), Douglas fir, loblolly pine and Norway spruce (Gupta 1992). Smaller trials with less than 100 micropropagules of larch, loblolly pine, radiata pine, redwood, sandalwood and eucalypts have also been established. Micropropagules of black spruce were more readily established in soil in the greenhouse than white spruce (Attree *et al.* 1990); 60% soil establishment frequency versus 18% respectively. The first large-scale evaluation of micropropagules produced from conifer somatic embryos, after one growing season in the field, has been reported for interior spruce (Webster *et al.* 1990). Growth rates, final height, shoot and root morphology, and frost hardiness were found to be similar for micropropagules from somatic embryos and seedlings. Indications from this study and from similar studies on Douglas fir, Norway spruce and loblolly pine (P. Gupta personal communication) are that no advanced maturation characteristics are evident with micropropagules from somatic embryos after one growing season in the field. This is encouraging, but not surprising since the original explants are from immature embryos and the multiplication phase is with embryogenic tissue which is juvenile. It is important to be cautious about conclusions from these initial trials because the maturation and germination conditions for somatic embryos are not optimized yet and micropropagules would have been produced from a variety of different experiments and treatments *in vitro* and *ex vitro*.

Further studies on the growth habits and genetic stability of germinated somatic embryos are being carried out, now that production of large numbers (1000–5000) of mature embryos is possible in some species (von Arnold 1991, Attree *et al.* 1990). This research is currently being carried out predominantly with spruces.

3. Micropropagation of Mature Trees

3.1. GENERAL

The micropropagation of material from mature forest trees has been much more difficult than juvenile trees. The maturation or transition from seedling to adult tree with flowers or cones is complex and varies from species to species. Readers are referred to recent reviews for thorough explanations of problems and possibilities for micropropagating mature trees and rejuvenation (Boulay 1987, Dunstan 1988, Pierik 1990, Thorpe *et al.* 1991). Mature trees can be micropropagated via the organogenesis and the embryogenesis routes, similar to juvenile trees. With propagation via organogenesis, two options are available in the choice of tissues for culture initiations in some species, namely (a) use of more juvenile tissues found in stump or sucker sprouts at the base of the tree and/or (b) use of more mature buds found near the top of the tree. The micropropagation of mature forest trees using juvenile explant tissues from the base of the tree has been successful for numerous hardwood species and for a few conifers (e.g. several species from the Taxodiaceae, redwood, *Cunninghamia*, and *Pinus rigida*), whereas micropropagation from more mature tissues has generally been less successful. Embryogenesis from mature trees is limited to only a few examples in hardwoods.

3.2. ORGANOGENESIS

3.2.1. *Pretreatment of Mother Plants.* Prior to placing tissues *in vitro*, pretreatments to the donor tissue (e.g., grafting, pruning, hedging, BAP treatment, serial grafting) are often made to improve its nutrition and health, arrest the maturation status, or partially rejuvenate it. Recently, micrografting *in vitro* has given encouraging results towards apparent rejuvenation for maritime pine (Dumas *et al.* 1989), redwood (Tranvan *et al.* 1991) and Western red cedar (*Thuja plicata* D. Don ex Lambert) (Mission *et al.* 1991). Micrografting for forest tree improvement is reviewed by Jonard (1986). The benefits are said to be less contamination, more accurate and reproducible control of the physiochemical environment, reduced correlative effects because of miniaturization and beneficial subculture effects (dilution of inhibitors, wounding effects etc.). Micrografting *in vitro* for 183-year-old Western red cedar was carried out with 6–7 mm buds which were successively micrografted 2 or 3 times onto juvenile shoot stocks, after which shoots appeared with juvenile morphology (Mission *et al.* 1991). These were multiplied on a modified MS medium containing 2-iP, elongated on a charcoal-containing medium and rooted (90%) using the same technique as is used for juvenile trees. Juvenile characteristics and an orthotropic growth habit were retained even after 4 years of growth in soil (Mission *et al.* 1991). Similar methods and results were obtained for micrografted buds from a 500-year-old redwood tree (Tranvan *et al.* 1991), but further improvements are still required for *in vitro* micrografting to be considered a reliable rejuvenation method.

3.2.2. *General.* Using either explants from more juvenile tissue or more mature tissue, the micropropagation of mature trees can be divided into the same four stages as micropropagation of juvenile trees, namely (1) initiation of shoots on explants, (2) shoot elongation and multiplication, (3) rooting, and (4) acclimatization of plantlets. The basic methods outlined for redwoods (*Sequoia sempervirens*) (Boulay 1974, Poissonier *et al.* 1980) and for eucalypts (de Fossard *et al.* 1977, Durand-Cresswell and Nitsche 1977) using more juvenile basal sprout explants have been used for other species e.g., blackwood (*Acacia melanoxylon* R. Br.) (Meyer and van Staden 1987) and *Cunninghamia lanceolata* (Bigot and Engelmann 1987). Using mature explants, reliable micropropagation methods have been developed for radiata pine (Horgan 1987, Horgan and Holland 1989) and giant sequoia (*Sequoiadendron giganteum* Buchholz) (Monteuuis *et al.* 1987) and for numerous broadleaved forest trees (Chalupa 1983, Mascarenhas *et al.* 1989).

The choice of nutrient media for micropropagation of mature trees is similar to that for micropropagation of juvenile trees. The addition of activated charcoal to the nutrient medium for elongation of shoots in coniferous species is often essential. To be successful with mature material, all steps of the procedure, the exact developmental state of the cultures, and environmental conditions within the culture vessel need to be defined and optimized to a much greater degree than is necessary for juvenile micropropagation (Horgan 1987). Mature material often shows more specific mineral and phytohormonal requirements and a lower capacity to be cloned (Monteuuis 1987).

3.2.3. *Initiation of Shoots on Explants.* As for juvenile trees, the choice of explant in the optimum developmental state is very important. The time of year can also be critical for some species (Thorpe *et al.* 1991). Buds or nodal microcuttings with young meristems are generally a prerequisite for obtaining axillary shoots *in vitro*. Needles also need to be in an immature state (approx. 1–2 mm long) for optimum adventitious shoot formation on their basal tissues (Selby and Harvey 1985). Endogenous cytokinin content was also correlated to the capability of clones to form shoots (von Arnold and Tillberg 1987). Selection of buds grown in full sunlight in the upper crown of the tree was important for optimum shoot initiation in radiata pine (Horgan 1987) and white spruce (Dunstan *et al.* 1987). The growth and multiplication of axillary shoots followed by rooting of these shoots is the most successful and reliable method employed to date. Adventitious shoots have been induced in many cases from buds, but these are often slow to elongate and do not form plantlets (Bonga 1981), with the recent exception of 17-year-old maritime pine (Dumas and Monteuuis 1991). The formation of adventitious shoots from immature female reproductive tissues of jarrah (*Eucalyptus marginata*) (Bennett and McComb 1982) and larch (Bonga 1984) has also occurred but rooting of shoots has been poor or non-existent.

Contamination is usually more of a problem with explants from mature trees than with explants from juvenile trees (McComb *et al.* 1990), although contamination of sprout explants is less than with mature branches (40% versus 100% respectively) (Warrag *et al.* 1990), and the use of enclosed buds avoids many problems (Horgan 1987). Delayed or latent contamination may appear during subsequent *in vitro* multiplication and this is still a problem in many laboratories. Media similar to those used to induce shoot formation from juvenile tissues are used for the production of shoots from mature tissue. Clonal differences and repeatability of results can be problematical.

3.2.4. *Shoot Elongation and Multiplication.* Shoot elongation occurs when cultures are transferred to a fresh gelled medium devoid of phytohormones and, for conifers often containing activated charcoal (0.5–2.0%). The requirement for charcoal, first demonstrated by Boulay (1979) is now shown to be widespread in micropropagation of mature conifers. When shoots have elongated they can be either rooted or remultiplied. For radiata pine, new cycles of axillary shoots are obtained by cutting nodal segments with small axillary meristems present and culturing them on a medium containing 5 mg/l BAP (22 μM) (Horgan 1987); however radiata pine shoots can die if they are placed on BAP medium in the wrong stage of development. For many hardwood species, shoots are either remultiplied by cutting them into nodal segments to stimulate new axillary shoots or by the division of shoot clusters and culturing them on a medium with BAP alone at low concentrations [(0.2–2.0 mg/l) (0.88–8.8 μM)] (Chalupa 1987), BAP and IBA (Meier-Dinkel 1991) or BAP, IBA and GA (Tricoli *et al.* 1985).

Repeated cycles on cytokinin medium followed by charcoal-containing medium or cytokinin-free medium have been shown to induce a degree of rejuvenation (Fouret *et al.* 1986, Monteuuis and Bon 1989). The degree to which this apparent rejuvenation is retained will show up in longer-term field trials. Continuous subculture on hormone-free media and a cold treatment have also induced more juvenile morphology in mature sitka spruce (Mac Ant-Saoir *et al.* 1991).

3.2.5. *Rooting and Acclimatization.* Rooting and acclimatization procedures for mature hardwoods have been successfully developed (Chalupa 1987, Meier-Dinkel 1991, Gupta and Mascarenhas 1987, Mascarenhas *et al.* 1987) using either more juvenile explants from the base of the tree or mature explants. An auxin treatment to induce roots has been found to be more necessary for mature shoots than juvenile shoots (Preece *et al.* 1991). Generally an IBA or IBA/NAA treatment is given to induce roots *in vitro* on gelled media, similar to methods for juvenile trees. Liquid MS medium containing IBA has also been used for rooting mature paulownia (*Paulownia taiwaniana*) shoots (Yang *et al.* 1989). The addition of rutin or quercitin (phenolic compounds and possibly auxin cofactors) has been found to enhance rooting of mature black cherry (*Prunus serotina* Ehrh.) under a 16 hour photoperiod (Tricoli *et al.* 1985). Large clonal differences in rooting are also found for birch (Meier-Dinkel 1991). Rooting of shoots produced from various coniferous species, particularly originating from mature explants, has been problematical (Evers *et al.* 1988, Thorpe *et al.* 1991). A reliable method for rooting *ex vitro* and acclimatizing shoots from mature 17- to 19-year-old radiata pine has been developed (Horgan and Holland 1989); 90–100% survival and 78% rooting were obtained. The main factors found to be important for root formation on mature shoots were: use of high quality shoots; a free-draining peat-pumice-perlite rooting mix; a well-managed watering regime and gradual conditioning from the day of planting; removal of basal callus; and a pretreatment on medium containing 6% sucrose. In addition, Boulay (1989) found that a 24-hour auxin dip containing benomyl fungicide produced 60–100% rooting in mature redwood shoots. These methods may be useful for improving the rooting and acclimatization of other mature conifer shoots.

3.3. EMBRYOGENESIS

True rejuvenation might be obtained if somatic embryos could be regenerated from mature trees. The establishment of clonal forests directly from field-proven superior trees would become possible and seed orchards might no longer be necessary. There have been no reports of somatic embryogenesis from conifer tissues of mature trees to date. However, encouraging results were obtained with the formation of embryogenic tissue and early stage somatic embryos from 4-month-old secondary needles of Caribbaean pine (*Pinus caribaea* Morelet) by Berlyn *et al.* (1991). Further work is required to mature and germinate the somatic embryos. Thus it may be possible to extend this technique to secondary needles from field-proven mature trees. Callus was also readily produced from male strobili tissues of jack pine (*Pinus banksiana* Lamb), collected after spring swelling and prior to acquiring a purplish hue (Chesick and Bergmann 1991). The medium used was GD medium containing 50 μm BAP and 1 μm NAA.

In some hardwood species, several major breakthroughs have been made and somatic embryos have been produced from a variety of male reproductive tissues. Male catkin tissues have been induced to form embryogenic tissue and somatic embryos in birch (*Betula pendula* Roth.) (Srivastava and Steinhauer 1981) and recently to produce germinated somatic embryos and regenerated plants in swamp white oak (*Quercus bicolor* Willd.) (Gingas 1991). Somatic embryogenesis in horse chesnut (*Aesculus hippocastanum* L.) was also achieved by culturing the filaments which bear the anthers and pollen grains (Jorgensen 1989). The stages of development when the male catkins were initiated into culture were the "semipendant stage" for oak and having "pollen mother cells" for birch; both of these stages are found in immature male catkins when the pollen is not ripe. The catkin tissues that produced embryogenic tissues were the diploid mother tissues in both cases — cut ends of the catkin axis and scales in birch, and from the basal ends of individual flowers on the peduncle in oak.

The media used for birch embryogenesis were firstly, a White's medium containing 2 mg/l IAA (11.4 μM), 5 mg/l kinetin (23.2 μM) and 1 gm/l casein hydrolysate for 4 weeks, followed by a medium containing the same levels of IAA and kinetin, but with 3.0 mg/l adenine (22.2 μM) (Srivastava and Steinhauer 1981). No follow-up publications have appeared on the birch system to optimize it further. A modified MS medium containing 200 mg/l casein hydrolysate and 1 mg/l 2,4-D (4.5 μM) was used for 4–6 weeks for initiation of embryogenic tissue in white oak for 4-6 weeks, followed by a MS medium containing 1 mg/l BAP (4.4 μM) for 6–8 weeks and then MS medium without growth regulators for the development of somatic embryos (Gingas 1991). Adventitious embryogenesis directly from the surface of the white oak embryos continued repetitively for 1.5 years. This is the first example of complete plantlet regeneration of somatic embryos produced from tissues of a mature forest tree. Somatic embryos have also been developed from callus derived from shoot segments of sandalwood (*Santalum album* L.) on MS medium containing 1 mg/l GA (2.9 μM) (Sita *et al.* 1979). These results, coupled with the earlier results of embryogenesis from grape reproductive tissue (Mullins and Srinivasan 1976), clearly show that the regenerative capacity of forest tree reproductive tissues for somatic embryogenesis should be further investigated.

3.4. FIELD EVALUATION OF MICROPROPAGULES FROM MATURE TREES

3.4.1. *Field Trials and Rejuvenation.* Micropropagules from mature trees have been even less extensively tested in the field than micropropagules from juvenile trees, but nevertheless there have been some reports. The most interesting question is whether or not rejuvenation has been achieved as a result of pretreatments to the mother plant, choice of the initial explant, *in vitro* culture and environment, or any other cultural and physical treatments. Depending on the end use of micropropagules, the goals can also be different, namely (1) rejuvenation or a degree of rejuvenation — trees having more juvenile morphological features may be required for fast growing plantation forestry, and (2) more mature micropropagules may be required for flowering seed orchards.

Field trials with micropropagules from mature trees produced via organogenesis have been established with maritime pine (Franclet *et al.* 1980), radiata pine (Horgan and Holland 1991), redwood (Boulay 1989), jarrah (*Eucalyptus marginata* Donn ex Smith) (Bennett *et al.* 1986), *Cunninghamia lanceolata* (Bigot and Engelmann 1987), teak (Mascarenhas *et al.* 1987), birch (McCown 1989) and numerous eucalypts (Gupta and Mascarenhas 1987). Most of these field trials have had small numbers of trees planted out (100–2000), similar to the early trials with micropropagules from juvenile trees, and we must be cautious about generalisations.

Early indications are that micropropagules from mature explants stay mature but may show some juvenile characteristics. Micropropagules of jarrah displayed mature foliage and leaf form (Bennett *et al.* 1986), flowering and seed set was earlier for *Eucalyptus citriodora* and teak (Mascarenhas *et al.* 1987) and micropropagules of *Cunninghamia lanceolata* stayed plagiotropic (Bigot and Engelmann 1987). Increased uniformity was found in birch micropropagules (McCown 1989). Shoots from epicormic buds of 60-year-old Douglas fir were also plagiotropic when collected from explants of shoots with a horizontal or downward growth habit; by contrast, upright growth of micropropagated shoots was observed in all explants from branches with an upright growth habit (Gupta and Durzan 1987). Sometimes orthotropic shoots are produced in culture with frequent transfer onto medium containing BAP and NAA and plagiotropic growth is observed after acclimatization *ex vitro* (Fouret *et al.* 1989). With radiata pine micropropagules from mature trees, some degree of rejuvenation was found after 2.5 years in the field, particularly with respect to growth rates (Horgan and Holland 1991). The long term growth rate and form of the trees has yet to be determined. The authors also state that maturation may occur as a result of sub-optimal treatments causing stress during the *in vitro* and *ex vitro* stages, similar to what has already been postulated for juvenile trees. Some of the developments in apparent rejuvenation of shoot growth *in vitro* and at the young plantlet stage using micrografting techniques and adventitious buds, appear promising and we await the results of field trials.

No micropropagules produced via embryogenesis from mature trees have been placed in field trials yet, but results to date are exciting and hold promise for the future.

3.4.2. *Biochemical Markers.* In order to avoid waiting months and more likely years to observe the effects of various treatments on the morphology(shape of leaves and buds), physiology (rooting ability, orthotropic versus plagiotropic habit etc.) and reproductive habit (flowering or vegetative) of micropropagules from mature forest trees, early detection methods using chemical and biochemical markers are being researched. For a review of this subject refer to Bon (1989). In mature Western red cedar the ratio of K/Ca in the tissue was influenced by cytokinin treatments *in vitro* and degree of juvenility in shoots in culture (de Canniere 1990). More juvenile material always has the highest K/Ca ratio. First use of a protein to confirm rejuvenation in meristems was reported by Bon (1989) where a membrane associated protein (J16), that was indicative of the juvenile phase, was detected in both juvenile and "rejuvenated" 100-year-old giant sequoia tissue. J16 was absent in mature tree and grafted tissues. The use of molecular biology methods to study the effect of maturation on morphological and photosynthetic characteristics in mature larch (*Larix lariciana*) (Hutchison *et al.* 1990) may also help find markers for rejuvenation.

4. Genetic Transformation of Forest Trees

4.1. GENERAL

Genetic improvement of forest trees by traditional breeding is a long process due to the considerable generation times, so there has been an increasing research emphasis on transformation of woody plants. This is in the hope that ultimately, genes of economic importance could be added to the breeding stock in a single tissue culture step. Micropropagation is an essential prerequisite for production of transformed trees through tissue-culture-based transformation systems. In recent years there has been a large number of reports of successful transient expression of genes introduced into various woody plant culture systems; while this may provide a useful means of assaying activity of promoters and constructs, the inability to regenerate stably transformed plants in most cases has been the roadblock to genetic improvement using the promoters that have been studied and genes that may be of functional or economic interest.

Demonstrable and heritable transformation requires that the transforming DNA must be detectably incorporated into the host genome, transcribed, and translated. This requires a DNA construct with (1) an appropriate promoter (which may be constitutive, or it may be desirable that it be expressed only in tissues of interest or under certain inducible conditions), (2) a desired coding region (which may be optimized for codon usage, start codons, organelle targeting, etc.), (3) a reporter gene and/or selectable marker to allow effective identification of transformed cells, (4) cells that are in a transformable, regenerable state (the two may not always coincide), and (5) an appropriate DNA delivery system.

4.2. TRANSFORMATION SYSTEMS

The promoters and coding region parameters that will work optimally with a given woody plant can only be determined through transformation experiments using reporter genes, so the application of DNA delivery systems reported to date has mostly been of this nature, though some genes of economic or functional interest have also been used for stable transformation in systems that had previously been explored with reporter genes.

DNA delivery systems may be divided into two groups — *Agrobacterium*-mediated insertion of DNA located on a tumor inducing (T_i) plasmid between the (transfer DNA) T-DNA borders, and direct DNA-mediated methods, which that do not necessarily use T_i plasmids.

4.2.1. *Agrobacterium.* The first major success in woody plant transformation that gave rise to regenerants was reported using *Agrobacterium tumefaciens* to infect leaf disks of a poplar (*Populus*) hybrid (Parsons *et al.* 1986, Fillatti *et al.* 1987). The efficiency of DNA delivery by *A. tumefaciens* has made it the method of choice for a number of plant species including woody plants, with walnut, apple and plum all having been stably transformed and regenerated (McGranahan *et al.* 1988, James *et al.* 1989, Mante *et al.* 1991), and many other woody dicot species have been transformed in possibly regenerable culture systems (e.g. *Alnus, Betula, Liquidambar, Prunus* sp. and *Salix* [Hanover and Keathley 1988]).

Agrobacterium infection has led to gall formation, and reporter genes and Southern blots have shown stable integration of the transforming DNA, in a number of conifers (Sederoff *et al.* 1986, Dandekar *et al.* 1987, Clapham and Ekberg 1988, Morris *et al.* 1988, Ellis *et al.* 1989, Stomp *et al.* 1990) but in all attempts made public to date the galls have not been regenerable into transformed plants; for the most part they have deteriorated or been sloughed off over time rather than even producing callus in culture (Dandekar *et al.* 1987, Stomp *et al.* 1990). Direct DNA-mediated transfer may be the better option for conifers.

Nearly all reports of optimizing woody plant transformation with *Agrobacterium* note differences in transformation frequency and stability by different strains of the bacterium. Genetic factors in the host woody plant cultivars or genotypes used may also come into play, as is being analyzed with clonally propagated shoots of various grapevine hybrids (Lowe and Krul 1991).

Tumors and roots induced by *A. rhizogenes* can sometimes be regenerated into plants more easily than can tumors induced by *A. tumefaciens*, although some regenerants and progeny show abnormal phenotypes such as wrinkled leaves, ageotropic roots, and increased lateral root initiation (Pythoud *et al.* 1987). Strain specificity is also a factor here. *A. rhizogenes*-transformed root tumors have recently been used to obtain stable transformants of the woody nitrogen-fixing angiosperm *Allocasuarina verticillata* (Phelep *et al.* 1991). Increased rooting of transformants has been used to advantage for rooting clonally propagated plants (Lambert and Tepfer 1991, MacRae and von Staden 1991).

4.2.2. *Direct DNA-Mediated Gene Transfer* methods, bypassing worries about *Agrobacterium* host range and strain specificities, have been much improved in recent years with the development of methods of bombarding intact plant cells with DNA-coated microprojectiles ("biolistic" methods; for a review, see Sanford 1988). This also bypasses the need to use protoplasts. Protoplast transformation using PEG (e.g. for *Citrus*, Schell *et al.* 1991) or electroporation has been reported in some woody plant systems, but the difficulty of regeneration from protoplasts for most woody species has precluded widespread application of protoplast-based methods for regeneration of transgenic plants (McCown 1988, though see Chand *et al.* 1988, in which this treatment was actually shown to stimulate regeneration in the woody plant *Solanum dulcamara*).

Despite the difficulty of using them in regeneration, one area where protoplast systems have been quite useful is quantitative testing of promoters for expression in heterologous systems by assaying a fused reporter gene such as the glucuronidase (GUS) gene (Jefferson *et al.* 1987) or others whose products are easily detectable in protoplasts (Bekkaoui *et al.*

1988 and 1990, Tautorus *et al.* 1989, Wilson *et al.* 1989, Gupta and Durzan 1991). Recently, however, microprojectile-based systems have allowed extensive and rapid transient assay testing of promoters in intact tissues. For example, a promoter deletion series is being used for identification of *cis*-acting DNA sequences affecting expression of a conifer ribolose bis-phosphate carboxylase, small subunit (*rbcS*) promoter in bombarded radiata pine suspension cells (Campbell and Neale 1990). Bombardment of various tissues has already allowed testing of promoter specificities in some tissues that might not be recoverable from transformed plants for quite a long time, such as developing xylem and resin ducts in loblolly pine (Loopstra and Sederoff 1990).

Various alterations of the basic microprojectile bombardment method have followed rapidly upon each other in the past two years to improve efficiency and decrease damage to the bombarded tissue, in the hope that better regeneration will thus be allowed (e.g. Christou *et al.* 1990, Iida *et al.* 1990, Sautter *et al.* 1991), and no time has been lost in exploiting this tremendous possibility for woody species that have been recalcitrant to regeneration from *Agrobacterium* or protoplast transformation methods. Transformants have been regenerated from bombarded *Liriodendron tulipifera* (Wilde *et al.* 1990), papaya (Fitch *et al.* 1990), and white spruce (Ellis *et al.* 1991).

4.3. TARGET TISSUES

As various transformation systems have resulted in transient expression, the target tissue for transformation is likely to be the major consideration for regeneration in the future. In addition, differences in transient expression have frequently been reported as associated with the tissue inoculated or bombarded. While it is likely that many of these reported differences stem from promoter specificity (and this points to the need for more knowledge of promoter parameters in conifers, in particular), there may also be differences in the biological receptivity of different target tissues to incorporation of the transforming DNA. Age or developmental stage of the target tissue is often important, for example, with juvenile or actively growing tissue usually reported as best, although in apple work with *A. rhizogenes*, burl formation in older trees enhanced transformability (Lambert and Tepfer 1991). In transformation of photoperiodic grape (*Vitis*) cultivars, a short-day treatment markedly lowered efficiency (Lowe and Krul 1991). Such effects are likely to be related to stages of maximal cell division and DNA replication (Stomp *et al.* 1990), while for *Agrobacterium* transformation the levels of inducing plant phenolics may also be important (Ellis *et al.* 1989).

Somatic or zygotic embryos, meristematic nodule cultures, or cultures which can be manipulated to give rise to either of these through *in vitro* manipulation of hormones, are attractive targets due to small cell numbers, ability to impose antibiotic or other selection systems, and regeneration potential. Such cultures have been used for biolistic transformation of yellow poplar (*Liriodendron tulipifera*), spruce, and pine (Wilde *et al.* 1990, Ellis *et al.* 1991, Robertson *et al.* 1991) and many more examples will undoubtedly be reported in the near future as such cultures have been developed for a variety of conifers and woody angiosperms as described in Sections 2 and 3 above.

4.4. IDENTIFICATION, REGENERATION AND MICROPROPAGATION OF TRANSFORMANTS

Micropropagation of transformed tissues has two aspects, the first being initial identification of transformed tissue for maximum efficiency of propagation effort, and the second being the selection of appropriate conditions for growth and development of that tissue.

Identification of *Agrobacterium*-transformed tissue was at first done simply by assay for a sort of reporter gene, i.e. opine synthesis and formation of galls or hairy root tumors. For armed strains of *Agrobacterium* this is still a useful means (e.g. Phelep *et al.* 1991), although it is sometimes forgotten that controls are necessary for both; opines are sometimes not detectable in transformed tissue if the precursor, arginine, is not present in sufficient amounts

(Ellis *et al.* 1989) and they may be found in wild-type tissue if arginine is present at high concentrations (Christou *et al.* 1986). Co-transformed reporter genes and selectable markers may also be used for identification of *Agrobacterium*-transformed tissue (Draper *et al.* 1988).

After disarmed or DNA-mediated transformation, identification usually depends upon expression of co-integrated selectable markers because this has the advantage of suppressing the growth of the untransformed tissue as well as identifying that which is putatively transformed. A variety of antibiotic selection markers can be used, with the most common, for historical reasons, being kanamycin resistance (Pythoud *et al.* 1987), but herbicide resistance can also be a good selectable marker and has the advantage of some possible economic utility in regenerants (Fillatti *et al.* 1988). These selective agents may be applied after transformation and sometimes through subsequent micropropagation steps.

The concentration of the selecting agent and the number of culture steps through which the selection must be applied is a matter for experimentation within each individual regeneration and micropropagation system. For example, in the first successful propagation example through somatic embryogenesis (McGranahan *et al.* 1988), kanamycin did not kill the untransformed tissue, but only inhibited embryo differentiation, so in order to avoid chimeric regenerants, kanamycin-resistant embryos needed to be subcultured onto fresh kanamycin-containing medium as they appeared, and selection was continued for several months. In many culture systems, more work is needed on parameters of maturation and rooting (Attree *et al.* 1990); it is likely that genotypes of target plants will differ to some extent with regard to regeneration potential under selection.

4.5. EVALUATION OF TRANSFORMED PLANT MICROPROPAGULES

At this stage, reporter genes under the control of a constitutive promoter are used to ascertain whether the transforming DNA has not been lost and is still capable of being expressed. However, GUS expression is not always a reliable guide to current gene expression, due to the stability of the GUS protein (Jefferson, personal communication); furthermore reporter gene expression has not been found to correlate well with levels of stable transformation of co-transformed genes in an extensive study of 2500 tobacco transformants (Drummond *et al.* 1991). Detection of the transforming DNA in the genome by polymerase chain reaction (PCR) (e.g. Ellis *et al.* 1991, Maynard 1991, Robertson *et al.* 1991, Schell *et al.* 1991), or more definitively (if there is enough tissue) by Southern blot, is important. We are still a long way from knowing much about the long-term stability of the transforming DNA in long-lived woody plants!

Whether the gene of interest will be expressed in the tissue of interest depends a great deal upon the specific promoter that was used and the DNA environment into which it was integrated, so there is no substitute for being able to assay the activity of the gene of interest directly using a reaction product or antiserum if at all possible. Transformed callus has been assayed directly for expression of incorporated insecticidal genes from *Bacillus thuringiensis* by feeding of Western tussock moth and spruce budworm larvae (Howe *et al.* 1990, Ellis *et al.* 1991). Morphological effects have allowed identification of a number of different transformed poplar clones expressing various levels of auxin and cytokinin biosynthesis transgenes in various tissues, which may be useful later as a transformation system for improving wood production (Olsson *et al.* 1990, Eklof *et al.* 1991).

Many of the traits that it would be desirable to engineer in woody plants, such as wood production (Timmis and Trotter 1989), and drought and salt resistance, are likely to be multigenic traits, and more work is needed to map quantitative trait loci and identify possible gene products that could be assayed and manipulated. Single genes controlling traits such as disease resistance and herbicide tolerance also have yet to be identified. It may be possible by studying the genes that are expressed in complex differentiation processes, such as reproduction, to engineer desirable traits such as male sterility. For conifers particularly, much research is also needed into the structure of promoters of native genes, so that efficient constructs can be designed.

With the many advances in conifer transformation for transient expression, and the simultaneous development of better regeneration systems, we can have high hopes that the two can be combined in several systems in the near future for testing the genes and promoters that are being identified in such basic molecular biology work.

5. Automation of Micropropagation

5.1. GENERAL

The production of planting-stock for reforestation by tissue culture has already begun. In New Zealand, 2.5 million plantlets are produced annually by the micropropagation of selected juvenile radiata pine (*Pinus radiata*) families and clones (Gleed 1991). Mature birch (*Betula pentula* Roth) plantlets are being produced on a commercial scale in the USA (McCown 1989) and Finland by Kemira, Oy Hortus AB and Enso-Gutzeit (Törmälä *et al.* 1990), eucalypts (*Eucalyptus*) are being produced in Australia (Rasmussen, personal communication) and Brazil (Purse, personal communication) and redwoods in France and the USA (Boulay 1989). These commercial units presently use the organogenesis route for plantlet production. The embryogenesis technique has not yet been developed to the same reliable degree.

Micropropagation is generally more expensive than other forms of propagation because of the high labour component (64%) (Kurtz *et al.* 1991). To reduce the cost per plantlet, new developments such as automation to reduce the labour input, via either the organogenic or embryogenic pathways, are necessary. Bioreactors or automated growing systems are required for the multiplication of meristems, shoots and somatic embryos, and separation (with or without cutting) and delivery systems are required to handle individual or multiple plant units. There are also other ways of reducing the cost, such as increasing the multiplication rate, decreasing losses due to contamination and even relocation of laboratories to countries where labour is cheaper.

5.2. ORGANOGENESIS

There are several stages during the micropropagation method that can be automated when shoots and plantlets are multiplied *in vitro* and rooted. The stages and methods used depend on the growth habits of the various species (Aitken-Christie 1991, Kozai *et al.* 1991). Automated and robotic systems have been investigated for the cutting and transfer of shoots/plantlets, multiplication and separation of meristematic nodules and handling plantlets *ex vitro*. More emphasis in future is needed on developing novel environmental control systems for automation (Kozai *et al.* this book), reducing the cost of automated and robotic systems and avoiding biological and physiological problems.

Automated methods have been developed for multiplying and separating meristematic nodules (Levin *et al.* 1988) which can now be produced in radiata pine (Aitken-Christie *et al.* 1988) eldarica pine (Phillips and Gladfelter 1991) *Eucalyptus grandis* (Warrag *et al.* 1991) and poplar (McCown *et al.* 1988). It appears that the automated system developed for lilies (Miwa 1991) could also be used for meristematic nodules with minor modifications.

The hedging system developed for growing radiata pine shoots (Aitken-Christie and Jones 1987) has also been automated for liquid nutrient replenishment (Aitken-Christie and Davies 1988). Sugar-free growth of shoots in an automated system to reduce contamination problems was recently demonstrated (Davies *et al.* 1992). The cost of this system is 20% cheaper than conventional micropropagation for eucalyptus plantlets produced on a commercial scale (Mascarenhas *et al.* 1989). There are numerous robotic and automated systems available for cutting and transplanting shoots with nodes and shoot clumps (Aitken-Christie 1991, Kozai *et al.* 1991). Robotic transplanting systems being developed for small seedlings in the greenhouse could also be adapted for handling tissue cultured plantlets (Ting 1991).

Encapsulated adventitious and axillary shoots have also been investigated as a delivery system similar to synthetic seeds for the woody plants mulberry (*Morus indica* L.) and *Valeriana wallichii* DC (Bapat *et al.* 1990, Mathur *et al.* 1989) and may be applied to a wider range of species.

5.3. EMBRYOGENESIS

Several stages of embryogenesis are being investigated for automation, namely (1) multiplication of embryogenic tissue, and (2) germination of somatic embryos.

Bioreactors have been tested for the growth and multiplication of embryogenic tissue of Douglas fir and Norway spruce (Gupta *et al.* 1991) and radiata pine (Smith *et al.* 1991). Embryogenic spruce cultures have also been cultured in liquid media in flasks (Hakman and von Arnold 1988). To date, cotyledonary embryos have not been produced in liquid shaker flasks or bioreactors, and transfer of embryogenic suspensions onto gelled media is necessary to allow further development of somatic embryos. Bioreactor systems being developed for other crop plants may also be useful for forest trees (Sondal and Noriega 1991). Low cost bioreactor units specifically designed for growth of forest-tree embryogenic tissues have yet to be developed.

The use of artificial seed as a delivery system for somatic embryos has been researched extensively for agricultural crop plants and is now being applied to both hardwood and conifer forest-tree species (Redenbaugh and Ruzin 1989, Bapat and Rao 1989). The most widely used method involves alginate beads containing somatic embryos germinated on gelled media rather than in soil. Germination rates of somatic embryos are still too low both under conventional conditions and in artificial seeds (see Table 1, Redenbaugh and Ruzin 1989 for review), and further work is necessary to develop optimum nutritional and environmental conditions for more synchronous germination. Only 1-2% germination has been achieved with encapsulated somatic embryos (Gupta *et al.* 1991). In addition to artificial seed, a range of other automated delivery systems for somatic embryos are being developed. A modified mini-plug transplanting system and hybrid systems, using aspects of both containerized and artificial seed technologies, is being developed by Weyerhaeuser Company, USA (Gupta *et al.* 1991). For example, crudely encapsulated embryos may be delivered via a greenhouse container system or into environmentally modified nursery beds (e.g., with plastic row covers) at lower overall cost. The use of robotics to retrieve and transplant spruce somatic embryos produced on gelled media (Bachop and Beaulac, unpublished results) may yield the most promising results earlier than other systems.

There is also potential to introduce imaging systems for the selection and screening of somatic embryos at the proper developmental stage but no work has been published on forest trees yet.

6. Concluding Remarks

Among forest trees, hardwoods such as poplar, birch, teak, eucalypts and sweetgum and conifers such as pines, spruces, Douglas fir and redwood have all proved amenable to large scale micropropagation, and numerous laboratory protocols have been developed. Micropropagation of tropical forest trees needed for reforestation of rain forests is also now under way and offers exciting prospects for many countries.

Tremendous advances have been made in the various areas of somatic embryogenesis from immature seed tissues, micropropagation of mature trees, and rejuvenation over the past five years. Clearly, however, more research is necessary on the manipulation of the microenvironment (physical, chemical, gaseous) to obtain high quality shoots, plantlets or somatic embryos at their various stages of development. The importance of field trials cannot be underestimated and careful evaluation of the performance of micropropagules produced using more optimal methods is necessary before the micropropagation method is used for large-scale propagation of forest trees.. Some of the factors involved in maturation status and early flowering of trees from micropropagules produced via organogenesis have now become apparent and lessons can be learned from these for future field trials of micropropagules produced using new methods.

Full exploitation of the potential of micropropagation biotechnology in forestry requires the combined effort of managers, tree breeders, nurserymen, engineers, biochemists and most importantly plant/tree physiologists for understanding and manipulating tree growth and development. Successful micropropagation systems are an essential prerequisite of tree improvement via genetic engineering, and though only a few model systems are currently workable, great advances may be expected in this area in the future.

7. Acknowledgements

The authors would like to thank Professor Warwick Silvester, Dr Derek White and Dr Michael Menzies for reviewing the manuscript, Helen Davies for referencing, David Blake for photography and Melanie Maika for word processing.

8. References

Abo El-Nil, M. and Wochok, Z.S. (1986) 'Seed weight and *in vitro* bud induction potential in *Pseudotsuga menziesii* cotyledons cultured *in vitro*', N.Z.J. For. Sci. 16, 283-288.

Ahuja, M.B. (1989) 'Storage of forest tree germplasm at sub-zero temperatures', in V. Dhawan (ed.), Applications of Biotechnology in Forestry and Horticulture, Plenum Press, New York and London, pp. 216-228.

Aitken-Christie, J. and Thorpe, T.A. (1984) 'Clonal propagation', Gymnosperms, in Cell Culture and Somatic Cell Genetics of Plants, Vol. 1, I.K. Vasil (ed.), Academic Press, Inc., Orlando, Florida, pp. 82-95.

Aitken-Christie, J. and Jones, C. (1987) 'Towards automation: radiata pine shoot hedges *in vitro*', Plant Cell Tissue and Organ Culture 8, 185-196.

Aitken-Christie, J. and Singh, A.P. (1987) 'Cold storage of tissue cultures', in J.M. Bonga and D.J. Durzan (eds.), Cell and Tissue Culture in Forestry, Vol. 2, Martinus Nijhoff Publishers, The Netherlands, pp. 285-304.

Aitken-Christie, J. and Davies, H.E. (1988) 'Development of a semi-automated micropropagation system', Acta Hortic. 230, 81-87.

Aitken-Christie, J., Singh, A.P. and Davies, H. (1988) 'Multiplication of Meristematic Tissue: A New Tissue Culture System for Radiata Pine', in James W. Hanover and Daniel E. Keathley, Genetic Manipulation of Woody Plants, Plenum Press, New York and London, pp. 413-432.

Aitken-Christie, J. (1991) 'Automation', P. Debergh and R.H. Zimmerman (eds.), Micropropagation: Technology and Application, Kluwer Academic Publishers, pp.

Amerson, H.V., Frampton, L.J. Jr., Mott, R.L. and Spaine, P.C. (1988) 'Tissue culture of conifers using Loblolly Pine as a model', in J.W. Hanover and D.E. Keathley (Eds.), Genetic Manipulation of Woody Plants, Plenum Press, New York and London, pp. 117-137.

Attree, S.M., Bekkaoui, F., Dunstan, D.I. and Fowke, L.C. (1987) 'Regeneration of somatic embryos from protoplasts isolated from an embryogenic suspension culture of white spruce (*Picea glauca*)'. Plant Cell. Rep. 6, 480-483.

Attree, S.M., Budimir, S. and Fowke, L.C. (1990) 'Somatic embryogenesis and plantlet regeneration from cultured shoots and cotyledons of seedlings from stored seeds of black and white spruce (*Picea mariana* and *Picea glauca*)', Can. J. Bot. 68, 30-34.

Attree, S.M., Tautorus, T.E., Dunstan, D.I. and Fowke, L.C. (1990) 'Somatic embryo maturation, germination and soil establishment of plants of black and white spruce (*Picea mariana* and *Picea glauca*)', Can. J. Bot. 68, 2583-2589.

Attree, S.M., Moore, D., Sawhney, U.K. and Fowke, L.C. (1991) 'Enhanced maturation and dessication tolerance of white spruce (*Picea glauca* [Moech] Voss) somatic embryos: effects of a non-plasmolysing water stress and abscisic acid', Ann. Bot. 68, 519-525.

Bajaj, Y.P.S. (1989) 'Biotechnology in Agriculture and Forestry 5., Trees II', Springer Verlag, Berlin.

Bajaj, Y.P.S. (1991) 'Biotechnology in Agriculture and Forestry 16, Trees III', Springer-Verlag, Berlin.

Bapat, V.A. and Rao, R.S. (1990) '*In vitro* growth of encapsulated axillary buds of mulberry (*Morus indica* L.)', Plant Cell, Tissue and Organ Culture 20, 69-70.

Becwar, M.R., Blush, T.D., Brown, D.W. and Chesick, E.E. (1991) 'Multiple paternal genotypes in embryogenic tissue derived from individual immature loblolly pine seeds'. Plant Cell, Tissue and Organ Culture 36, 37-44.

Bekkaoui, G., Pilon, M., Laine, E., Raju, D.S.S., Crosby, W.L. and Dunstan, D.I. (1988) 'Transient gene expression in electroplated *Picea glauca* protoplasts', Plant Cell Rep. 6, 481-484.

Bekkaoui, F., Datla, R.S.S., Pilon, M., Tautorus, T.E., Crosby, W.L. and Dunstan, D.I. (1990) 'The effects of promoter on transient expression in conifer cell lines', Theor. Appl. Genet. 79, 353-359.

Bender, L., Harry, I.S., Yeung, E.C. and Thorpe, T.A. (1987) 'Root histology, and nutrient uptake and translocation in tissue culture plantlets and seedlings of *Thuja occidentalis* L.', Trees 1, 232-237.

Bennett, I., Tonkin, C.M., Wroth, M., Davison, E. and McComb, J.A. (1986) 'A comparison of growth of seedlings and micropropagated *Eucalyptus marginata* (jarrah)', I. Early growth to 2 years, Forest Ecol. Manag. 14, 1-12.

Bennett, I.J. and McComb (1982) 'Propagation of jarrah (*Eucalyptus marginata*) by organ and tissue culture', Comb. Proc. Int. Plant Prop. 40, 86-90.

Berlyn, G.P., Kohls, S.J. and Anoruo, A.O. (1991) 2. 'Caribbean Pine (*Pinus caribaea* Morelet.)', in Y.P.S. Bajaj (ed.), Biotechnology in Agriculture and Forestry, Vol. 16, Springer-Verlag, Berlin Heidelberg, pp. 253-268.

Berthon, J.Y., Tahar, S.B., Gaspar, T. and Boyer, N. (1990) 'Rooting phases of shoots of *Sequoiadendron giganteum in vitro* and their requirements', Plant Physiol. Biochem. 28, 631-638.

Bigot, C. and Engelmann (1987) 9. 'Vegetative propagation *in vitro* of *cunninghamia lanceolata* (Lamb.) Hook', in J.M. Bonga and D.J. Durzan (eds), Cell and Tissue Culture in Forestry Vol. 3 Case Histories: Gymnosperms, Angiosperms and Palms, pp. 114-127.

Biondi, S. and Thorpe, T.A. (1982) 'Growth regulator effects, metabolite changes, and respiration during shoot vitrification in cultured cotyledon explants of *Pinus radiata*', Bot. Gaz. 143, 20-25.

Bon, M-C. (1989) 'Concerning some biochemical markers of juvenility: review', in 1988-1989, Ann. AFOCEL, pp. 41-42.

Bonga, J.M. (1981) 'Organogenesis *in vitro* of tissues from mature conifers', *In vitro* 17 (6), 511-518.

Bonga, J.M. (1984) 'Adventitious shoot formation in cultures of immature female strobili of *Larix decidua*', Physiol. Plant 62, 416-421.

Bonga, J.M. and Durzan, D.J. (1987) 'Cell and Tissue Culture in Forestry, Vol. 2, Specific Principles and Methods: Growth and Developments', Martinus Nijhoff Publishers, The Netherlands.

Bonga, J.M. and Durzan, D.J. (1987) 'Cell and Tissue Culture in Forestry, Vol. 3, Case Histories: Gymnosperms, Angiosperms and Palms', Martinus Nijhoff Publishers, The Netherlands.

Boulay, M. (1979) 'Multiplication et clonage rapide de *Sequoia sempervirens* par la culture *in vitro*', Ann. AFOCEL, 12, 49-56.

Boulay, M. (1987) '*In vitro* Propagation of Tree Species', in C.E. Green, D.A. Somers, W.P. Hackett and D.D. Biesboer (eds.), Plant Biology, Vol. 3, Plant Tissue and Cell Culture, Alan R. Liss, Inc., New York, pp. 367-382.

Boulay, M. (1987) 15. 'Conifer micropropagation: Applied research and commercial aspects', in J.M. Bonga and D.J. Durzan (eds), Cell and Tissue Culture in Forestry Vol. 3, Case Histories: Gymnosperms, Angiosperms and Palms, pp. 185-206

Boulay, M. (1989) 4. 'Redwood (*Sequoia sempervirens*)', in Y.P.S. Bajaj (ed.), Biotechnology in Agriculture and Forestry, Vol. 5, Trees II, Springer-Verlag Berlin Heidelberg pp. 549-573.

Campbell, M. and Neale, D. (1991) 'Production of chimeric plasmids for deletion analysis of a larch *rbcS* promoter', Molecular Genetics Working Party Abstracts, International Union of Forestry Research Organizations.

de Canniére (1990) 'Evolution de statut mineral de microboutures de *Thuja plicata* Lamb. en culture *in vitro*', Thesis, Univ. Catholique, Louvain la Neuve, Belgium, 186 pp.

Carman, J.G. (1990) 'Embryogenic cells in plant tissue cultures: occurrence and behaviour'. *In Vitro* Cell. Dev. Biol. 26, 746-753.

Chalupa, V. (1983) 'Micropropagation of conifer and broadleaved forest trees', Communicationes Instituti Forestalis, Czechosloveniae 13, 7-39.

Chalupa, V. (1985) 'Somatic embryogenesis and plantlet regeneration from cultured immature and mature embryos of *Picea abies* (L.) Karst'. Communi. Inst. For. Cech. 14, 57-63.

Chalupa, V. (1987) 17. 'European hardwoods', in J.M. Bonga and D.J. Durzan (eds), Cell and Tissue Culture in Forestry Vol. 3 Case Histories: Gymnosperms, Angiosperms and Palms, pp. 224-260.

Chalupa, V. (1991) 12. 'Larch (*Larix decidua* Mill.)', in Y.P.S. Bajaj (ed.), Biotechnology in Agriculture and Forestry, Vol. 16, Springer-Verlag, Berlin Heidelberg, pp. 446-470.

Chand, P.K., Ochatt, S.J., Rech, E.L., Power, J.B. and Davey, M.R. (1988) 'Electroporation stimulates plant regeneration from protoplasts of the woody medicinal species *Solanum dulcamara* L.', J. Exp. Bot. 39, 1267-1274.

Chang, S., Sen, S., McKinley, C.R., Aimers-Halliday, J. and Newton, R.J. (1991) 'Clonal propagation of Virginia Pine (*Pinus virginiana* Mill.) by organogenesis', Plant Cell Rep. 10, 131-134.

Cheliak, W.M. and Rogers, D.L. (1990) 'Integrating biotechnology into tree improvement programs', Can. J. For. Res. 20, 452-464.

Chesick, E.E. and Bergmann, B.A. (1991) 1. 'Jack Pine (*Pinus banksiana* Lamb.)', in Y.P.S. Bajaj (ed.), Biotechnology in Agriculture and Forestry, Vol. 16, Springer-Verlag, Berlin Heidelberg, pp. 241-253.

Christou, P., McCabe, D.E., Martinell, B.J. and Swain, W.F. (1990) 'Soybean genetic engineering-commercial production of transgenic plants', Trends in Biotech. 8, 145-151.

Christou, P., Platt, S.G. and Ackerman, M.C. (1986) 'Opine synthesis in wild-type plant tissue', Plant Physiol. 82, 218-221.

Clapham, D.H. and Ekberg, I. (1988) 'Induction of tumors by various strains of *Agrobacterium tumefaciens* on *Abies nordmaniana* and *Picea abies*', in Hanover, J.W. and Keathley, D.E. (eds.), Genetic Manipulation of Woody Plants, Plenum Press, New York, p. 463.

Dandekar, A.M., Gupta, P.K., Durzan, D.J. and Knauf, V. (1987) 'Transformation and foreign gene expression in micropropagated Douglas-fir (*Pseudotsuga menziesii*)', Bio/Technology 5, 587-590.

Davies, H.E., Kubota, C., Kozai, T. and Aitken-Christie, J. (1992) 'Sugar-free growth of *Pinus radiata* shoots *in vitro* for conventional and automated micropropagation', Australian Journal of Botany (Special Issue, in press).

Draper, J., Scott, R., Armitage, P. and Walden, R. (1988) 'Plant Genetic Transformation and Gene Expression', Blackwell Scientific Publications, Oxford.

Drummond, B., Beck, P., Martich, Lane, S., Ross, M., Blair, D., Fox, T. and Bowen, B. (1991) 'The amount of input DNA affects phenotypic variation in tobacco by the particle gun', in Molecular Biology of Plant Growth and Development, The International Society for Plant Molecular Biology Third International Congress, Abstract #1809.

Dumas, E. and Monteuuis, O. (1991) 'Régénération *in vitro* de pins maritimes agés par bourgeonnement adventif sur euphylles', Ann. AFOCEL, pp. 43-58.

Dumas, E., Franclet, A. and Monteuuis, O. (1989) 'Microgreffage de méristemes primarires caulinaires de Pins maritimes (*Pinus pinaster* Ait.) âgés sur jeunes semis cultives *in vitro*. C.R. Acad. Sci. 3, 309, 723-728.

Dunstan, D.I. and Thorpe, T.A. (1986) 'Regeneration in forest trees', I.K. Vasil (ed.), Cell Culture and Somatic Cell Genetics in Plants, Academic Press, New York, pp. 223-241.

Dunstan, D.I. (1988) 'Prospects and progress in conifer biotechnology', Can. J. For. Res. 18, 1497-1506.

Dunstan, D.I., Mohammed, G.H. and Thorpe, T.A. (1987) 'Morphogenetic response of vegetative bud explants of adolescent and mature *Picea glauca* (Moench) Voss *in vitro*', New Phytol. 106, 225-236.

Durand-Cresswell, R. and Nitsch, C. (1977) 'Factors influencing the regeneration of *Eucalyptus grandis* by organ culture', Acta Hortic. 78, 149-155.

Eklof, S., Imbault, N., Sandberg, G. and Olsen, O. (1991) '*ipt*-transformed poplar and tobacco plants as a model to study IAA metabolism', in Molecular Biology of Plant Growth and Development, The International Society for Plant Molecular Biology Third International Congress, Abstract #917.

El Maâtaoui, M., Espagnac, H. and Michaux-Ferriére, N. (1990) 'Histology of Callogenesis and somatic embryogenesis induced in stem fragments of cork oak (*Quercus suber*) cultured *in vitro*', Annals of Botany 66, 183-190.

Ellis, D., McCabe, D., McInnis, S., Martinell, B., Roberts, D. and McCown, B. (1991) 'Transformation of white spruce by electrical discharge particle acceleration', in Molecular Biology of Plant Growth and Development, The International Society for Plant Molecular Biology Third International Congress, Supplemental Abstract.

Ellis, D., Roberts, D., Sutton, B., Lazaroff, W., Webb, D. and Flinn, B. (1989) 'Transformation of white spruce and other conifer species by *Agrobacterium tumefaciens*', Plant Cell Rep. 8, 16-20.

Ellis, D.D. and Bilderback, D.E. (1991) 7. 'Ponderosa Pine (*Pinus ponderosa* Laws.)', in Y.P.S. Bajaj (ed.), Biotechnology in Agriculture and Forestry, Vol. 16, Springer-Verlag, Berlin Heidelberg, pp. 339-357.

Evers, D.W., Donkers, J., Prat, A. and Vermeer, E. (1988) 4. '*Pseudostuga menziesii*', in Micropropagation of forest trees through tissue culture Pudoc Wageningen pp. 19-25.

Fillatti, J.J., McCown, B.H., Sellmer, J., Haissig, B.E. and Comai, L. (1987) '*Agrobacterium*-mediated transformation and regeneration of poplar', Mol. Gen. Genet. 206, 192-199.

Fitch, M.M.M., Manshardt, R.M., Gonsalves, D., Slighton, J.L. and Sanford, J. (1990) 'Stable transformation of papaya via microprojectile bombardment', Plant Cell Rep. 9, 189-194.

de Fossard, R.A., Barker, P.K. and Bourne, R.A. (1977) 'The organ culture of nodes of four species of *Eucalyptus*', Acta Hortic. 78, 157-165.

Fouret, Y., Arnaud, Y., Larrieu, C. and Miginiac, E. (1986) '*Sequoia sempervirens* as an *in vitro* rejuvenation model,' New Zealand Journal of Forestry Science 16, (3) 319-327.

Fouret, Y., Larrieu, C. and Arnaud, Y. (1989) '*In vitro* rejuvenation of *Sequoia sempervirens* (Endl.): Research of the most efficient protocol', Ann. AFOCEL, 55-56.

Frampton, L.J. Jr. and Isik, K. (1987) 'Comparison of field growth among Loblolly pine seedlings and three plant types produced *in vitro*', Tappi 70, 119-123.

Franclet, A., David, A., David, H., Boulay, M. (1980) 'Premiere mise en evidence morphologique d'un rajeunessement de meristemes primaires caulinaires de pin maritime age (*Pinus pinaster* Sol.)', C.R. Acad. Sci. Paris D., pp. 927-930.

Gates, J.C. and Greenwood, M.S. (1991) 'The physical and chemical environment of the developing embryo of *Pinus resinosa*', American Journal of Botany 78 (7), 1002-1009.

Gingas, V.M. (1991) 'Asexual Embryogenesis and Plant Regeneration from Male Catkins of *Quercus*', Hort. Sci. 26 (9), 1217-1218.

Gingas, V.M. and Lineberger, R.D. (1989) 'Asexual embryogenesis and plant regeneration in *Quercus*', Plant Cell, Tissue and Organ Culture 17, 191-203.

Gleed, J.A. (1991) 'Toward Clonal Afforestation', in J.T. Miller (ed.), Proceedings, FRI/NZFP Forests Ltd, Clonal Forestry Workshop, Forest Research Institute, New Zealand, pp. 61-62.

Goldfarb, B., Howe, G.T., Bailey, L.M., Strauss, S.H., Zaerr, J.B. (1991) 'A liquid cytokinin pulse induces adventitious shoot formation from Douglas fir cotyledons - propagation; effect of benzyladenine', Plant Cell Rep. 10, 156-160.

Gresshoff, P.M. and Doy, C.H. (1972) 'Development and differentiation of haploid *Lycopersicon esculentum* (tomato)', Planta. 107, 161-170.

Gupta, P.K. and Durzan, D.J. (1985) 'Biotechnology of somatic polyembryogenesis and plantlet regeneration in loblolly pine', Bio/Technology 5, 147-151.

Gupta, P.K. and Durzan, D.J. (1987) 'Micropropagation of phase specificity immature, elite Douglas fir', J. Amer. Soc. Hort. Sci. 112 (6), 969-971.

Gupta, P.K. and Durzan, D.J. (1991) 9. 'Loblolly pine (*Pinus taeda* L.)', in Y.P.S. Bajaj (ed.), Biotechnology in Agriculture and Forestry, Vol. 16, Springer-Verlag, Berlin, Heidelberg, pp. 383-407.

Gupta, P.K. and Mascarenhas, A.F. (1987) 29. 'Eucalyptus', in J.M. Bonga and D.J. Durzan (eds), Cell and Tissue Culture in Forestry Vol. 3 , Martinus Nijhoff, The Netherlands, pp. 385-399.

Gupta, P.K. and Pullman, G.S. (1991) 'Method for reproducing coniferous plants by somatic embryogenesis', United States Patent 4 957 866.

Gupta, P.K., Timmis, R. and Mascarenhas, A.F. (1991) 'Field performance of micropropagated forestry species, *In Vitro* Cell. Dev. Biol. 27, 159-164.

Gupta, P.K., Timmis, R., Pullman, G., Yancy, M., Kreitinger, M., Carlson, W. and Carpenter, C. (1991) 'Development of an Embryogenic System for Automated Propagation of Forest Trees', in Indra K. Vasil (ed.), Scale-up and Automation in Plant Propagation, Academic Press, New York, pp. 75-93.

Hakman, I. and von Arnold, S. (1985) 'Plantlet regeneration through somatic embryogenesis in *Picea abies* (Norway spruce)', J. Plant. Physiol. 121, 149-158.

Hakman, I. and von Arnold, S. (1988) 'Somatic embryogenesis and plant regeneration from suspension cultures of *Picea glauca* (white spruce)', Physiol. Plant. 72, 579-587.

Hanover, J.W. and Keathley, D.E. (1988) 'Genetic Manipulation of Woody Plants', Plenum Press, New York.

Harry, I.S. and Thorpe, T.A. (1991) 10. 'Engelmann Spruce (*Picea engelmannii* Parry ex. Engelm.)', in Y.P.S. Bajaj (ed.), Biotechnology in Agriculture and Forestry, Vol. 16, Springer-Verlag, Berlin, Heidelberg, pp. 408-422.

Harry, I.S., Thompson, M.R., Lu, C.-Y. and Thorpe, T.A. (1987) '*In vitro* plantlet formation from embryonic explants of eastern white cedar (*Thuja occidentalis* L.)', Tree Phys. 3, 273-283.

Hartney, V.J. (1980) 'Vegetative propagation of Eucalypts', Aust. For. Res. 10, 191-211.

Horgan, K. and Aitken, J. (1981) 'Reliable plantlet formation from embryos and seedling shoot tips of radiata pine', Physiol. Plant. 53, 170-175.

Horgan, K. (1987) '*Pinus radiata*', in Tissue Culture in Forestry, ed. by J.M. Bonga and D.J. Durzan, Vol. 3, Martinus Nijhoff Publishers, The Netherlands, 128-145.

Horgan, K. and Holland, L. (1991) 'Micropropagating mature radiata pine - potential for clonal forestry', in J.T. Miller (ed.), Proceedings of the FRI/NZFP Forests Ltd Clonal Forestry Workshop 1-2 May, 1989, pp. 78-80.

Horgan, K. and Holland, L. (1989) 'Rooting micropropagated shoots from mature radiata pine', Can. J. For. Res. 19, 1309-1315.

Howe, G.T., Strauss, S.H., Beckwith, R. and Goldfarb, B. (1990) 'Transgenic poplar callus containing an insect toxin gene from *Bacillus thuringiensis* inhibits survival and growth of Western tussock moth', Molecular Genetics Working Party Abstracts, International Union of Forestry Research Organizations.

Iida, A., Seki, M., Kamada, M., Yamada, Y. and Morikawa, H. (1990) 'Gene delivery into cultured plant cells by DNA-coated gold particles accelerated by a pneumatic particle gun', Theor. Appl. Genet. 80, 813-816.

Ishii, K. (1991) 13. 'Hinoki Cypress (*Chamaecyparis obtusu*)', Y.P.S. Bajaj (ed.), in Biotechnology in Agriculture and Forestry, Vol. 16, Springer-Verlag, Berlin, Heidelberg, pp. 471-478.

James, D.J., Passey, A.J., Barbara, D.J. and Bevan, M. (1989) 'Genetic transformation of apple (*Malus pumila* Mill.) using a disarmed Ti-binary vector', Plant Cell Rep. 7, 658-661.

Jang, J.C. and Tainter, F.H. (1991) 'Micropropagation of short leaf, Virginia and loblolly x shortleaf pine hybrids via organogenesis', Plant Cell, Tissue and Organ Culture 25, 61-67.

Jefferson, R.A., Kavanagh, T.A. and Bevan, M.W. (1987) 'GUS fusions: glucuronidase as a sensitive and versatile gene fusion marker in higher plants', EMBO J. 6, 3901-3907.

John, A. and Pearson, D.J. (1986) 'Induction of vitrification in *Picea sitchensis* cultures', N.Z. J. For. Sci. 16(3), 328-342.

Jokinen, K. and Törmälä, T. (1991) 'Micropropagation of Silver Birch (*Betula pendula* Roth.) and Clonal Fidelity of Mass Propagated Birch Plants' (in press).

Jonard, R. (1986) 'Micrografting and its applications to tree improvement', in Y.P.S. Bajaj (ed.), Biotechnology in Agriculture and Forestry Vol. I Trees, I Springer, Berlin Heidelburg New York, pp. 31-48.

Jorgensen, J. (1989) 'Somatic embryogenesis in *Aesculus hippocastanum* L. by culture of filament callus', J. Plant Physiol. 135, 240-241.

Joy IV, R.W., Yeung, E.C., Kong, L. and Thorpe, T.A. (1991) 'Development of white spruce somatic embryos: I. Storage Product Deposition', *In Vitro* Cell. Dev. Biol., 27 p. 32-41.

Khuspe, S.S., Gupta, P.K., Kulkarni, D.K., Mehta, U. and Mascarenhas, A.F. (1987) 'Increased biomass production by tissue culture of eucalyptus, Can. J. For. Res. 17, 1361-1363.

Klimaszewska, K. (1989) 'Recovery of somatic embryos and plantlets from protoplast cultures of *Larix* x *eurolepsis*', Plant Cell Rep. 8, 440-444.

Klimaszewska, K., Ward, C. and Cheliak, W.M. (1991) 'Cryopreservation of embryogenic cultures of larch (*Larix eurolepsis*) and black spruce (*Picea mariana*), J. Exp. Bot. 43, 73-79.

Komamine, A., Matsumoto, M. Tsukahara, M., Fujiwara, A., Kawahara, R., Ito, M., Smith, J., Nomura, K. and Fujimura, T. (1990) 'Mechanisms of somatic embryogenesis in cell cultures - physiology, biochemistry and molecular biology', in H.J.J. Nijkump, L.H.W. van der Plas, J. van Aartrijk (eds.), Progress in Plant Cellular and Molecular Biology, Kluwer Academic Publishers, pp. 307-313.

Kozai, T. (1991) 'Micropropagation under photoautotrophic conditions', in P. Debergh and R.H. Zimmerman (eds.), Micropropagation: Technology and Application, Kluwer Academic Publishers, pp. 447-469.

Kozai, T., Ting, K.C. and Aitken-Christie, J. (1991) 'Considerations for automation of micropropagation systems', Trans. of the ASAE 35, 503-517.

Kurtz, S.L., Hartman, R.D. and Chu, I.Y.E. (1991) 'Current methods of commercial micropropagation', in 1. K. Vasil (ed.), Scale and Automation of plant propagation. Academic press New York, pp. 7-34.

Kvaalen, H. and von Arnold, S. (1991) 'Effects of various partial pressures of oxygen and carbon dioxide on different stages of somatic embryogenesis in *Picea abies*', Plant Cell, Tissue and Organ Culture 27 (1), 49-57.

Laine, E. and David, A. (1990) 'Somatic embryogenesis in immature embryos and protoplasts of *Pinus caribaea*', Plant Sci. 69, 215-224.

Lakshmi Sita, G., Raghava Ram, N.V. and Vaidyanathan, C.S. (1979) 'Differentiation of embryoids and plantlets from shoot callus of sandalwood', Plant Science Letters 15, 265-270.

Lambert, C. and Tepfer, D. (1991) 'Use of *Agrobacterium rhizogenes* to create chimeric apple trees through genetic grafting', Bio/Technology 9, 80-82.

Lelu, KM.A. (1988) 'Preliminary study of the morphological and genetic variations obtained after somatic embryogenesis in *Picea abies*', Ann. AFOCEL, pp. 35-36.

Lelu, M.P., Boulay, M.P. and Bornman, C.H. (1990) 'Somatic embryogenesis in cotyledons of *Picea abies* is enhanced by an adventitious bud-inducing treatment', New Forests 4, 125-135.

Levin, R., Gaba, V., Tal, B., Hirsch, S., Nola, D. and Vasil, I.K. (1988) 'Automated plant tissue culture for mass propagation', Biotechnology 6, 1035-1040.

Lloyd, G. and McCown, B. (1980) 'Commercially-feasible micropropagation of mountain laurel, *Kalmia latifolia*, by use of shoot tip culture', Comb. Proc. Int. Plant Prop. 30, 421-427.

Loopstra, C.A. and Sederoff, R.R. (1990) 'Gene expression in xylem of loblolly pine', Molecular Genetics Working Party Abstracts, International Union of Forestry Research Organizations.

Lowe, B.A. and Krul, W.R. (1991) 'Physical, chemical, developmental, and genetic factors that modulate the *Agrobacterium-Vitis* interaction', Plant Physiol. 96, 121-129.

Lu, C.-Y. and Thorpe, T.A. (1988) 'Shoot-bud regeneration in subcultured callus of Engelmann Spruce', *In Vitro* Cell. Dev. Biol. 24, 239-242.

Lu, C.-Y. and Thorpe, T.A. (1987) 'Somatic embryogenesis and plantlet regeneration in cultured immature embryos of *Picea glauca*", J. Plant. Physiol. 128, 297-302.

Mac Ant-Saoir, S., O'Brian and Selby, C. (1991) 'Apparent rejuvenation of mature Sitka spruce in culture', Acta Hortic. 289, 337-338.

MacRae, S. and von Staden, J. (1991) 'Use of *Agrobacterium rhizogenes*-mediated transformation to improve rooting of eucalypts', in Molecular Biology of Plant Growth and Development, The International Society for Plant Molecular Biology Third International Congress, Abstract #903.

Malajczuk, N. and Hartney, V.J. (1986) 'Procedure for inoculation of micropropagated plantlets of *Eucalyptus camaldulensis* with ectomycorrhizal fungi, and comparison with seedling inoculation using inoculum contained in peat/vermiculite carrier', Aust. For. Res. 16, 199-206.

Mante, S., Morgens, P.H., Scorza, R., Cordts, J.M. and Callahan, A.M. (1991) '*Agrobacterium*-mediated transformation of plum (*Prunus domestica* L.) hypocotyl slices and regeneration of transgenic plants', Bio/Technology 9, 853-857.

Mascarenhas, A.F., Kendurkar, S.V., Gupta, P.K., Khuspe, S.S. and Agrawal, D.C. (1987) 'Teak', in J.M. Bonga and Don J. Durzan (eds.), Cell and Tissue Culture in Forestry, Vol. 3, Martinus Nijhoff Publishers, Dordrecht, Boston, Lancaster, pp. 300-315.

Mascarenhas, A.F., Khuspe, S.S., Nadgauda, R.S., Gupta, P.K. and Khan, B.M. (1988) 'Potential of cell culture in plantation forestry programs', in J.W. Hanover and D.E. Keathley (eds.), Genetic Manipulation of Woody Plants, Plenum Press, New York and London, pp. 391-412.

Mascarenhas, A.F., Khuspe, S.S., Nadgauda, R.S., Gupta, P.K., Muralidharen, E.M. and Khan, B.M. (1989) 'Biotechnological application of plant tissue culture to forestry in India', in V. Dhawan (ed.), Applications of Biotechnology in Forestry and Horticulture, Plenum Press, New York and London, pp. 73-86.

Mathur, J., Ahuja, P.S., Lal, N. and Mathur, A.K. (1989) 'Propagation of *valeriana wallichi* DC using encapsulated apical and axial shoot buds', Plant Sci. 60, 111-116.

McCown, B.H. (1988) 'Recent advances in protoplast culture of horticultural crops: ornamental trees and shrubs', Scientia Horticulturae 37, 257-265.

McCown, B.H., Zeldin, E., Pinkalla, H.A. and Dedolph, R.R. (1988) 'Nodule culture: A developmental pathway with high potential for regeneration automated micropropagation and plant metabolite production from woody plants', in J.W. Hanover and D.E. Keathley (eds.), Genetic manipulation of wood plants, Plenum Press, New York, pp. 149-166.

McCown, B.H. (1989) 16. 'Birch (Betula spp.)', in Y.P.S. Bajaj (ed.), Biotechnology in Agriculture and Forestry, Vol. 5 Trees II. Springer-Verlag Berlin Heidelberg, pp. 324-341.

McGranahan, G.H., Leslie, C.A., Uratsu, S.L., Martin, L.A. and Dandekar, A.M. (1988) '*Agrobacterium*-mediated transformation of walnut somatic embryos and regeneration of transgenic plants', Bio/Technology 6, 800-804.

McKeand, S.E. (1985) 'Expression of mature characteristics by tissue culture plantlets derived from embryos of loblolly pine', J. Amer. Soc. Hort. Sci. 110 (5), 615-623.

McKeand, S.E. and Allen, H.L. (1984) 'Nutritional and root development factors affecting growth of tissue culture plantlets of loblolly pine', Physiol. Plant 61, 523-528.

Mehra-Palta, A. (1982) 'Clonal propagation of *Eucalyptus* by tissue culture', Plant Sci. Lett. 26, 1-11.

Meier-Dinkel, A. (1991) 'Micropropagation of Birches (Betula spp.)', in Y.P.S. Bajaj (ed.), Biotechnology in Agriculture and Forestry Trees III. Vol. 16. Springer, Berlin New York pp.

Menzies, M.I., Klomp, B.K. and Holden, D.J. (1991) 'Optimal physiological age of propagules for use in clonal forestry', in J.T. Miller (ed.), Proceedings FRI/NZFP Forests Ltd, Clonal Forestry Workshop, Forest Research Institute, New Zealand, pp. 142-145.

Merkle, S.A. and Sommer, H.E. (1986) 'Somatic embryogenesis in tissue cultures of *Liriodendron tulipifera* ', Can. J. For. Res. 16, 420-422.

Merkle, S.A., Wiecko, A.T., Sotak, R.J. and Sommer, H.E. (1990) 'Maturation and conversion of *Liriodendron tulipifera* somatic embryos', *In Vitro* Cell. Dev. Biol. 26, 1086-1093.

Meyer, H.J. and van Stoden, J. (1987) 'Regeneration of *Acacia melanoxylon* plantlets *in vitro*', S. Afri. J. Bot. 53(3) 206-209.

Mission, J.P., De Cannière, C.H. and André, P. (1991) 14. 'Western Red Cedar (*Thuja plicata* D. Don ex. Lambert)', in Y.P.S. Bajaj (ed.), Biotechnology in Agriculture and Forestry, Vol. 16, Springer-Verlag, Berlin Heidelberg, pp. 479-490.

Miwa, Y. (1990) 'Automation of plant tissue culture process', in I. Karube (ed.), Automation in Biotechnology, Elsevier Science Publishers B.V., pp. 217-233.

Mohammed, G.H. and Vidaver, W.E. (1988) 'Root production and plantlet development in tissue-cultured conifers', Plant Cell, Tissue and Organ Culture 14, 137-160.

Mohammed, G.H. and Vidaver, W.E. (1990) 'The influence of acclimatization treatment and plantlet morphology on early greenhouse-performance of tissue-cultured Douglas fir (*Pseudostuga menziesii* [Mirb.] Franco)', Plant Cell, Tissue and Organ Culture 21, 111-117.

Monteuuis, O. (1987) '*In vitro* meristem culture of juvenile and mature *Sequoiadendron giganieum*', Tree Phys. 3, 265-272.

Monteuuis, O. (1991) 'Rejuvenation of a 100-year-old *Sequoidendron giganteum* through *in vitro* meristem culture, 1. Organogenic and morphological arguments', Physiol. Plant. 81, 111-115.

Monteuuis, O. and Bon, M.C. (1989) 'Rejuvenation of a 100-year old giant sequoia (*Sequoiadendron giganteum* Buchholz) through *in vitro* meristem culture', Ann. Sci. For. 46 suppl. 1835-1865.

Monteuuis, O. Bon, M.C. and Berthan, J.Y. (1987) 'Micropropagation aspects of *Sequoiadendron giganteum* juvenile and mature clones', Acta Hortic. 212, 489-497.

Morris, J.W., Castel, L.A. and Morris, R.O. (1988) 'Transformation of *pinaceous* gymnosperms by *Agrobacterium*' in Hanover, J.W. and Keathley, D.E. (eds.), Genetic Manipulation of Woody Plants, Plenum Press, New York, p. 481.

Mullins, M.G. and Srinivasan, C. (1976) 'Somatic embryos and plantlets from an ancient clone of the grapevine (cv. *Cabarnet sauvignon*) by apomixis *in vitro*', J. Exp. Bot. 27, 1022-1030.

Murashige, T. and Skoog, F. (1962) 'A revised medium for rapid growth and bioassays with tobacco tissue cultures', Physiol. Plant 15, 473-497.

Muralidharan, E.M. and Mascarenhas, A.F. (1987) '*In vitro* plantlet formation by organogenesis in *E. camaldulensis* and by somatic embryogenesis in *Eucalyptus citriodora*', Plant Cell Rep. 6, 256-259.

Nairn, B.J. (1987) "Significance of gelling agents in a production tissue culture laboratory', Comb. Proc. I.P.P.S. 37, 200-205.

Norgaard, J.V. and Krogstrup, P. (1991) 'Cytokinin induced somatic embryogenesis from immature embryos of *Abies nordmanniana* Lk.', Plant Cell Rep. 9, 509-513.

Olsson, O., Nilsson, O., Sitbon, F., Alden, T., Little, C.H.A., Sundberg, B., Koncz, C. and Sandberg, G. (1990) 'Manipulation of the endogenous IAA level to study plant development', Molecular Genetics Working Party Abstracts, International Union of Forestry Research Organizations.

Paranjothy, K., Saxena, S., Banerjee, M., Jaiswal, V.S. and Bhojwani, S.S. (1991) 'Chapter 8, Clonal Multiplication of Woody Perennials', in S.S. Bhojwani (ed.), Plant Tissue Culture: Applications and Limitations, pp. 190-219.

Parsons, T.J., Sinkar, V.P., Stettler, R.F., Nester, E.W. and Gordon, M.P. (1986) 'Transformation of poplar by *Agrobacterium tumefaciens*', Bio/Technology 4, 533-536.

Patel, K.R., Rumary, C. and Thorpe, T.A. (1986) 'Plantlet formation in black and white spruce, III. histological analysis of *in vitro* root formation and the root-shoot union', N.Z. J. For. Sci. 16 (3), 289-296.

Phelep, M., Petit, A., Martin, L., Duboux, E. and Tempe, J. (1991) 'Transformation and regeneration of a nitrogen-fixing tree, *Allocasuarina verticillata* Lam.', Bio/Technology 9, 461-466.

Phillips, G.C. and Gladfelter, H.J. (1991) 3. 'Eldarica Pine, Afghan Pine (*Pinus eldarica* Medw.)', in Y.P.S. Bajaj (ed.), Biotechnology in Agriculture and Forestry, Vol. 16, Springer-Verlag, Berlin Heidelberg, pp. 269-287.

Pierik, R.L.M. (1990) 'Rejuvenation and Micropropagation', in H.J.J.Nijkamp, L.H.W. Van der Plas, J. van Aartrijk (eds), Progress in Plant Cellular and Molecular Biology, Kluwer Academic Publishers, pp. 91-101.

Poissonier, M., Franclet, A., Dumant, M.J. and Gauntrey, J.Y. (1980) 'Enracinement de tigelles *in vitro* de *Sequoia sempervirens*', Ann AFOCEL, 231-253.

Preece, J.E., Huettemann, C.A., Ashby, W.C., Roth, P.L. (1991) 'Micro- and cutting propagation of silver maple, I. Results with adult and juvenile propagules', J. Am. Soc. Hortic. Sci. 116, 142-148.

Pythoud, F., Sinkar, V.P., Nester, E.W. and Gordon, M.P. (1987) 'Increased virulence of *Agrobacterium rhizogenes* conferred by the *vir* region of pTiBo542: application to genetic engineering of poplar', Bio/Technology 5, 1323-1327.

Quoirin, M. and Lepoivre, P. (1977) 'Etudes de milieux adaptes aux cultures *in vitro* de *Prunus*', Acta Hortic. 78, 437-442.

Radogevic, L. (1980) 'Somatic and androgenic embryogenesis in some woody species', Bulletin de la Société Botanique de France, Actualités Botaniques, 127 (3/4), 99-107.

Rancillac, M. (1991) 6. 'Maritime Pine (*Pinus pinaster* Sol.)', in Y.P.S. Bajaj (ed.), Biotechnology in Agriculture and Forestry, Vol. 16, Springer-Verlag, Berlin Heidelberg, pp. 317-338.

Rao, P.S. and Ozias-Akins, P. (1985) 'Plant regeneration through somatic embryogenesis in protoplast cultures of sandalwood (*Santalum album* L.)', Protoplasma 124, 80-86.

Redenbaugh, K. and Ruzin, S.E. (1989) 6. 'Artificial seed production and forestry', in V. Dhawan (ed.), Applications of Biotechnology in Forestry and Horticulture, Plenum Press New York and London, pp. 57-71.

Ritchie, G.A. and Long, A.J. (1986) 'Field performance of micropropagated Douglas fir', N.Z. J. For. Sci. 16(3) 343-356.

Roberts, D.R., Flinn, B.S., Webb, D.T., Webster, F.B. and Sutton, B.C.S. (1990a) 'Abscisic acid and indole-3-butyric acid regulation of maturation and accumulation of storage proteins in somatic embryos of interior spruce', Physiol. Plant. 78, 355-360.

Roberts, D.R., Sutton, B.C.S. and Flinn, B.S. (1990b) 'Synchronous and high frequency germination of interior spruce somatic embryos following partial drying at high relative humidity', Can. J. Bot. 68, 1086-1090.

Roberts, D.R., Webster, F.W., Flinn, B.S. and Lazaroff, D.R. (1991) 'Propagation of interior spruce through somatic embryogenesis', in B.E. Haissig, T. Kent-Kirk, W.L. Olsen, K.F. Raffa, J.M. Slavicek (Eds.), Abstracts of papers presented at the International Symposium on Applications of Biotechnology to Tree Culture, Protection and Utilization, p. 127.

Robertson, D., Ackley, R., Weissinger, A.K., Stomp, A.-M. and Sederoff, R.R. (1991) 'Stable transformation of Norway spruce embryogenic callus through microprojectile bombardment', in Molecular Biology of Plant Growth and Development, The International Society for Plant Molecular Biology Third International Congress, Abstract #1036.

Ruaud, J.N. (1991) 'Recherche d'une relation entre bourgeonnement adventif et embryogenése somatique chez des cotylédons de jeunes semis de *Picea abies*, Premiers résultats', AFOCEL, pp. 23-41.

Sanford, J. (1988) 'The biolistic process', Trends in Biotech. 6, 299-302.

Sautter, C., Waldner, H., Neuhaus-Url, G., Galli, A., Neuhaus, G. and Potrykus, I. (1991) 'Micro-targeting: high efficiency gene transfer using a novel approach for the acceleration of micro-projectiles', in Bio/Technology 9, 1080-1085.

Schell, J.L., Barthe, G.A.S., Grosser, J.W. (1991) 'Genetic transformation of citrus protoplasts by PEG-mediated direct DNA uptake and the regeneration of transgenic plants', in Molecular Biology of Plant Growth and Development, The International Society for Plant Molecular Biology Third International Congress, Abstract #870.

Schenk, R.U. and Hildebrandt, A.C. (1972) 'Medium and techniques for induction and growth of monocotyledonous and dicotyledonous plant cell cultures', Can. J. Bot. 50, 199-204.

Schwartz, O.J., Beaty, R.M. and Franco, E.O. (1991) 5. 'Egg-Cone Pine (*Pinus oocarpa* Schieole.)', in Y.P.S. Bajaj (ed.), Biotechnology in Agriculture and Forestry, Vol. 16, Springer-Verlag, Berlin Heidelberg, pp. 304-316.

Sederoff, R.R., Stomp, A.-M., Chilton, W.S. and Moore, L.M. (1986) 'Gene transfer into loblolly pine by *Agrobacterium tumefaciens*', Bio/Technology 4, 647-649.

Selby, C. and Harvey, B.M.R. (1985) 'The influence of natural and *in vitro* bud flushing on adventitious bud production in Sitka spruce (*Picea sitchensis* [Bong.] [Carr.]) bud and needle cultures, New Phytol. 100, 549-562.

Sen, S., Newton, R.J., Fong, F. and Neuman, P. (1989) 'Abscisic acid: a role in shoot enhancement from loblolly pine (*Pinus taeda* L.) cotyledon explants', Plant Cell Rep. 8, 191-194.

Smith, D.L. and Krikorian, A.D. (1989) 'Release of somatic embryogenic potential from excised zygotic embryos of carrot and maintenance of proembryonic cultures in hormone-free medium', Amer. J. Bot. 76 (12), 1832-1843.

Smith, D.R., Donaldson, L.A. and Warr, A. (1991) 'Bioreactor technology for mass propagation of *Pinus radiata* by somatic embryogenesis', *In Vitro* 27 (3), Abstract 43A.

Smith, M.A.L. and McClelland, M.T. (1991) 'Gauging the influence of *in vitro* conditions on *in vivo* quality and performance of woody plants', *In Vitro* Cell. Dev. Biol., 27, 52-56.

Sommer, H.E. and Brown, C.L. (1980) 'Embryogenesis in tissue cultures of sweetgum', Forest Sci. 26 (2), 257-260.

Sotak, R.J., Sommer, H.E. and Merkle, S.A. (1991) 'Relation of the developmental stage of zygotic embryos of yellow-poplar to their somatic embryogenic potential, Plant Cell Rep. 10, 175-178.

Söndahl, M.R. and Noriega, C. (1991) 'Bioreactor Systems for plant micropropagation', Trans. of the ASAE 35, 497-501.

Srivastava, P.S. and Steinhauer, A. (1981) 'Regeneration of birch plants from catkin tissue cultures', Plant Sci. Lett. 22, 379-386.

Stomp, A.-M., Loopstra, C., Chilton, W.S., Sederoff, R.R. and Moore, L.W. (1990) 'Extended host range of *Agrobacterium tumefaciens* in the genus *Pinus*', Plant Physiol. 92, 1226-1232.

Tautorus, T.E., Attree, S.M., Fowke, L.C. and Dunstan, D.I. (1990) 'Somatic embryogenesis from immature and mature zygotic embryos, and embryo regeneration from protoplasts in black spruce (*Picea mariana* Mill.)', Plant Sci. 67, 115-124.

Tautorus, T.E., Bekkaoui, F., Pilon, M. Datla, R.S.S., Crosby, W.L., Fowke, L.C. and Dunstan, D.I. (1990) 'Factors affecting transient gene expression in electroporated black spruce (*Picea mariana*) and jack pine *Pinus banksiana*) protoplasts', Theor. Appl. Genet. 78, 531-536.

Tautorus, T.E., Fowke, L.C. and Dunstan, D.I. (1991) 'Somatic embryogenesis in conifers', Can. J. Bot. 69, 1873-1899.

Thorpe, T.A. (1988) 'Physiology of bud induction in conifers *in vitro*', in J.W. Hanover and D.E. Keathley (eds.), Genetic Manipulation of Woody Plants, Plenum Press, New York, pp. 167-184.

Thorpe, T.A. and Patel, K.R. (1986) 'Comparative morpho-histological studies on the sites of shoot initiation in various conifer explants', N.Z. J. For. Sci. 16, 257-68.

Thorpe, T.A., Harry, I.S. and Kumar, P.P. (1991) 'Application of micropropagation to forestry', in P.C. Debergh and R.H. Zimmerman (eds.), Micropropagation Technology and Application, Kluwer Academic Publishers, Dordrecht, Boston, London, pp. 311-336.

Timmis, R. and Trotter, P.C. (1989) 'Biological and economic feasibility of genetically engineered trees for lignin properties and carbon allocation', Applications of Biotechnology in Forestry and Horticulture, Plenum Press New York and London, pp. 349-367.

Ting, K.C. (1991) 'Robotic workcell for flexibly automated handling of young transplants', in I. Karube (ed.), Automation in Biotechnology, Elsevier Science Publishers, The Netherlands, pp. 261-278.

Törmälä, T. (1991) 'Genotype-Phenotype Interplay in Micropropagation', in H.J.J. Nijkamp, L.H.W. van der Plas and J. van Aartrijk (eds.), Progress in Plant Cellular and Molecular Biology, Kluwer Academic Publishers, Dordrecht, Boston, London, pp. 102-107.

Tranvan, H., Bardat, F., Jacques, M. and Arnaud, Y. 'Rajeunissement chez le *Sequoia sempervirens*: effe du microgreffage *in vitro*', Can. J. Bot. 69, 1772-1779.

Tricoli, D.M., Maynard, C.A. and Drew, A.P. (1985) 'Tissue culture propagation of mature trees of *Prunus serotina* Ehrh. I. Establishment multiplication and rooting *in vitro*', Forest Sci. 30, (1) 201-205.

van der Moezel, P.G., Bell, D.T., Bennett, I.J., Strawbridge, M. and McComb, J.A. (1990) 'Development of Salt Tolerant Clonal Trees in Australia', Comb. Proc. Int. Plant Prop. Soc. 40, 73-78.

von Aderkas, P., Bonga, J.M. and Nagmani, R. (1987) 'Promotion of embryogenesis in cultured megagametophytes of *Larix decidua*'. Can. J. For. Res. 17, 1293-1296.

von Arnold, S. (1987) 'Improved efficiency of somatic embryogenesis in mature embryos of *Picea abies* (L.) Karst'. J. Plant Physiol. 128, 233-234.

von Arnold, S. (1991) 'Current status of tissue culture for vegetative regeneration of forest trees', in B.E. Haissig, T. Kent-Kirk, W.L. Olsen, K.F. Raffa, J.M. Slavicek (Eds.), Abstracts of papers presented at the International Symposium on Applications of Biotechnology to Tree Culture, Protection and Utilization, pp. 9-13.

von Arnold, S. and Hakman, I. (1986) 'Effect of sucrose on initiation of embryogenic callus cultures from mature zygotic embryos of *Picea abies* (L.) Karst. (Norway spruce)', J. Plant Physiol. 122, 261-265.

von Arnold, S. and Tillberg, E. (1987) 'The influence of cytokinin pulse treatments on adventitious bud formation on vegetative buds of *Picea abies*. Plant Cell, Tissue and Organ Culture 9, 253-261.

Warrag, E., Lesney, M.S. and Rockwood, D.J. (1991) 'Nodule culture and regeneration of *Eucalyptus grandis* hybrids', Plant Cell Rep. 9, 586-589.

Warrag, E.I., Lesney, M.S. and Rockwood, D.L. (1990) 'Micropropagation of field tested superior *Eucalyptus grandis* hybrids', New Forests 4, 67-79.

Webb, D.T. and Flinn, B.S. (1991) 8. 'Eastern White Pine (*Pinus strobus* L.)', in Y.P.S. Bajaj (ed.), Biotechnology in Agriculture and Forestry, Vol. 16, Springer-Verlag, Berlin Heidelberg, pp. 358-382.

Webster, F.B., Roberts, D.R., McInnis, S.M. and Sutton, B.C.S. (1990) 'Propagation of interior spruce by somatic embryogenesis', Can. J. For. Res. 20, 1759-1765.

Wilde, D., Merkle, S. and Meagher, R. (1990) 'Transformation and regeneration of yellow-poplar', Molecular Genetics Working Party Abstracts, International Union of Forestry Research Organizations.

Wilson, S.M., Thorpe, T.A. and Moloney, M.M. (1989) 'PEG-mediated expression of GUS and CAT gene expression in protoplasts from embryogenic suspension cultures of *Picea glauca*', Plant Cell Rep. 7, 704-707.

Wisniewski, L.A., Brandon, D.L., McKeand, S.E. and Amerson, H.V. (1991) 'The effect of root pruning on the maturation of Loblolly pine (*Pinus taeda*) plantlets, rooted hypocotyls and seedlings', Can. J. For. Res. 21, 1073-1079.

Wisniewski, L.A., Frampton, L.J. Jr., and McKeand, S.E. (1986) 'Early shoot and root quality effects on nursery and field development of tissue-cultured loblolly pine, Hort. Science 21 (5), 1185-1186.

Yang, J.C., Chang, S.H. and Ho, C.K. (1989) 'Micropropagation of *Paulownia taiwaniana* from mature tissues. Ann. Sci. For. 46 suppl. 1655-1675.

Ziv, M. and Ariel, T. (1991) 'Bud proliferation and plant regeneration in liquid-cultured philodendron treated with ancymidol and paclobutrazol', J. Plant Growth Regul. 10, 53-57.

IN VITRO MASS PROPAGATION OF RICE

T. HIROSAWA
Nursery Technology Inc.
3377 Kitsuregawa, Tochigi 329-14
Japan

ABSTRACT. Factors which affect the overall *in vitro* production efficiency of rice plantlets were studied. Highly efficient plant regeneration systems utilizing liquid medium (>5000 plantlets / g callus / 6 weeks) have been developed. Field performance trials of the regenerated plants were carried out, and changes in agronomic features (i.e. culm length, heading date, fertility) were observed.

1.GENERAL INTRODUCTION

Recent advances in plant tissue culture technology make the *in vitro* mass propagation of cereal crops a possibility. At the same time, production of F1 hybrids and genetically engineered rice plants have been reported. The production of F1 hybrid seeds via cytoplasmic male sterility, requires complex procedures to match heading date of seed parents to that of pollinators. In addition, substantial amount of time is also required to produce a large quantity of seeds from a few genetically engineered plants. Tissue culture provides alternative methods for the rapid, large-scale production of plantlets of various types of rice cultivars including breeding lines.

A culture system for the above purposes was required which met the following prerequisites:

1)High regeneration efficiency
2)Stable, reproducible regeneration
3)Liquid regeneration medium
4)Possibility for scaling-up
5)Maintenance of genetic uniformity.

The aim of our research was to establish an efficient, high-yielding culture system for the large-scale commercial production of rice plantlets. This paper reports on the major results obtained to date.

K. Kurata and T. Kozai (eds.), Transplant Production Systems, 195–212.

2. CULTURE SYSTEM

2.1 PLANT MATERIALS AND CALLUS INITIATION

Several types of mass propagation systems have been reported for rice. Tanaka (1983) reported a system using the shoot primordium, and claimed that 240 million plantlets could be produced from single shoot primordium in a year. Hisajima *et al.*(1987) reported a system in which multiple shoot buds were induced from seeds in a medium containing benzylaminopurine (BAP). Theoretically, 1.6 million clumps consisting of multiple shoots could be produced in a year by successive shoot multiplication using this system. These two types of culture system were claimed to be efficient and free from somatic variation avoiding callus culture. However, these systems require a great amount of manual labor and are not applicable to a large-scale industrial operation. Therefore, we chose callus culture as propagation system.

Regeneration of plants from rice callus was first reported by several Japanese researchers in 1968. They used either mature seeds (Nishi *et al.*1968; Maeda 1968; Tamura 1968), anthers(Niizeki and Oono 1968), or roots (Kawata and Ishihara 1968) as explants. Calli derived from immature embryos were reported to be highly competent for regeneration, especially via somatic embryogenesis. However, there is difficulty in supplying the immature embryos at the exact time needed and this fact posed serious limitations for further research and practical application of the system. Therefore, we chose mature seeds of japonica rice as an explant.

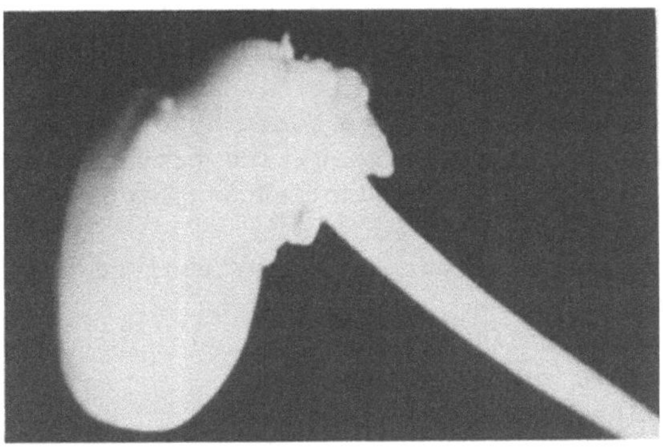

Figure 1. Callus induced around scutellar region of rice after 8 days of culture on callus induction medium.

Callus was initiated from the scutellar region (Fig. 1) with a medium containing inorganic salts of N6 medium (Chu *et al.* 1975) supplemented with 12mM proline, 0.1g/l casein acid hydrolysate, 4mg/l 2,4-dichlorophenoxyacetic acid (2,4-D), 10g/l sucrose, 30g/l sorbitol, 5mM 2-(N-morpholino)ethanesulfonic acids (MES) and 2g/l gellan gum (Matsuno *et al.*1990a,c; Tsukahara and Hirosawa 1992). After 8-10 days of incubation, callus which formed at the scutellar region of the seeds was removed, transferred onto fresh initiation medium, and incubated another 14 days. This subculture from the initial induction medium to fresh medium in 8-10 days appeared to be critical in obtaining a highly competent callus culture. Suspension cultures were initiated by transferring approximately 1g of callus to 500ml Erlenmeyer flasks containing 100ml of subculture medium (callus initiation medium devoid of gellan gum). The cell suspensions were cultured on a rotary shaker at 80rpm in the dark and was subcultured at 7-day-intervals. Highly competent callus suspension cultures were obtained with this method irrespective of genotype such as japonica, indica, javanica, F1 hybrids and African wild rice (Matsuno *et al.* 1990b). Unless otherwise stated, callus thus obtained from *Oryza sativa* L. cv. sasanishiki was used in the following experiments.

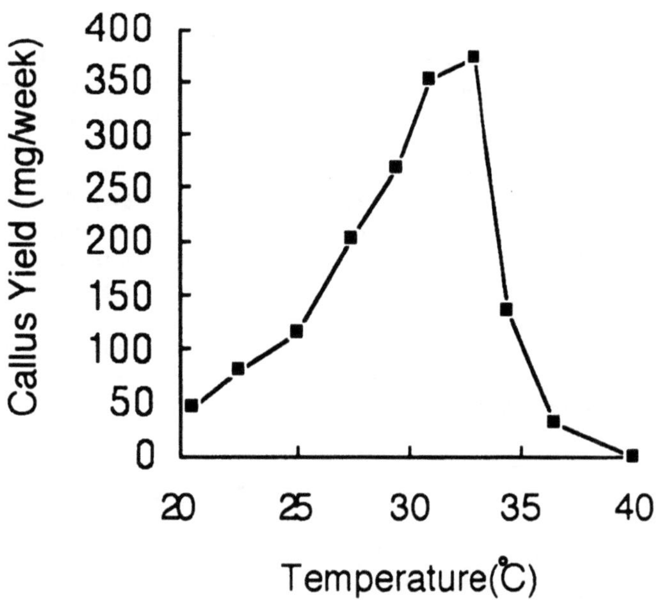

Figure 2. Effect of incubation temperature on callus yield. Calli were incubated at respective temperatures for one week in a temperature gradient incubator (Hirosawa unpublished).

2.2 OPTIMIZATION OF CALLUS GROWTH

2.1.1. *Temperature*. In order to increase total production efficiency, conditions for optimum callus growth were studied. Figure 2 indicates callus yield of rice suspension (*Oryza sativa* L. cv. sasanishiki) at respective culture temperatures. The callus yield at 33 ^{o}C was two-fold as high as that obtained at 27 ^{o}C, a widely used culture temperature. A notable feature of the curve is the narrow range of optimum temperature at 30-33°C.

2.1.2. *Nitrogen source*. An analysis of the components of the growth medium during course of callus growth revealed that NH_4^+ and PO_4^- are depleted from the culture medium 5 to 9 days respectively, after the onset of culture (Okamoto *et al.* 1991). Proline was also depleted by the 10th day. This rapid depletion of NH_4^+ was followed by the gradual decline of NO_3^- concentration which lasted until the 13th day. In addition, this rapid change in NH_4^+ concentration was accompanied by a rapid decline of medium pH from 5.8 to 4.0. This depletion of NH_4^+ from the growth medium also coincided with the transition of the cultures from exponential growth phase to a linear growth phase. It appeared that the proper supply of nitrogen sources might improve callus yield. Table 1 shows the effects of nitrogen feeding on callus growth. Elevation of the initial NH_4^+ concentration by the addition of 3.5mM ammonium sulfate did not rise any significant effects. Instead, a fed-batch culture by nitrogen feeding has developed to supply nitrogen along callus growth. Ammonium hydroxide (5 %) and nitric acid (1 N) were fed in a way to keep the medium pH 5.8. During the initial stage of the culture, NH_4OH was fed, and during later stage HNO_3 was fed. With these nitrogen-fed batch cultures, callus yield was increased by 100%. Relative growth constants of nitrogen fed culture and control culture were 0.5/day and 0.32/day, respectively. Since pH adjustment with 1N HCl and 1N NaOH did not show such a promotion in callus yield, the positive effects of nitrogen feeding is not likely

TABLE 1. Effects of nitrogen feeding on callus growth rate.

Feeding methods	Growth rate (X / 2 weeks)
pH adjustment with 1N HNO_3 and 5% NH_4OH	21.2
elevation of initial NH_4^+concentration by 100% with $(NH_4)_2SO_4$	12.7
pH adjustment with 1N HCl and 1N NaOH	10.8
control	11.7

due to the effects of pH adjustment. Callus thus obtained was characterized by higher regeneration frequency than those obtained from ordinary batch culture (Okamoto *et al.* 1991). Supplementation of phosphate and proline did not support the stimulation of callus growth. With the combination of above optimum conditions, a growth rate of 4X per week can be obtained on a routine basis.

2.3 MARKERS FOR THE QUALITY OF CALLUS

The successful commercialization of tissue culture plant production schemes relies on the efficient separation of competent cells from other cells. Even in the highly competent cultures, only a small fraction of the cells may be capable of regeneration. Thus, the early chemical determination of the ability of cultured cells to produce plants would be extremely valuable as a screening tool to obtain an efficient regeneration system. Presently, visual criteria for the distinction of embryogenic callus from non-embryogenic callus are being used (Nabors *et al.* 1983).

TABLE 2. Some biochemical and physiological differences between various types of callus.

Callus type	Starch (mg/g DW)	GABA (μ moles/g FW)	Oxygen uptake (μ moles/g FW/hr)
R-callus	7.99	0.97	16.36
	8.44	0.84	19.13
NR-callus	0.25	2.24	88.81
	0.32	1.39	56.28
R - NR callus	0.29	—	—
	0.39	—	—

Starch was detected enzymatically using a commercial kit.
GABA content was measured with an amino acid analyzer.
Oxygen uptake was measured polarographically.

Table 2 shows the differences observed in biochemical and physiological parameters between highly regenerating callus (R-callus), non-regenerating callus (NR-callus), as well as NR-callus which had been converted from R-callus by prolonged culture without subculture. The most distinct difference was observed in starch content. R-callus yielded a starch content more than 30 times as much as that associated with NR-calli. Wurtele *et al.*(1988) reported similar results in which carrot embryogenic callus contained 15-40-fold greater amount of starch as compared to non-embryogenic callus. Another significant difference was detected in amino acid composition, particularly in regard to gamma-aminobutyric acid (GABA) content was higher in NR-callus compared to that of R-callus. The significance of this difference is not clear, but it has been reported that undifferentiated callus tissue and actively growing meristematic tissues accumulated more GABA compare to other tissues (Kishinami and Ojima 1980;Kishinami 1987, 1988). This may suggests that highly regenerative callus are somehow differentiated to regenerate, or are not completely dedifferentiated. The higher respiration rate observed in NR-callus supports the idea that NR-callus was highly dedifferentiated and actively growing, and that they are thereby accumulating a relatively higher amount of GABA compared to that in R-callus. Although starch content seems to be the best marker for the quality of callus, quantitative correlations between starch content and regeneration ability remain to be established.

2.4 REGENERATION OF PLANTS

2.4.1. *Plantlet Regeneration on Solid Medium.* We first studied the regeneration on solid media to identify the essential factors affecting plantlet regeneration. Among the factors studied, concentrations of gelling agents, osmoticum, and combinations of growth regulators in the regeneration medium were found to be important (Tsukahara and Hirosawa 1992). High concentrations (4-6g/l gellan gum,10-16g/l agar) of gelling agents promoted regeneration frequency.

Figure 3 shows that the addition of 15-75g/l sorbitol to the regeneration medium greatly stimulated regeneration frequency. In contrast to sorbitol, the addition more than 30g/l of mannitol was inhibitory to regeneration, while no inhibition was observed in callus growth. In the previous reports, several authors showed the importance of osmolarity of medium (Kavi Kishor and Reddy 1986a,b; Kavi Kishor *et al.* 1989; Liu and Lai,1991), however, such a discrimination between sorbitol and mannitol was not mentioned. Preliminary results obtained in our lab indicate that sorbitol is absorbed by calls, while the mannitol concentration of the medium was unchanged. This suggests that sorbitol is not a mere osmoticum and has to be metabolized in order to stimulate the regeneration.

Liu and Lai (1991) showed that water status of rice callus is closely correlated with regeneration ability. Conditions which reduce water content of callus such as high osmotic pressure in the growth medium and higher agar content promoted regeneration frequency. It was found that a brief dehydration treatment of callus promoted regeneration rate (Tsukahara and Hirosawa 1992). Callus was placed in a Petri dish with a single layer of filter paper and kept for respective time periods at room temperature. With a 24hr dehydration treatment, water content was reduced to 56% of that of control. At the same time,

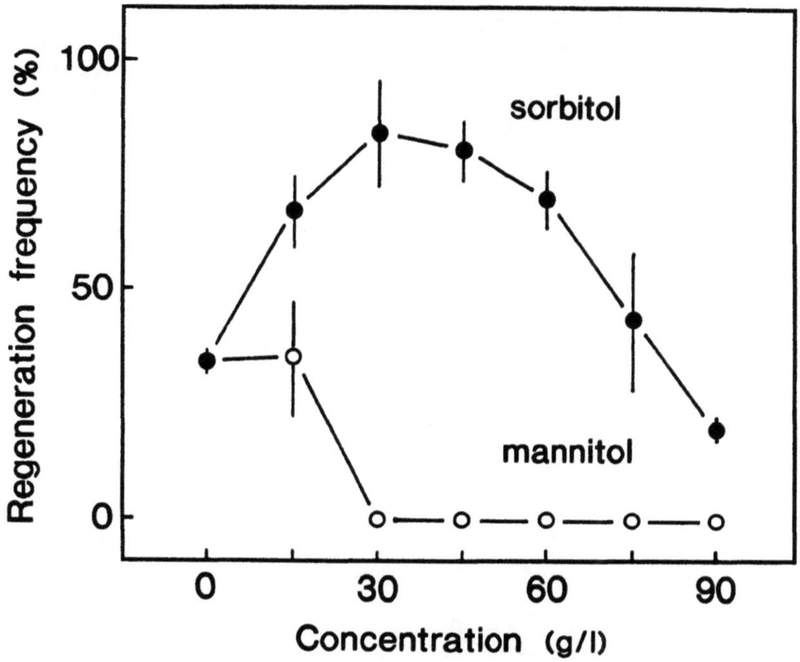

Figure 3. Effects of sorbitol and mannitol concentration on regeneration frequency (Tsukahara and Hirosawa 1992).

regeneration rate was elevated from 2.5% to more than 50%. This simple method provides an alternative method for improving regeneration efficiency and stability in regeneration efficiency.

Direct morphological observations with scanning electron microscope were made on the development of regenerated plantlets (Fig.4 a-d). Unfixed fresh tissues were directly observed with scanning electron microscope. Undifferentiated callus without epidermal tissue shows disruption of surface cells during observation under high vacuum. Within one week of incubation on the regeneration medium, smooth parts which could tolerate observation condition become visible, indicating epidermal tissue differentiation has started. In two weeks, shoot primordia-like structures become discernible and in four weeks, green plantlets could be observed with the naked eye.

2.4.2. *Plantlet Regeneration in Liquid Media*. For the large-scale industrial propagation of rice plantlets, it is required to produce large quantities of plants in a synchronous fashion. For this reason, a regeneration system in a liquid medium is required. Among monocotyledonous species, however, only a few liquid regeneration systems have been reported for rice. Abe and Futsuhara (1986) reported that formation of globular embryo-like structure in root derived callus

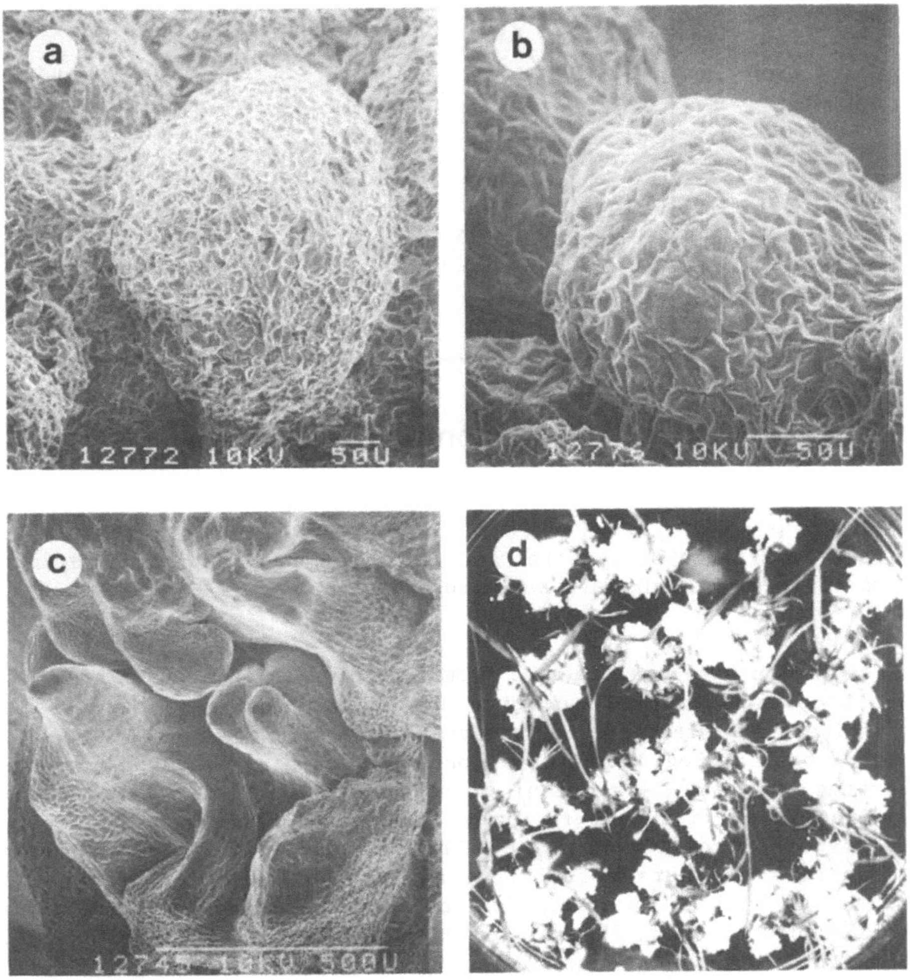

Figure 4 a-d. Scanning electron micrographs of the plantlet regeneration process on solid medium. a.Callus without epidermal tissue. b.Callus with developing epidermal tissue after 1 week of culture. c.Differentiation of shoot,2 weeks after culture. c.Regenerated plantlets,4 weeks after culture(Hirosawa and Matsushima, unpublished).

culture. Yoshida *et al.*(1990) reported a rotary culture system for rice F1 hybrids. Recently, Ozawa and Komamine (1989) reported successful embryogenesis in a liquid culture with a particular rice variety (*O. sativa* L. cv. konansou). Components of regeneration medium such as salt concentration, concentration and combination of growth regulators, sucrose concentration, callus inoculation density were reported to be important. Reported regeneration efficiency was 48% which was relatively lower than that obtained on solid medium (93%). They also reported that callus subculture at 3day-intervals is essential for high regeneration rates. These previously described methods were not directly applicable to commercial production because of their difficulty in scaling-up.

Two different types of regeneration systems have now been developed using callus produced and maintained by methods described in the previous section.

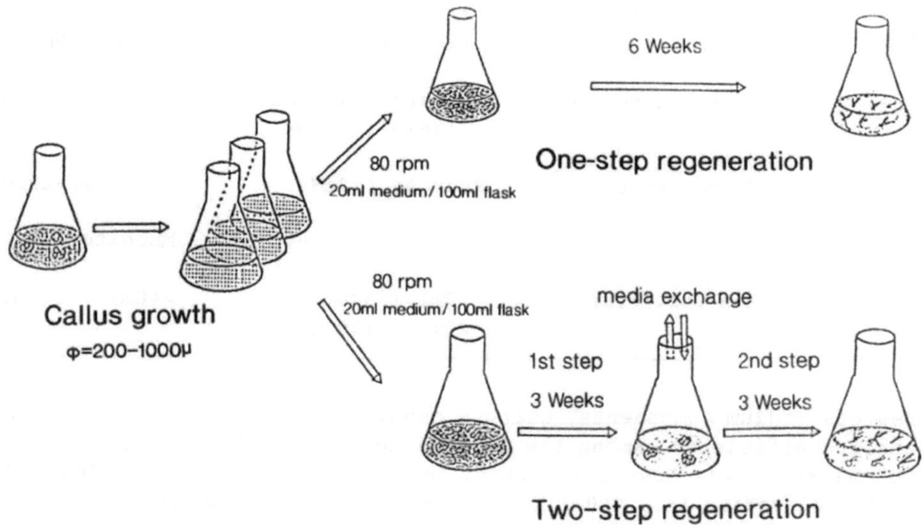

Figure 5. Schematic presentation of two types of regeneration systems using liquid media.

One step regeneration method. Figure 5 illustrates the procedure for regeneration using liquid media. After 7 days of suspension culture in a fresh callus growth medium, callus was collected by using a nylon mesh with a pore size of 200u and 1000u. Thus, callus with a diameter between 200-1000u was inoculated into a 100ml Erlenmeyer flask containing 20ml of respective media. Flasks were cultured on a rotary shaker at 80 rpm for 4-6 weeks.

Table 3 shows the summary of factors tested for the optimization of the regeneration system. The auxin 2,4-D was inhibitory at any concentration tested, and the combination of 0.1mg/l NAA and 0.1mg/l kinetin was found to be best. Nitrate ion was preferable over ammonium ion for regeneration. Higher concentration,more than 29mM, of sucrose was inhibitory and only callus growth was observed. As is the case with solid regeneration medium, species and concentration of osmotica were found to be very important. Addition of 30g/l sorbitol promoted regeneration rate ,while the addition of mannitol was inhibitory.

TABLE 3. Summary of factors affecting plantlet regeneration in liquid regeneration medium.

Factors	Promotive	Inhibitory
Growth regulators	NAA , Kinetin (0.1mg/l)(0.1mg/l)	2,4-D (at any conc.)
Nitrogen sources	Nitrate / Ammonium (56mM / 3.5mM)	Ammonium alone
Carbon source	Sucrose (20mM)	Sucrose(>29mM)
Osmotica	Sorbitol (30-60g/l)	Mannitol
Med.vol./Flask vol.	30ml / 300ml 20ml / 100ml	>40ml / 300ml

Besides medium components, culture conditions greatly affected regeneration efficiency. Among the factors tested, medium volume to vessel volume ratio was most influential over regeneration rate (Table 3). Thirty ml medium in a 300ml Erlenmeyer flask was found to be the optimum medium volume to vessel ratio.

Upon the combination of the optimum conditions of each factors tested, a regeneration medium composed of N6 salts supplemented with 56mM KNO_3, 7g/l sucrose, 60g/l sorbitol, 12 mM Proline, 0.1 g/l casein acid hydrolysate, 5 mM MES, 0.1 mg/l NAA, and 0.1 mg/l kinetin was established and resulted in the regeneration of a maximum of 200 plantlets (3-10mm in shoot length) from 20mg of callus, after 6 weeks of culture using above described liquid medium.

General morphological features of the differentiation process resembled those observed with solid medium (Fig. 4). After 3 weeks of incubation, roots and shoots became visible, and in 6 weeks, plantlets with root(s) were formed. Scanning electron microscopy revealed that as early as two days after the initiation of regeneration culture, organization of callus surface had started. Figure 6 shows plantlets which regenerated via the outlined one step regeneration method.

With this one-step regeneration culture, more than 5,000 plantlets were produced from 1g of callus inoculated in 6 weeks, thus exceeding that possible on solid regeneration medium which would be 5,000 plantlets.

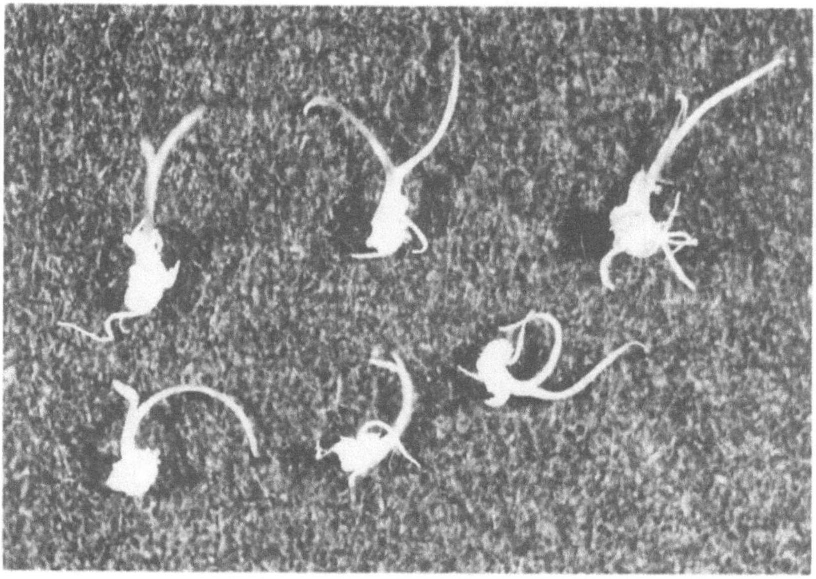

Figure 6. Plantlets regenerated after 6 weeks of incubation in a liquid regeneration medium (one step regeneration method).

Two-step regeneration method. A similar, but even more efficient system has also been developed (Kobayashi and Hirosawa 1991a,b). In a control culture, callus were inoculated into 100ml Erlenmeyer flasks containing 20ml of liquid regeneration medium consisting of MS basal medium supplemented with 12mM Proline, 2mg/l Casein acid hydrolysate, 5mM MES, 10g/l sucrose, 30g/l sorbitol, 0.4mg/l NAA, and 0.5mg/l Kinetin. After 6 weeks of culture on a rotary shaker, an average of 10 plantlets were obtained per flask. A substantial increase in regeneration efficiency was observed by replacing the initial regeneration medium during the culture with 40ml of replacing medium consisting of half-strength MS basal medium supplemented with 2g/l casein acid hydrolysate, 2.5mM MES, 15g/l sucrose, 1.5% sorbitol, 1.0mg/l NAA, and 0.5mg/l Kinetin. The timing for the replacement was critical and the beginning of the 4th week of culture was found to be optimum. With this medium replacement, 100-150 plantlets were obtained. Moreover, 75% of total plants obtained during culture were regenerated within 10 days after the replacement. The increase in the number of regenerated plantlets was also observed by the simple addition of 20ml of the replacing medium instead of medium replacement. This suggested that the cause for the enhancement of regeneration is at least partially due to the supply of depleted medium component(s). Another possible reason for the promotion is removal of inhibitory substance(s) excreted into the culture medium as suggested by Ozawa and Komamine (1989).

The relatively rapid decline of regeneration potential with culture age has been reported (Oono 1975; Henke *et al.*1978; Inoue and Maeda 1980; Nabors *et al.* 1983). Figure 7 shows that rapid loss of potential for the one-step regeneration in liquid medium. Although, several methods have been reported to overcome this problem, the ultimate solution must be established.

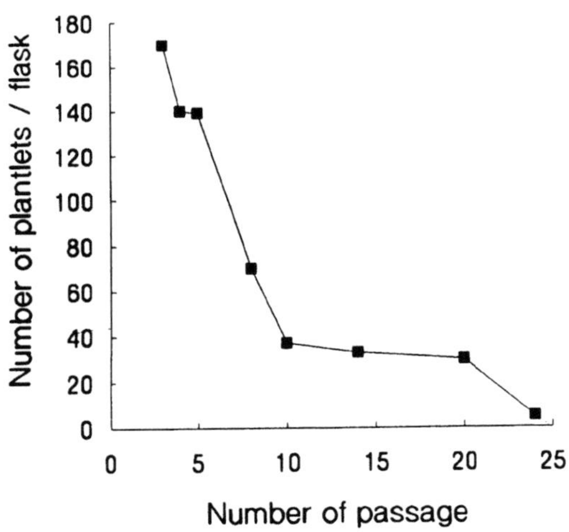

Figure 7. Effects of culture passage of callus on the regeneration frequency. Callus was subcultured every week. Regeneration frequency is expressed as the number of plantlets regenerated in a 100ml flask containing 20 ml medium.

3. FIELD PERFORMANCE OF REGENERATED PLANTS

Field performance of regenerated plants (more than 1,000 plants of each regeneration method) have been observed in the years 1990 and 1991 (Matsuno *et al.* 1991). Table 4 shows the results obtained in 1991. Regenerated plants were characterized by shorter culm, reduced fertility, and earlier heading-date, irrespective of the culture method used. However, the magnitude of variation in those traits were significantly different between regeneration conditions with which the tested plants were obtained. In general, plants regenerated in liquid medium showed values closer to that of control than those regenerated on solid medium. Non-fertile plants showed curled anthers and very few well developed pollen grains (Fig.8 a-f). At least major reason for the lower fertility seemed to be a decrease in pollen fertility. Another factor which closely correlated to the variations was found to be the number of culture passage. After 24 passages with one week subculture intervals, fertility became almost lost.

It appears that these changes pose serious limitations for the commercial application of this culture system. However, the differences observed between plants derived from liquid and solid media suggests prevention of variation in agronomic characters may be possible by selecting proper culture conditions.

TABLE 4. Comparison of agronomic characters between plants regenerated under different regeneration conditions.

		Regenerated plants	
	Seedings	Solid medium	Liquid medium
Culm length	88.3 ± 7.5	68.4 ± 8.2	73.7 ± 9.6
Panicle length	21.0 ± 1.7	20.3 ± 2.1	20.9 ± 1.9
Panicle number	11.3 ± 3.6	12.8 ± 5.4	13.2 ± 4.4
Fertility (%)	93.8 ± 3.4	48.7 ± 35.0	64.8 ± 30.8
Heading date	8/2 ± 2.1	7/26 ± 2.8	7/29 ± 2.9

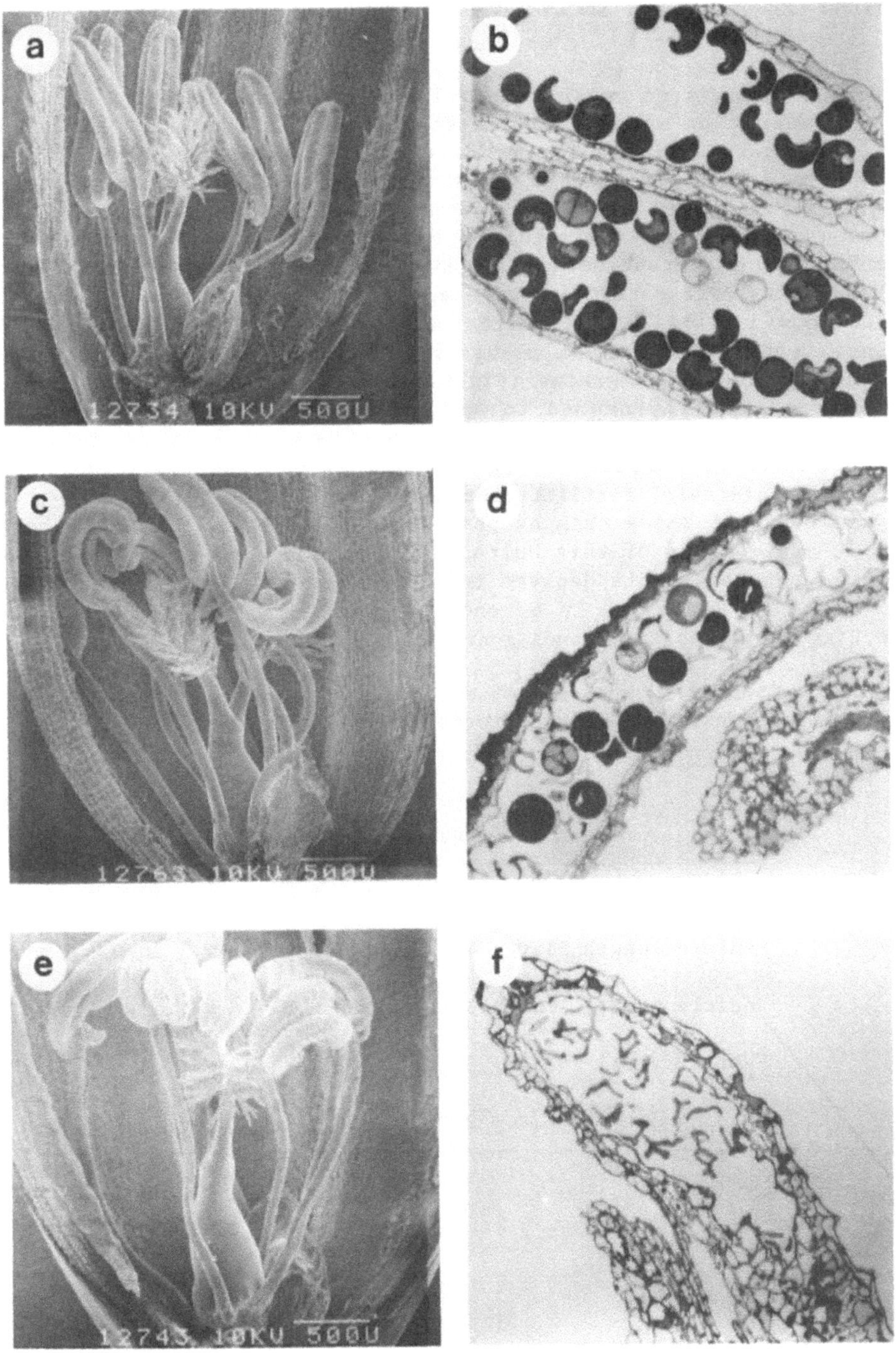
a
12734 10KV 500U
b
c
12763 10KV 500U
d
e
12743 10KV 500U
f

4. CONCLUDING REMARKS

Novel tissue culture methods developed in this study offer a great opportunity for the industrial large-scale production of rice plantlets. As far as quantity is concerned, the regeneration reached a level high enough for practical commercialization. However, the following technologies must be established before commercial production of rice seedling becomes a reality:

1)Prevention of somaclonal variation
2)Synchronous regeneration
3)Storage of regenerated plants
4)Culture system using a bioreactor type vessel
5)Automated planting system for plantlets

Research on the establishment of the above technologies are underway.

Acknowledgements. The author is grateful to all researchers in Nursery Technology Inc. for their cooperation and to Dr.Sharon Kurtz, Twyford International,Inc., for correcting English grammar.

REFERENCES

Chu, C.C., Wang, C.C., Sun, C.S., Hsu, C., Yin, K.C., Chu, C.Y., and Bi, F.Y. (1975) 'Establishment of an efficient medium for anther culture of rice through comparative experiments on the nitrogen sources', Sci. Sin., 659-668.

Henke, R.R., Mansas, M.A., Constantin, M.J. (1978) 'Organogenesis and plantlet formation from organ and seedling derived calli of rice (*Oryza sativa*)', Physiol. Plant. 44,11-13.

Heyser, J.W., Dykes, T.A., DeMott, K.J., and Nabors, N.W. (1983) 'High frequency, long term regeneration of rice from callus culture', Plant Sci. Lett. 29, 175-182.

Inoue, M. and Maeda, E. (1980) 'Effects of auxins and cytokinins on the occurrence of green regions in rice callus cultures', Jpn. J. Crop Sci. 49, 167-174.

Kavi Kishor, P.B. (1987) 'Energy and osmotic requirement for high frequency regeneration of rice plants from long-term cultures', Plant Science 48, 189-194.

Kavi Kishor, P.B. and Reddy, G.M. (1986a) 'Regeneration of plants from long-term cultures of *Oryza sativa* L.'. Plant Cell Reports 5, 391-393.

Kavi Kishor, P.B. and Reddy, G.M. (1986b) 'Retention and revival of regenerating ability by osmotic adjustment in long-term culture of four varieties of rice', J.Plant Physiol. 126,49-54.

Kawata, S. and Ishihara, A. (1968) 'The regeneration of rice plant, *Oryza sativa* L., in the callus derived from the seminal root' Proc. Jpn. Acad. 44, 549-553.

Kishinami, I. (1987) 'Accumulation of gamma-aminobutyric acid due to ammonium in various cultured plant cell', Plant Cell Physiol. 28,1459-1464.

Kishinami, I.(1988) 'Effects of auxins (2,4-dichlorophenoxyacetic acid, indole-3-acetic acid and naphthaleneacetic acid) on the accumulation of gamma-aminobutyric acid in excised rice root tips', Plant Cell Physiol. 29, 581-585.

Kishinami, I. and Ojima, K. (1980) 'Accumulation of gamma-aminobutyric acid due to adding ammonium or glutamine to cultured rice cells', Plant Cell Physiol. 21,581-589.

Kobayashi, H. and Hirosawa, T. (1991a) 'Improved plantlet regeneration in the rice (*Oryza sativa* L.) suspension culture', Jpn. J. Breed. 41 suppl. 1, 22-23.

Kobayashi, H. and Hirosawa, T. (1991b) 'Effect of osmotica on the plantlet regeneration in rice suspension culture', Jpn. J. Breed. suppl. 2, 254-255.

Lai, K.L. and Liu, L.F. (1988) 'Increased plant regeneration frequency in water-stressed rice tissue culture', Jpn. J. Crop Sci. 57,553-557.

Maeda, E. (1968) 'Subculture and organ formation in the callus derived from rice embryos in vitro', Proc. Crop Sci. Soc. Jpn. 37, 51-58.

Matsuno, T., Ishizaki, K., and Horn M.E. (1990a) 'Establishment of mass production systems by somatic embryogenesis in rice I' Jpn. J. Breed. 40 suppl. 1,266-267.

Matsuno, T., Maruyama, K. and Hirosawa, T. (1990b) 'Applicability of rice scutellar embryonic callus methods among different types of rice', Proc. 2nd Plant Tissue Culture Colloquium, 100-101.

Matsuno, T. and Murayama, H. (1990c) 'Effects of osmotic potentials and gel strength on SE conversion in rice', Jpn. J. Breed. 40 suppl. 2, 50 - 51.

Matsuno, T., Tsukahara, M., Shimizu, A.,Osanai, E.,and Kobayashi, O. (1991) 'Variations observed among R-0 regenerants from rice tissue culture', Jpn. J. Breed. 41 suppl. 2,312-313.

Nabors, M.W., Heyser, J.W., Dykes, T.A. and DeMott, K.J. (1983) 'Long-duration,high-frequency plant regeneration from cereal tissue cultures', Planta 157, 385-391.

Niizeki, H. and Oono, K. 'Induction of haploid rice plant from anther culture', Proc. Jpn. Acad. 44, 544-557.

Nishi, T., Yamada, Y., and Takahashi, E. (1968) 'Organ redifferentiation and plant restoration in rice callus', Nature 219, 508-509.

Okamoto, A. Sakurazawa, H. and Hirosawa, T.(1991) 'Optimization of rice cell culture by nitrogen feeding', Nippon Nogeikagaku

Kaishi 65, 607.
Oono, K. (1975) 'Production of haploid plants of rice(*Oryza sativa*) by anther culture and their use for breeding', Bul. Natl. Inst. Sci. Jpn. 26, 139-222.
Ozawa, K. and Komamine, A. (1989) 'Establishment of a system of high-frequency embryogenesis from long-term cell suspension culture of rice (*Oryza sativa* L.)', Theor. Appl. Genet. 77,205-211.
Tanaka, R. (1983) 'Cloning of useful plants' in Seedling industries and new breeding techniques, CMC Press,Tokyo.
Tamura, S. (1968) 'Shoot formation in calli originated from rice embryo', Proc. Jpn. Acad. 44, 544-548.
Tsukahara, M. and Hirosawa, T. (1992) 'Characterization of factors affecting plantlet regeneration from rice (Oryza sativa L.) callus' Bot. Mag. Tokyo 1078, in press.
Tsukahara, M. and Hirosawa, T. (1991) 'Characterization of factors affecting the regeneration of plantlets in the rice suspension culture', Proc. Ann. Met. Jpn. Soc. Plant. Physiol.,90.
Yoshida T., Otsuki Y., Oda F. and Ko J. (1989) '*In vitro* propagation of rice F1. 1. Application of liquid culture and efficient plant regeneration, Jpn. J. Breed. 38, suppl 2, 140-141.
Wurtele, E.S, Keller G.L., Nikolau, B.J., Ulrich, T.H.(1988) 'Quantitation of starch and ADP-glucose pyrophosphorylase in non-embryogenic cells and embryogenic cell clusters from carrot suspension cultures', J. Plant Physiol. 132,683-689.

TOWARDS A MODEL OF MINERAL NUTRITION IN VITRO

R. R. WILLIAMS
Department of Agronomy and Soil Science
University of New England
Armidale, NSW 2351
Australia

ABSTRACT. The mineral nutrition of plants in vitro is discussed with an emphasis on the mechanisms involved. The process includes the initial supply of minerals, changes which take place in the medium, transport of the minerals to the plant, uptake by the plant and transport through the plant. Based on this outline, the concept of a mechanistic mathematical model of mineral nutrition in vitro is developed providing a focus for future research and possibly the development of improved control systems for automated micropropagation production systems.

1. Introduction

The formulation of nutrient media remains an important part of the development of protocols for plant tissue culture and micropropagation. When a new species or cultivar is being initiated into in vitro culture it is usually necessary to adapt a culture medium formulation to obtain satisfactory growth and multiplication. Manipulation of the growth regulator component is probably critical in obtaining shoot multiplication and rooting. Sometimes varying the overall concentration of minerals in the medium is also part of the procedure for regulating the developmental stages. However, less consideration is given to the changing relationship between the plant and the mineral component of the medium over time and from sub-culture to sub-culture.

I propose to discuss some aspects of this relationship between the mineral component of in vitro media and the mineral nutrition of the explants. Most importantly, I will show that this is a dynamic relationship suggesting that more frequent variation of the mineral component may be necessary for sustained, optimum culture performance. Development of a model can help us not only to study and understand this complex system but can also provide a tool for the 'dynamic' management which will be necessary in future production systems.

A practical limitation to optimizing the supply of minerals, particularly for micropropagation, has been the time and cost involved in changing the medium frequently. Cell cultures have the advantage because bioreactors can enable continuous liquid medium circulation and replacement. The emergence of new micropropagation systems such as double layer (agar-liquid) media (Maene and Debergh, 1987) and recirculation liquid culture, combined with automation (Aitken-Christie and Jones, 1987; Aitken-Christie, 1990), is largely overcoming this labour-cost limitation so it is opportune to reassess our approach to mineral supply.

As well as making ongoing regulation of the mineral supply possible, automation itself demands renewed attention to culture medium management. The conflicting factors of the high capital cost of automation and the need to lower unit plant costs mean that optimum culture performance will be necessary. Optimization of the mineral supply throughout the productive life of a culture will be critical both in terms of sustained yield and, particularly for automation of cutting and explant transfer, uniformity of culture vigour and morphology (Johnson, 1989).

K. Kurata and T. Kozai (eds.), Transplant Production Systems, 213–229.

In this paper I will be less concerned about the actual formulation used and more with the rationale on which the development of media is based, the important factors, and the notion that more dynamic approach to mineral supply will be necessary.

Typically, the medium is only varied when a change in development stage is required; and then it is usually only the growth regulator component which is changed. It is generally assumed that a medium found to be satisfactory over one or two subcultures, at any particular stage of the culture cycle, will be suitable indefinitely. It is also often assumed that what is supplied in the initial medium is what is available to the plant. These assumptions are quite wrong.

One of the problems often encountered in commercial micropropagation laboratories is a periodic decline in culture vigour. Progressive accumulation of growth regulators in the plant tissues is one cause, particularly where cytokinins are used to promote shoot multiplication. Accumulation of inhibitory substances in the medium may be a problem during extended periods between transfers to fresh medium. The gaseous environment, rising ethylene or depleted CO_2 levels, is another aspect receiving considerable attention. But what about the mineral component?

2. The Mechanistic Approach

Much of the work published on mineral nutrition of in vitro- grown plants has involved empirical manipulation of media components with the response measured in terms of plant growth and development. This approach has made possible the widespread application of in vitro culture today but it suffers from one major drawback, it does not provide a reliable means of determining the nutrient requirements for previously untried species or cultivars. We still have a limited understanding of the processes involved and how the factors interact. A more mechanistic approach to media research may increase our understanding of the system and ultimately provide a basis for a predictive model (Williams, 1992). Such a model could provide the algorithms necessary for future automated continuous adjustment of the medium components.

Heller and colleagues (Heller, 1965) have used a mechanistic approach to study the nutrition of callus cultures. There has also been considerable work on in vivo plants and root uptake (e.g. Marschner, 1986). The task is to find the middle road which applies to micropropagation i.e. typically shoot cultures, usually without roots at least during the multiplication stage.

The essence of this approach is to define the sequence of processes involved in the supply and uptake of minerals in vitro then investigate the underlying mechanisms and critical factors. Once the key pieces of the model have been identified the relationships between the pieces may be described mathematically to establish a quantitative, mechanistic, predictive model.

The overall process of mineral nutrition of plants involves a number of steps: the input of minerals; the interaction between the ions as supplied; interaction between these ions and the medium substrate (liquid or semisolid gel); movement of ions through the substrate to the plant surface; uptake (and excretion) by the plant; transport through the plant; and finally assimilation by the plant. Remobilization of minerals from the plant tissues is also a possibility but is not considered part of the initial system.

The mere process of identifying these steps focuses our attention on key parts of the system and raises questions about what might be the critical factors.

3. The Initial Supply

The first parameter of the mineral supply model is the composition of minerals provided in the initial medium. Much research has been done in developing the basic formula to supply a balanced regime of essential elements (George et al., 1987). There are a number of standard formulations now widely used, the most common being adaptions of Murashige and Skoog's (1962) medium developed specifically for tobacco callus culture. Others contributing to the refinement of the basic mineral component include Heller (1953) and White (1963). Media have also been developed for specific groups of plants e.g. Gamborg et al., (1968) soybean root cells ; Schenk and Hildebrandt (1972) for monocots; Lloyd and McCown (1981) for woody plants.

To date the concentrations and proportions of minerals initially provided in the medium have been mostly determined by measuring the plant response. Ideally we would like to be able to determine what the plant needs for optimum performance then calculate from this what to supply in the medium. While we may be able to accurately measure the mineral content of healthy plants, in vitro or in vivo, the process of interpretation of such data and extrapolation to the supply, is fraught with difficulty.

For a start, what we measure in a healthy plant or tissue does not necessarily indicate the optimum regime. Plants have wide tolerance for most minerals. An associated problem is the unequal distribution of elements within a plant over time and space. However, maybe we can determine the range of tolerance for individual elements consistent with optimum growth at different stages.

Even if we can identify what mineral regime the plant needs, we clearly cannot simply put this 'ideal' regime in the medium. We now must consider the relationship between what is supplied and what the plant finally assimilates.

Mineral uptake into plants grown in vitro is dependent on the supply in the medium. However, the supply depends not only on what is added to the medium at the time of preparation but also on the availability of those minerals to the plant. Availability will depend on the solubility of the ions, after any chemical changes that may occur in the medium. Where a gelling agent is used, mineral availability will also depend on the extent of bonding of the ions with the gel and the rate of diffusion of ions through the gel to the plant-gel interface. Water relations between the medium, the plant and the atmosphere within the culture vessel are also important (Murashige and Skoog, 1962, Debergh, 1983).

When considering the nutrient supply in the medium we must remember that the plant probably does not have access to the entire volume of the medium. Romberger and Tabor (1971) introduced the term "Effective Media Volume" referring to the volume of medium in the vicinity of the plant to or from which ions or organic substances might move. This is analogous to the situation in soil where much of the total mineral supply is not readily available. In soil there is a limited pool of minerals in solution within the soil water adjacent the roots but there can be continuous exchange between this pool and the insoluble or bound minerals in the surrounding soil. Similarly water may be freely available or bound by chemical and physical forces.

The total mineral pool of the in vitro medium is on the one hand finite, determined by the initial medium concentration and volume. On the other hand, it is at least initially in a soluble, readily available form. The ratio of the available ions in vitro is presumably configured to suit the plant. However, the examples discussed elsewhere in this paper illustrate that both the initial total available supply and the ratio of ions can be substantially modified within the medium. The actual size and composition of the unavailable pool in vitro is a pertinent area for research.

Heller (1965) reported that the inclusion of a colloidal layer beneath a liquid medium enhanced growth. This may be akin to the soil situation where there is a reservoir of minerals attached to the colloidal particles and exchangeable with the free ion pool and is similar to dual-phase media. It may thus be possible to provide higher levels of minerals in the initial medium without the problem of supraoptimal concentrations in the liquid phase. This could both extend the duration of culture and moderate changes in the availability of individual ions over the culture period. This principal has been applied to Radiata pine (Aitken-Christie and Jones, 1987)

4. Chemical Changes to Media

The actual ionic (mineral and organic) composition of the medium may undergo marked change even before the plant is introduced. The effect of pH is discussed later. Procedures and conditions commonly used during media preparation, particularly autoclaving and exposure to light, are major contributors to such changes (Dalton et al., 1983; Skirvin et al., 1986; Elmore et al., 1990; Williams et al., 1990; Sarma et al., 1990). The result may be reduced availability or loss of components and sometimes the production of toxic byproducts.

EDTA is commonly used as a chelating agent to maintain the solubility of iron in culture media but it also contributes to the precipitation of minerals from standard MS medium. Dalton et al. (1983) found that a significant proportion of the supplied Fe, Zn, P, S and Ca (45%, 20%, 13%, 11% and 5% respectively) was precipitated within a few days. This was due to an imbalance in the ratio of EDTA chelate to Fe and could be overcome by reducing the Fe to one third. Higher levels of EDTA may be toxic.

Photochemical oxidation of EDTA produces formaldehyde which can accumulate to inhibitory levels (Hangarter and Stasinopoulos, 1991). This may also lead to iron deficiency due to precipitation of the unchelated iron. Hangarter and Stasinopoulos (1991) showed that this photo-oxidation could be reduced by filtering out ultraviolet and blue wavelengths from the light source.

In the broader context, the organic components of media are also subject to changes during preparation or in standing medium. Ascorbic acid is included in many media because of its antioxidant properties. However we now know that ascorbic acid added to culture media is completely oxidized to dehydroascorbic acid within a few hours (Elmore et.al., 1990). Even the dehydroascorbic acid was itself destroyed within 12 hrs. This process is also accelerated by autoclaving, shaking, high light intensity and increasing pH even before plants are inserted.

Sucrose, the usual carbohydrate source, is inherently unstable. Firstly sucrose is partially hydrolyzed to fructose and glucose when dissolved in water. Subsequent autoclaving results in further breakdown of fructose and glucose to phenolics and other potentially toxic substances (Ball, 1953). This process is catalyzed by EDTA (Schenk et.al., 1991). To minimize this effect, these authors recommend autoclaving the FeNa-EDTA separately. In addition, they suggest that KH_2PO_4 also be autoclaved with the FeNa-EDTA to avoid possible precipitation of other ions by the phosphate.

Other recent studies have also investigated the instability of organic components in culture media; vitamins (Hagen et.al., 1991), auxins (Nissen and Sutter, 1990; Ebert and Taylor, 1990).

5. pH Changes

In many laboratories pH of the medium is largely taken for granted. Most tissue culture reports specify the pH set point. Standard practice is to adjust the pH during preparation of the medium then ignore it. However it is well established that the pH changes both during media preparation, particularly autoclaving, and in the presence of living plant material; cell or callus cultures (Gautheret, 1959; Gamborg and Shyluk, 1981), organ cultures (Singha et al., 1987; Williams et al., 1990) or germinating seeds (Skirvin et al., 1986).

While the pH change is rapid and equilibrium is reached within 24-48hrs, there can be lasting effects on plant growth (Adams et al., 1979) or hormone activity (Williams et al., 1985).

This change in pH when plant material is added is not dependent on wounding at the time of transfer but appears to involve an ion imbalance between the medium and the plant (Williams et al., 1990). There can be a substantial efflux of minerals out of the plant into the medium following transfer to fresh medium (Williams 1992) and this may be due in part to this pH imbalance. The equilibrium point is dependent on the species but is similar irrespective of the initial set pH.

In most studies of pH it is only the setpoint that is compared. Since the pH rapidly moves toward the same equilibrium point there is soon no pH difference between such treatments. We do not know what the effect of sustained pH differences will be. In view of the adjustment process discussed below, the plant would probably be stressed if the pH were maintained out of the normal equilibrium range.

A major factor in the pH change is the balance between NO_3 and NH_4 supply and uptake (Dougall, 1980; Congard et al., 1986). Where NO_3 is preferentially taken up there is a corresponding efflux of anions to balance the charge leading to an increase in pH and conversely, acidification occurs with NH_4 uptake. As this balance changes over the culture period there may be an initial decrease in pH followed by an increase (Gamborg and Shyluk, 1981; Singha et al., 1987).

An alternative hypothesis is that acidification of the medium is due to the accumulation of metabolic byproducts particularly CO_2 (Leva et al. 1984, cited by Singha et al., 1987).

Congard et al. (1986) showed that the effect of the plant on pH also depended on the developmental stage with acidification during formation of the primary root then an increase in pH with the secondary roots.

The distribution of these pH changes through the medium is unclear. In soil the pH of the rhizosphere can vary $\pm$ 2 units compared to the surrounding bulk soil (Marschner et al., 1986). Constable (1963) reported a localized change at the site of contact with the in vitro medium. Hush et al. (unpublished) have measured small currents around roots for 24 hr after wounding. They detected Ca^{++} as a carrier as well as OH^- and H^+. However others have reported non-localized pH changes around organ cultures in gelled media (Skirvin et al., 1986; Williams et al., 1990) and liquid (Singha et al., 1987).

If cultures tend toward an equilibrium pH, then it would be logical to adjust the initial medium to this pH to avoid the equilibration problem. The equilibrium point for the plant may well change over time so the process would have to be dynamic. Flowing liquid culture systems offer the opportunity to regulate pH in the bulk medium; we need to understand how the equilibrium point changes with plant development.

Another approach to regulation of pH is the use of buffers. The role of phosphates in the buffering of nutrient media has been discussed (Butenko, 1968; Dougall, 1980) but there does not appear to be any definitive research on their use for micropropagation as opposed to cell culture.

Parfitt et al. (1988) tested some organic buffers and found MES to be non-toxic and effective for pH 5-6. Buffering of their media was not generally beneficial but they were buffering at pH 5-6 whereas the equilibrium point was less than 5! We need to investigate buffers effective in the right range.

Apart from the direct effects on the plant, pH will affect the solubility and dissociation of ions in the medium. It may also affect the physical characteristic of gelled media. Organic components of the system can also be affected.

6. Physical Characteristics of Gels

The use of semi-solid media is still the rule for micropropagation because of its convenience in supporting the plants but gelling agents, principally agars or gelrite, cannot be considered as inert components of the system. Both their chemical and physical properties can influence culture performance.

An important aspect of gelling agents is their effect on the water balance within the culture. The higher the concentration of gel the higher the water potential. Reducing the water potential promotes the occurrence of vitrification under inducing conditions e.g. high cytokinin levels, rapidly growing tissues (Debergh et al., 1981). Increased water potential can also be expected to influence mineral availability, by increasing the concentration of ions in the free water, and mineral uptake by directly reducing water uptake by the plants.

A major disadvantage of semi-solid media compared to liquid systems is the difficulty of manipulating the medium. Liquid can be readily recirculated for automation. Gels may still be useful for holding plants in place where this is required by robotic systems.

These limitations of gels are to some extent overcome by the use of double layer systems, particularly when control of vitrification is achieved by ventilation or control of humidity within the culture vessel (Meane and Debergh, 1987) or periodic flushing with the liquid phase (Aitken-Christie and Jones, 1987).

A largely overlooked aspect of gels is their direct effect on availability of nutrients to the plant. The limiting factor to mineral nutrition in vitro is not usually the total supply in the medium because there are still adequate amounts of the elements remaining in the medium after typical culture periods (Singha et al., 1987; Williams, 1992). Availability may be a function of binding or ion exchange with the medium. Most important may be the rate at which ions can move to the plant-gel interface.

The medium may also have a direct osmotic effect on plant cells (Dougall, 1980) but this is likely to be less important in micropropagation where organs are cultured rather than cells or callus. Osmotic forces at the medium-plant interface could affect the uptake of minerals.

7. Ion Diffusion

If there is to be diffusion there must be concentration gradients. Gradients do exist across the medium toward the plant. For example, as P is absorbed the supply in the medium adjacent the tissues declines (Heller, 1965). This should promote the movement of P towards the tissues; the localized decrease in concentration shows that the P in the medium is not freely mobile.

Romberger and Tabor (1971) argue that the influence of gel type and concentration on plant growth is largely attributable to the rate of diffusion through the gel. In particular they suggest plant growth is limited by the rate of diffusion of invertase from the plant. Using an equation for estimating diffusion of macromolecules through gels (Ackers and Steere, 1962) they estimated the available pool (or Effective Medium Volume) for diffusion of invertase based on the gel pore size and the size of the diffusing molecules. In their case the EMV was about 50% of the total media volume. These calculations do not take into account particle charges and gel binding sites but the general approach could be applicable to a model of mineral ion movement and availability.

Scherer et al. (1988) used a range of anionic and cationic dyes to compare diffusion rates through agar versus gelrite. Cationic dyes diffused faster in gel than agar but there was little difference for anionic dyes.

This type of data may be useful for estimation of optimal plant : medium volume ratios and the theoretical limitation in the rate and duration of mineral supply from gelled media.

8. Uptake by the Plant

In intact plants in vivo we recognize three main mechanisms of mineral uptake; mass flow, driven mainly by the transpiration stream; diffusion along a concentration gradient; and active (energy dependent, selective) uptake against a concentration gradient and across a semipermeable barrier. To what extent are these mechanisms involved in mineral uptake in vitro?

There are some important differences between in vivo and in vitro plants. The in vitro plant often lacks a root system and where roots are present there is doubt about their effectiveness. This difference could effect mass flow because of the reduced water uptake. Roots also have the capacity for selective uptake. The in vitro root environment is also different. The size and nature of the available pool of minerals was discussed in the section on the initial supply.

There have been numerous studies of uptake of minerals by cell cultures (e.g reviews by Dougall, 1980; Heller 1965) but these are only partly applicable to the plant or organ cultures of micropropagation. In cell cultures the cells are usually dispersed within a liquid medium giving direct contact with the nutrient solution. In addition suspended cells may be maintained largely undifferentiated in contrast to tissue or organ explants. The mechanism and kinetics of uptake into cell cultures may be comparable with the final step of assimilation into the tissues of plant or organ cultures.

Even uptake into suspended cells involves more than simple diffusion and appears to be biphasic depending on the availability of low or high concentrations of the ion (Dougall, 1980). The rate of uptake may be correlated with growth e.g phosphate uptake by tobacco cells varied from 17.3 during exponential growth to 135 nmol/g/hr during the stationary phase (Bellamy and Bieleski, 1966).

At this point it is worth emphasizing that minerals and organic molecules, move out of the plant as well as in. This may be an important factor for mass production programmes in that explants can have a significant loss of dry weight following transfer to fresh medium and may take several weeks to regain their initial weight e.g. a 33% decrease in mineral dry weight after 2 weeks for *Ptilotus* shoot cultures and 18% for *Lechenaultia* (Williams, 1992; Williams et.al. unpublished). This suggests an imbalance between the plants and the medium. The observed improvement in growth of cultures where transfer of plants is avoided by the addition of supplementary liquid medium (Maene and Debergh, 1985) may well be due to avoidance of this transitory loss of dry weight.

If diffusion into the plant plays a role in the uptake process there must be gradients from the adjacent medium into the plant. The effective concentration of ions in the medium has been discussed but the magnitude of the gradient also depends on the concentration in the tissues. From the outset it must be remembered that the mineral content of the plant tissues will not be uniformly distributed through the plant. There is also compartmentalization; extra cellular, intracellular; bound and unbound. Thus data from tissue analyses must be interpreted with caution.

Dougall (1980) suggested that if diffusion were important in uptake then the concentration of ions in the medium should be relatively high compared with in the tissues. In our experiments with *Ptilotus* on MS medium a crude calculation of elemental concentrations indicated that the concentration in the medium was 1.2-2.1 times that in the plant except for P where the figure was only about 0.3. These figures do not indicate a strong concentration gradient but they do not take into account compartmentalization within the plant or possible localized depletion of the medium.. What is needed is a measure of the concentration within the free space. Generally ion concentrations are much higher within root cells (the cytoplasm or symplast) compared to the surround spaces. An exception is Ca where the apoplastic Ca level may be about 1000μM while the cytoplasm has only 0.1-1.0μM (Eklund and Eliasson, 1990).

The very low relative concentration of P in the medium, compared to other ions, may partly explain why this element becomes the limiting factor for plant growth in these cultures and typically in other cell (Heller, 1965; Dougall, 1980) and organ cultures (Butcher, pers. com.; Lumsden 1990).

If diffusion through the medium is a limiting factor, uptake and hence growth, should be better on liquid media. However hypohydration or vitrification can nullify the benefit of improved mineral supply.

For *Ptilotus* shoot cultures on agar, the concentration of minerals in the plant and the net uptake, generally increased with the concentration of minerals supplied in the medium (Winney, 1988). Growth was also proportional to the supply of minerals initially but by the eighth week growth had stopped on the high mineral medium (equivalent to MSx2). This inhibition of growth at high supply levels appears to be due to cessation of P uptake by the sixth week (Williams, 1992) even though there were still appreciable supplies of all minerals in the medium at this stage (Williams, 1991). Clearly, it is the availability of the minerals which is limiting growth not the total supply. It is not known why P uptake stopped in the high minerals medium first. Perhaps there is precipitation at the higher supply level?

Lumsden et al (1990) varied the number of plants per container and measured total uptake (N, P, K) per container or uptake per plant. There was an initial increase in uptake of each element with plant number then uptake levelled off progressively; P after 5 plants, N after 10 while K uptake was maintained with 20 plants. Uptake per plant generally declined with increased number again suggesting that the rate of supply of ions was limiting. Is this a reflection of the rate of diffusion of ions through the medium to the plant?

The gradient from the medium to the plant also depends on the concentration in the tissues. Increasing the concentration of minerals in the plant tissues should reduce the concentration gradient from the medium into the plant and so reduce uptake. This was tested with *Ptilotus* by applying a pretreatment to establish high mineral levels in the tissues then subculturing onto fresh medium with different levels of minerals (Kibbler, 1990). High mineral levels in the tissues did reduce subsequent mineral uptake at normal levels of supply. Subculture onto medium with a high supply of minerals overcame this inhibition presumably because an adequate gradient was once again established.

Thus, as suggested by Dougall (1986), a concentration gradient from the medium to the plant appears necessary. In addition, it is not the total supply of minerals in the medium that counts; there must be localized depletion.

Another consideration is the cumulative effect of uptake from one sub-culture to the next. Even where *Ptilotus* shoots were sub-cultured onto the same medium there was a further increase in the concentration of ions in the tissues. Heller (1953) showed decreasing tolerance of carrot callus to higher levels of P over successive subcultures. This could offer an explanation for the decline in culture vigour often encountered after several transfers onto the same, initially favorable medium.

9. Transport Through the Plant

In terms of the net uptake of minerals by the plant, transport within the plant is important because it removes ions from the point of entry, thereby maintaining the relative concentration in the base of the plant i.e. on the tissue side of the plant-medium interface. Presumably once the ions are in the plant they are distributed through the plant by normal transport mechanisms but in vitro conditions may modify the relative activity of these mechanisms, particularly where mass flow is involved.

It is likely that mass flow, driven by the transpiration stream, would be limited by the high humidity within the closed culture vessel. A reduction in humidity in the vessel has been found to increase uptake of calcium, the transport of which is dependent on mass flow due to its low mobility in plants (Geerts et al., 1987). However, this response cannot be simply due to mass flow of ions into the plant because it is a selective process; there is no corresponding increase in other ions which are also available in the medium (Williams, 1992).

For diffusion to contribute to mineral transport through the plant there must a concentration gradient between the base of the plant and the assimilating tissues. This may be maintained by growth creating a sink effect; by assimilation of minerals into bound forms; by compartmentalization of minerals in the tissues; or by loading of the transport system at the base of the plant. The loading of the vascular system occurs in roots but does it occur in rootless shoot cultures? Also, to what extent does the epidermis around shoots in vitro restrict diffusion? Is this part of the reason woody plants have been more difficult in vitro?

10. Empirical Models

There are many questions to be answered regarding the process of mineral nutrition in vitro. Development of a model may facilitate a rational approach to research on the problem. The simplest approach to quantifying the process of mineral nutrition is to look for gross empirical relationships between the concentration of minerals supplied and uptake or growth of the plants. For example, the general correlation between the concentration of minerals supplied in the medium and the concentration found in the plant after 4-8 weeks can be expressed mathematically. Linear regression of crude concentrations in the plant versus the medium (μg/g dwt) for *Ptilotus* produced the following relationship (Williams et al, unpublished):

$$C_p = b + 0.9C_m \quad r^2 = 0.953$$

where

C_p and C_m = concentrations in the plant and medium respectively

This relationship changed slightly over the duration of a culture cycle with the slope = 0.8 and 1.0 for the first and second 4wk periods respectively. The interpretation is that the concentration of minerals in the plant is controlled passively by the supply to the plant rather than by the plant itself. However this gross generalization masks wide differences between individual elements where the slope varied from approximately 20 for K to <0.007 for Zn. Also, in these experiments the concentrations of all elements were varied together.

If the mineral status of the plant (and hence growth) is limited by the supply of one or more minerals from the medium then net mineral uptake (µg) rather than concentration may be a better parameter. A similar empirical equation could be derived for the relationship between mineral concentration and net uptake

There are two main limitations to this empirical approach. Firstly the relationship is only valid for the specific conditions under which it was derived. Testing the relationship under a wide range of conditions may produce 'average' values of the parameters but these would be less accurate in specific cases and insensitive to changing conditions. The second limitation is that such empirical models provide little insight to the underlying mechanisms. A far more useful tool is a mechanistic model.

11. The Mechanistic Model

A mechanistic model has the advantage of providing a tool for developing our understanding of the system. What follows is the outline of a possible mechanistic approach with some examples of the type of mathematical relationships that could be applied to mineral nutrition in vitro.

The process of mineral supply in vitro and the relationships between the main factors are summarized as schematic model in Figure 1. To what extent can we quantify these relationships between the factors or parameters involved? While there are no known comprehensive studies of this type of model, there are some studies directly or indirectly relevant to steps in the overall process.

In order to quantify the passage of minerals from the source to the point of assimilation we need to identify the entities which are moving through the system. These entities are the mineral ions (or ion complexes) in solution. At any point in the system there will be a pool of these available ions. The composition of the pools may change because of inputs, deletions or recombinations enroute. The end fate of the ions is their incorporation into the plant tissues where they no longer participate in the 'available' pools.

As a first step in the evolution of a quantitative model the process is represented as a series of mineral pools at different steps along the pathway (Figure 2). For each ion there will be a concentration S_{1x}, S_{2x} ... in each pool. Three notional sites of potential pool modification are identified, the medium at large, the medium-plant interface, and in the plant tissues where assimilation occurs. At each site the two pools represent different associations of the ions; these associations will generally be reversible. Pools at different sites (i.e. S_2-S_3 and S_4-S_5) will differ primarily in the concentration of the ions. Transport of ions occurs between the sites, firstly through the medium and then through the plant.

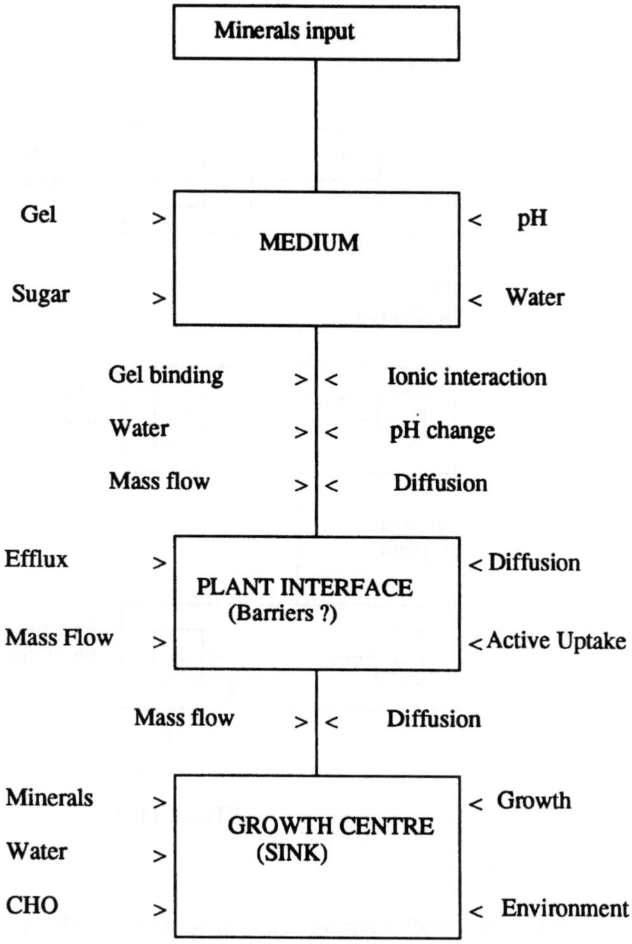

Figure 1. The mineral supply pathway

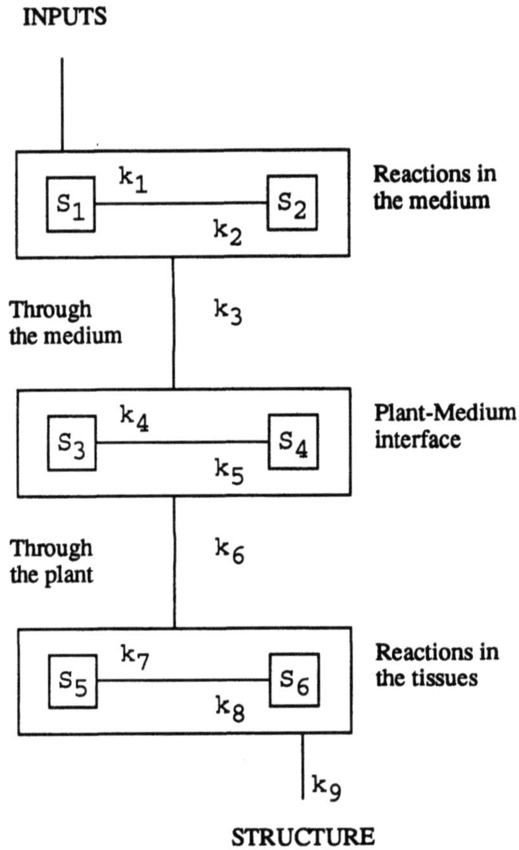

Figure 2. Schematic diagram of the transport path with the ion pools and the resistances (after Thornley and Johnson, 1990)

In this simplistic model there are single resistances or coefficients (for each constituent ion) for conversion between pools. In reality there may be more than one interconversion e.g. within the medium between ion forms of differing solubility or between free and gel-bound complexes. Similarly k_3 and k_6 represent the net effect of various resistances to transport.

With this broad model as a basis we can begin to identify mathematical expressions which may be useful in describing the interactions and responses of component processes. This is illustrated by the following examples of expressions relevant to the first stage, reactions in the medium.

The input, i.e. the relative concentrations of the ions supplied, is known for defined media. This must be adjusted for the mineral components in the gel used. Once in the medium the original composition of minerals may be changed by differences in the degree of dissociation or interaction of ions in solution. To quantify this we can call upon the basic chemistry of solutes.

For each ion the degree of dissociation can be described in terms of the effective concentration γ of the dissociated ion viz:

$$\gamma = a/m$$

where

m = the molality of the ion

a = activity coefficient of the ion

A temperature dependant equilibrium constant k_T can be determined for each ionic reaction in the medium viz:

For the general reaction

$$aA + bB <-> cC + dD$$

$$k_T = [A]^a[B]^b / [C]^c[D]^d$$

These coefficients can be summed for a given medium formulation, at a defined temperature

$$K_T = k_1 + k_2 + k_3 \ldots + k_n$$

where k_1, k_2, ... k_n are the dissociation constants for each of the ionic components of the medium.

Adsorption of ions by a gelling agent may be represented by the Freundlich equation:

$$y = kc^n$$

where

y = g of ion adsorbed per g of medium

c = equilibrium concentration of the ion

k & n are constants (dependent on temperature)

The values of k_1 and k_2 in Figure 2 would be a function of the three above parameters, the effective concentration γ, the equilibrium constant K_T and the adsorption coefficient y.

An expression for k_3, the summation of the transport coefficients for the medium, may be derived in a similar manner. Thornley and Johnson (1990) have defined a number of expressions covering a range of transport processes from simple diffusion to various options of aided diffusion or active transport across membrane barriers. To illustrate this take the case of simple concentration-dependent diffusion following Frick's Law:

$$J = -D\ (dS/dX)$$

where

J = flux density ($kgm^{-2}s^{-1}$) of substrate S in the direction of x
D = the diffusion coefficient (m^2s^{-1})
dS/dX = the gradient

From this:

$$J = -D\ (S_2-S_1)/(X_2-X_1)$$

For a particular culture situation (X_2-X_1) is essentially fixed leaving the concentration difference as the main variable. The diffusion coefficient can be estimated experimentally by measuring the flux down a known gradient.

Another approach to the diffusion process is that taken by Romberger and Tabor (1971) discussed above in the section on ion diffusion. While the equation they used was specifically derived for diffusion of macromolecules through an agar membrane (Ackers and Steere, 1962) the principal of relating diffusion to the physical properties of the gel could be applied to mineral ion transport. Pore size may be a useful parameter to characterize gelled media relative to larger molecules but mineral ions would require incorporation of factors for the electrochemical charges and the degree of dissociation of ions as well as the contribution of mass flow.

These examples are provided to illustrate the mathematical process involved in developing a quantitative model. Space does not permit an exhaustive coverage. There are undoubtedly other expressions developed for mineral nutrient supply from soils which could be utilized.

Like any biological system, the whole process of mineral nutrition is very complex and a model to cope with all the possibilities is impractical. In this treatment I have largely ignored the organic components of the system although their contribution will generally be accounted for in the derived values of the constants or parameters. However, by defining expressions for parts of the overall process we can begin to identify those factors or parameters likely to have a significant effect on the overall outcome. A practical, simplified model based on these can provide a rational basis for predicting responses of the system and hence the development of dynamic control systems.

12. Conclusions

This overview of nutrition in vitro serves several purposes. Firstly it illustrates some of the critical aspects of the procedures used in preparing and using media and suggests some precautions. It also illustrates the complexity of the system. Never-the-less, having identified the factors involved it should be possible to quantify them and therefore allow for them in a mathematical model. Finally, I hope it demonstrates the need for continued research in this field and its potential to contribute to efficient in vitro transplant production systems in the future.

Acknowledgements

I thank my students who have contributed to the experimental work behind this paper and my colleagues Drs Graeme Blair, Ian Johnson and Acram Taji for their stimulating discussions of the topic.

References

Ackers, G. K. and Steele, R. L. (1962) Restricted diffusion of macromolecules through agar-gel membranes. Biochim. Biophys. Acta 59,137-149.

Adams, R. R., Koenigsberg, S. S. and Langhans, R. W. (1979) In vitro propagation of *Cephalotus follicularis* (Australian Pitcher Plants). HortScience 14,512-513.

Aitken-Christie, J. (1990) Automation. in Zimmerman, R. H. and Debergh, P. C. (eds.) Micropropagation, Kluwer Scientific, Dordrecht.

Aitken-Christie, J. and Jones, C. (1987) Towards automation: Radiata pine shoot hedges in vitro. Plant Cell Tissue and Organ Culture 8,185-196.

Ball, E. (1953) Hydrolysis of sucrose by autoclaving media, a neglected aspect in the technique of culture of plant tissue. Bulletin of the Torrey Botanical Club 80,409-411.

Bellamy, A. R. and Bieleski, R. L. (1966) Some-salt uptake and tissue-aging phenomena studied with cultured tobacco cells. Australian J. Biological Science 19,23.

Butenko, R. G. (1968) Plant tissue culture and plant morphogenesis. Israel Program for Scientific Translations, Israel.

Constable, F. (1963) Untersuchungen uber extracellulare hydrolyse von kohlenhygraten. Planta 59,330-337.

Congard, B., Beaujard, F. and Viemont, J-D. (1986) Les bruyeres in vitro VI Croissance de Calluna vulgaris sur milieu strictement nitrique ammoniacal et cinetique du pH en fonction du developpement des plantes. Canadian J Botany 64,959-964.

Dalton, C. C., Iqbal, K. and Turner, D. A. (1983) Iron phosphate precipitation in Murashige and Skoog media. Physiologia Plantarum 57,472-476.

Debergh, P. C. (1983) Effects of agar brand and concentration on the tissue culture medium. Physiologia Plantarum 59,270-276.

Debergh, P., Harbaoui, Y. and Lemeur, R. (1981) Mass propagation of globe artichoke (*Cynara scolymus*) : Evaluation of different hypotheses to overcome vitrification with special reference to water potential. Physiol Plantarum 53,181-187.

Dougall, K. D. (1980) Nutrition and metabolism, in, E. J. Staba, (ed.), Plant Tissue Culture as a Source of Biochemicals, CRC Press, Florida, pp. 21-58.

Ebert, A. and Taylor, H. F. (1990) Assessment of the changes of 2,4-dichlorophenoxyacetic acid concentrations in plant tissue culture media in the presence of activated charcoal. Plant Cell Tissue and Organ Culture 20,165-172.

Elmore, W. H., Samples, B., Sharma, S. and Harrison, M. (1990) Influence of cultural and physiological factors on ascorbate stability in plant tissue cultures. Plant Cell Tissue and Organ Culture 20,131-135.

Eklund, L. and Eliasson, L. (1990) Effects of calcium ion concentration on cell wall synthesis. J. Experimental Botany 41,863-867.

Gamborg, O. L. Miller, R. A. and Ojima, K. (1968) Nutrient requirements of suspension cultures of soybean root cells. Experimental Cell Research 50,151-158.

Gamborg, O. L., Shyluk, J. L. (1981) Nutrition media and characteristics of plant cell and tissue cultures, in, T. A. Thorpe (ed.), Plant Tissue Culture Methods and Applications in Agriculture, Acad Press, pp. 21,44.

Gautheret, R. J. (1959) La culture des tissues vegetaux. Masson et Cie., Paris.

Geerts, M. Van Cleemput, O. and Debergh, P. C. (1987) Translocation of calcium in micropropagated Prunus avium L. Med. Fac. Landbouww. Rijksuniv. Gent 52,1449-1456.

George, E. F., Puttock, D. J. M. and George, H. J. (1987) Plant culture media. Vol 1. Formulations and Uses. Exergetics Limited, Westbury, England.

Hagen, S. R., Muneta, P., Augustin, J. A. and LeTourneau, D. (1991) Stability and utilization of picloram, vitamins, and sucrose in a tissue culture medium. Plant Cell Tissue and Organ Culture 25,45-48.

Hangarter, R. P. and Stasinopoulos, T. C. (1991) Effect of Fe-catalized photooxidation of EDTA on root growth on plant culture media. Plant Physiology 96,843-847.

Heller, R. (1965) Some aspects of the inorganic nutrition of plant tissue cultures, in, White, P. R. (Ed.), Proceedings of the International Conference on Plant Tissue Culture, McCutchan, Berkley, pp. 1-17.

Heller, R. (1953) Recherches sur la nutrition minerale des tissus vegetaux cultives in vitro. Ann. Sc. Nat. Bot., 11th Ser, 1-223.

Kibbler, H. (1990) Nutritional problems of establishing Sturt's Desert Pea in vitro and investigation of the carryover effect using *Ptilotus exaltatus*. Dip.Hort.Sci. thesis, University of New England, Australia.

Johnson, B. J. (1989) Towards automation of crop tissue culture. Australian J Biotechnology 3,278-280.

Lloyd, G. and McCown, B. (1981) Commercially-feasible micropropagation of Mountain laurel, *Kalmia latifolia*, by use of shoot tip culture. Proc. International Plant Propagators Society 30,421-427.

Lumsden, P. J., Pryce, S. and Leifert, C. (1990) Effect of mineral nutrition on the growth and multiplication of in vitro cultured plants, in, Nijkamp, H. J. J., van de Pals, L. H. W. and Aartrijk, J. (eds.), Progress in Plant Cellular and Molecular Biology, Kluwer, pp. 108-113.

Maene, L. J. and Debergh, P. C. (1985) Liquid medium additions to established cultures to improve elongation and rooting in vivo. Plant Cell Tissue and Organ Culture 5,23-33.

Maene, L. J. and Debergh, P. C. (1987) Optimalization of the transfer of tissue cultured shoots to in vitro conditions. Acta Horticulturae 212,335-348.

Marschner, H. (1986) Mineral nutrition of higher plants. Academic Press, London.

Marschner, H., Romheld, V., Horst, W. J. and Martin, P. (1986) Root induced changes in the rhizosphere. Importance for the mineral nutrition of plants. Z. Pflanzenernaehr Bodenk 149,441-456.

Murashige, T., Skoog, F. (1962) A revised medium for rapid growth and bioassays with tobacco tissue cultures. Physiol Plantarum 15,473-497.

Nissen, S. J. and Sutter, E. G. (1990) Stability of IAA and IBA in nutrient medium to several tissue culture procedures. HortScience 25,800-802.

Parfitt, D. E., Almedhi, A. A. and Bloksberg, L. N. (1988) Use of organic buffers in plant tissue-culture systems. Scientia Horticulturae 36,157-163.

Romberger, J. A., Tabor, C. A. (1971) The *Picea abies* shoot apical meristem in culture. I Agar and autoclaving effects. American J Botany 58,131-140.

Sarma, K. S., Maesato, K., Hara, T. and Sonoda, Y. (1990) Effect of method of agar addition on autoclave pH of the tissue culture media. Annals of Botany 65,37-40.

Schenk, N., Hsiao, K. C. and Bornman, C. H. (1991) Avoidance of precipitation and carbohydrate breakdown in autoclaved plant tissue culture media. Plant Cell Reports 10,115-119.

Schenk, R. U. and Hildebrandt, A. C. (1972) Medium and techniques for induction and growth of monocotyledonous plant cell structure. Canadian Journal of Botany 50,199-204.
Scherer, P. A., Muller, E., Loppert, H. and Wolff, G. (1988) Multielement analysis of agar and gelrite impurities investigated by inductively coupled plasma emission spectrometry as well as physical properties of tissue culture media prepared with agar or the gellan gum gelrite. Acta Horticulturae 226,655-658.
Singha, S., Oberly, G. H. and Townsend, E. C. (1987) Changes in nutrient composition and pH of the culture medium during in vitro shoot proliferation of crab apple and pear. Plant Cell Tissue and Organ Culture 11,209-220.
Skirvin, R. M., Chu, M. C., Mann, M. L., Young, H., Sullivan, J. and Fermanian, T. (1986) Stability of tissue culture medium as a function of autoclaving time and cultured plant material. Plant Cell Reports 5,292-294.
Thornley, J. M. and Johnson, I. R. (1990) Plant Crop Modelling. A mathematical approach to plant and crop physiology, Clarendon Press, Oxford.
White, P. R. (1963) The cultivation of animal and plant cells. 2nd edit., Ronald Press, New York.
Williams, R. R. (1991) Factors determining mineral uptake in vitro. Acta Horticulturae 289,165-166.
Williams, R. R. (1992) Mineral nutrition in vitro - A mechanistic approach. Australian J Botany, in press.
Williams, R.R., Taji, A. M., Bolton, J. A. (1985) Specificity and interaction among auxins, light and pH in rooting of Australian woody species in vitro. HortScience 20,1052-1053.
Williams, R. R., Taji, A. M. and Winney, K. A. (1990) The effect of *Ptilotus* plant tissue on pH of in vitro media. Plant Cell Tissue and Organ Culture 22,153-158.
Winney, K. A. (1988) The effect of nutrient concentration and media pH on nutrient balance and uptake in *Ptilotus exaltatus* in vitro. B.Sc.Hons thesis, University of New England, Australia.

ENVIRONMENTAL AND HORMONAL EFFECTS IN MICROPROPAGATION

PAUL E. READ
Department of Horticulture
University of Nebraska - Lincoln
Lincoln, NE 68583-0724
USA

ABSTRACT. Micropropagation has become important in the latter quarter of the Twentieth Century, both in a commercial sense and as an exceedingly valuable research tool for the study of plant physiology, growth and development. A myriad of papers have been published on a wide variety of subjects related to micropropagation including the critically important areas of environment and hormones. In this paper, special attention is given to the influence of environment and hormones, and their interactions as they affect the stock plant and the subsequent performance of the explant when cultured in vitro. Explant selection (genotype, organ chosen, age and physiological stage, health), stock plant management (nutrition; light intensity, quality and photoperiod; temperature; chemical and physical manipulations), gaseous atmosphere (RH, CO_2, O_2, ethylene) and both conventional and novel plant growth regulating chemicals are topics presented. These topics are discussed in relation to future challenges and opportunities, such as automation and robotization of micropropagation technology.

1. Introduction

Ever since Haberlandt first explored plant cell culture in 1902, scientists have assiduously sought to define those conditions requisite to successful production of viable plant cell, tissue and organ cultures. Although Haberlandt probably intended to use plant cell culture as a means to study the concept of totipotency and to explore morphogenesis, it is doubtful that he could have envisioned the dramatic progress achieved in the ninety years since his historic publication. The most practical and commercialized impact of the many possible uses of plant tissue culture is the use of micropropagation to achieve rapid and economically feasible mass propagation of desirable genotypes. An excellent early review of plant micropropagation was written by Murashige (1974) in which he provides eloquent descriptions of the stages of the process and details of appropriate medium constituents. Indeed, he lists information relevant for micropropagation of over 100 genera from more than 50 families. Many factors contribute to the successful micropropagation. Most of these critical factors can be addressed under the headings of environmental factors and hormonal or plant growth regulant (PGR) factors. Both environment and PGR contribute greatly to the success of the plant tissue culture system and much progress has been made toward a better understanding of their roles since the successful culture of callus by Nobecourt and Gautheret in France and by White in the U.S. (White, 1963). It is impossible to give proper credit to the myriad dedicated scientists who have helped achieve the great strides that have contributed to the growth and application of the technology of micropropagation. However, some of the most significant contributions will be presented in the following categories: The Stock Plant, Treatment of the Explant, The culture Environment and Hormones and Plant Growth Regulators.

K. Kurata and T. Kozai (eds.), Transplant Production Systems, 231–246.

2. The Stock Plant

Discussion of the influence of the stock plant (donor plant, mother plant) can be divided into two general categories, explant selection, and stock plant treatment, or perhaps more correctly, stock plant management. Asexual methods of plant propagation have been practiced for millennia. The propagator, especially in recent history has learned to select specific plant parts and to manage stock plants to facilitate effective plant propagation by cuttings, grafting and layering. It is therefore quite logical that many of the principles and practices that enhance the success of micropropagation techniques could therefore be applied to a better understanding and more effective practices in micropropagation (Read, 1990).

Indeed, cutting propagation has been demonstrated to be influenced by many different stock plant factors including nutrition, light, temperature and physical and chemical treatments. Impetus was given to development of understanding the influence of stock plant nutrition by the pioneering work of Kraus and Kraybill (1918). Their demonstration that carbohydrate-to-nitrogen ratio was important was corroborated by subsequent research, including that of Starring (1924) and Brandon (1939). Interaction of nutrients and light (Dhillon, 1963), macro-and micronutrients (Haun and Cornell, 1951; Samish and Speigel, 1957, Weiser and Blaney, 1960), etiolation and reduced light intensities (Gardner, 1937; Kawase, 1965; Biran and Halevy, 1973; Hanson, 1976; Moe, 1976), photoperiod (Bachelard and Stowe, 1963; Henny and Read, 1971; Heins and Wilkins, 1979), temperature and growth-retardant treatments (Heide, 1969) and physical manipulations (Stoltz and Hess, 1966) are all stock plant treatments that have been reported to modify the rate of propagation success.

2.1 EXPLANT SELECTION

In his frequently referenced treatise on the use of plant tissue culture for propagation purposes, Murashige (1974) refers to the importance of proper explant selection and indicates that almost all types of organs can be potentially successful explants. It is important to consider a number of factors when selecting explants, including the genotype, organ chosen, chronological and physiological age, season of year, stage of growth, and overall health of the stock plant serving as the source of explants.

2.1.1. Genotype. It is clear that even in species considered to be relatively easy to culture in tissue culture, the genotype of the plant may exert a significant influence on culture results. For example in "model" species such as Nicotiana, Cheng and Smith (1973) found noticeable differences among genomes for autonomous growth of pith cultures. Similarly dramatic differences in shoot numbers were produced from leaf segment cultures of a range of petunia cultivars (Read, et al, 1979). In more recent work with Brussel sprouts (Brassica oleracea var. gemnifera), Williams, et al (1990) found differing levels of shoot proliferation among three genotypes tested. In petiole callus cultures of cotton (Gossypium hirsutum L.), distinctly different embryogenic abilities were found among genotypes (Gawel and Robacker, 1990). Furthermore, Jaramillo and Summers (1990) have shown that genotype affected both number and diameter of calli produced by tomato (Lycopersicon esculenturn Mill.) anther cultures.

2.1.2. Organ Chosen. As mentioned earlier, Murashige (1974) indicates that a wide range of organs and tissues can be used as explants for in vitro culture. He also points out that the explant may vary from species to species, suggesting that stem tips were best for petunia, for example and that the cotyledon has been the most likely part of seedlings to regenerate (see also Doerschung and Miller, 1967 and Cade, et al, 1990). Inflorescences, or parts of inflorescences are often good explants, as in the case for Festuca arundinacea (Eizenga and Dahleen, 1990). Similarly, receptacle tissues of garlic (Allium sativum L.) proved to be the preferred explant for production of bulbs in vitro (Matsubara, et al, 1990). In the case of embryoid formation from protoplasts of Helianthus annuus L., only protoplasts derived from hypocotyl tissue divided, and that even the part of the hypocotyl was important for embryogenesis to occur (Dupuis, et al, 1990). In elegant studies with thin cell layer techniques with tobacco, Tran Thanh Van (1973a, 1973b) demonstrated that the production of different organs such as flowers, shoots and roots was dependent on the part of the plant from which the thin cell layer was taken and the hormone concentrations. However, it is well to remember that the explant may also lead to formation of plantlets distinctly different in phenotype from that of the donor plant, as seen in work by Bush, et al (1976) in which they obtained plants with different flower types and colors when plantlets were regenerated from ray flower segments of chrysanthemum.

2.1.3 Age and Physiological Stage. Chronological age of the potential explant tissue and season of the year that the explant is obtained are important only insofar as they influence the extent of differentiation of the cells and their physiological age. It is clear, for example, that in woody plants, cells which have cell walls that have fully lignified are virtually impossible to culture in vitro. In this regard, Geier, et al (1979) have commented that in general a preexisting low organogenic competence cannot be completely compensated for by manipulating the composition of the medium. Many researchers have demonstrated an "age" effect, such as Pierik's (1970) report that rooting would occur readily in young Lunaria leaves, but not in older leaves. Similarly, Hackett (1969) showed that regeneration ability of bulb scales of Lilium was highest for the older (outer) scales and decreased for scales taken from the inner, younger part of the bulb. Kane and Albert (1989) also found inherently greater regenerability in adult leaves of Myriophyllum heterophyllum than for juvenile leaves. This is in contrast with the frequent observations that parallel the phenomena demonstrated for Hedera helix cuttings that juvenile material will root more readily than mature cuttings and that tissue cultures of juvenile English ivy are much more organogenic than are cultures of mature tissue. Hackett (1983) has discussed such phase change factors in detail, so no further discussion will be provided here.

2.1.4. In Vitro-Derived Explants. In recent years, many investigators have reported that in-vitro derived explants often are more regenerative than explants taken from field- or greenhouse-grown stock plants. This was shown for the micropropagation ability of azalea (Economou and Read, 1986), strawberry (Liu and Sanford, 1988) and grape (Stamps, et al, 1990). Furthermore, in blueberry (Vaccinium spp., Grout and Read, 1986), tissue culture-derived plants had a "carry-over" of ability to enhance propagation ability of both cuttings and in vitro cultures. For protoplast cultures, aseptically cultured plantlets were used as source materials for a number of vegetable species, including members of the Solanaceae, Brassicaceae and Cucurbitaceae (Nishio, et al, 1989). It should be noted, however, that Lu, et al (1991) reported that their best success with wheat anther cultures was attained when the anthers to be cultured were taken from field-grown plants. Nevertheless, it can be generally accepted that in most cases, in-vitro derived explants have undergone apparent physiological changes conducive to better in vitro performance.

2.1.5. Explant Health. It is generally accepted that, as is true for stock plants used as donors for propagules to be used for macro-propagation, disease-free stock plants are preferred for explants to be used for in vitro culture. Morel's classic work on use of meristem culture to produce virus-free plants (Morel, 1952, 1955) may be helpful in some cases, but the micropropagator otherwise must be careful to select explants from disease-free stock plants. If culture of diseased explants is successful, the result is large numbers of diseased plants. Furthermore, as Matsubara, et al (1990) have shown for garlic, greater production results from explants taken from disease free plants. In their work, a larger number and heavier bulblets were produced by explants taken from virus-free plants than for those taken from virus-infected stock plants.

2.2. STOCK PLANT MANAGEMENT

Treatments and manipulations applied to the stock plant allow the micropropagator to exert a purposeful control over the potential outcome of culture of the explants taken from such stock plants. In addition to the obvious requirement that the explant material be thoroughly disinfested in order to prevent contamination and loss of the cultures, several other factors can be managed to control and modify the physiological and physical condition of the stock plant. These include nutrition of the stock plant, temperature and light conditions under which the stock plant is grown, and direct physical and chemical treatments of the stock plant and of the explant itself.

2.2.1. Stock Plant Nutrition. It is logical that the use of healthy plant material is important and that the mineral nutrition of the stock plant contributes to this health. As was found by Kraus and Kraybill (1918), tomato cuttings taken from plants grown under a high nitrogen regime rooted less well than those taken from plants that received a lower nitrogen level. Similarly, Read, et al (1979) found lesser shoot proliferation from leaf segments taken from tomato plants grown under a high level of nitrogen fertilizer. A very low level of nitrogen fertilizer was also deleterious to shoot production and the leaf nitrogen levels seemed to be related to the ability of leaf segments to produce shoots in vitro. Nitrogen and potassium levels in the nutrient solution used for growing stock plants of several taxa of Salix were found to exert a strong effect on organogenic response of explants cultured in vitro from such stock plants (Garton, et al, 1983). However, manipulation of the stock plant nutrition was not sufficient to overcome the propensity of certain taxa for poor shoot proliferation.

2.2.2. Temperature Under Which The Stock Plant is Grown. Although culture temperature conditions have been frequently studied, little literature is available on the effects on cultures resulting from stock plant temperature treatments. An interaction was shown by Heide (1965) between stock plant temperature and growth regulator levels in begonia leaf segment cultures and Appelgren and Heide (1972) found that adventitious bud and root formation was stimulated when leaf segments of streptocarpus were cultured from stock plants grown at 12°C. According to Pierik and Ruibing (1973), the number of hyacinth bulblets was not affected by storage of the source bulbs at 25°C, but bulblet weight was increased if the bulbs were stored for 90 days. Ko (1986) found a more profound effect when storing hyacinth bulbs used as explant sources at various temperatures and lengths of storage. These limited number of reports on the effects of stock plant temperature on subsequent performance in vitro suggests that there is potential for further research in this area of study.

2.2.3. Stock Plant Light Conditions. Concern for light intensity applied to the stock plant was given little and infrequent attention in the early years of plant tissue culture work. Dunwell and

Perry (1973) found that a greater yield of haploid plantlets could be obtained from anthers taken from tobacco plants grown under high light intensities, while Hughes, et al (1978) found that a high yield of protoplasts could be produced from barley stock plants grown under low light intensities. Both of these reports were with plants grown under short-day conditions, but Heide (1968) illustrated a positive effect of long days for begonia leaf cultures and Paterson, et al (1979) showed an increased regeneration ability for explants from jade plants held in darkness on short days.

Economou and Read (1986a, 1986b) demonstrated profound effects of light duration, intensity and quality applied to stock cultures on subsequent performance of explants taken from such cultures. For example, if stock cultures were held under low light intensities (10μM/(s·m^2)] rooting of microcuttings was better than for microcuttings taken from stock cultures held at 30 or 75μM/(s·m^2). The intermediate light intensity led to a greater number of shoots produced that rooted satisfactorily, so it was deemed to be the overall better treatment. In grapes, Yu and Meredith (1986) found a similar effect, since explants from shaded shoot tips proliferated shoots better than those from full sun. As early as 1951, Mes suggested that loss of endogenous auxin could be caused by high light intensity, thus providing a possible explanation for the reduced rooting ability of microcuttings produced at high light intensities.

Rooting of azalea microcuttings was promoted when they were produced under a stock culture regimen of two weeks of far red light followed by two weeks of red light. (Read and Economou, 1982). Read et al (1978) had earlier demonstrated an increase in shoot proliferation from petunia leaf discs taken from stock plants receiving 30 minutes of red light as end-of-day treatments. A possible explanation of these phenomena as well as reports of shaded leaves or plant parts having an enhanced ability to produce more shoots could be that red light increases cytokinin availability and/or reduces auxin level (Fridborg and Eriksson, 1975; Read et al, 1978). These reports and others relating in vitro responses to light levels, duration and quality suggest that stock plant light treatments may be a fertile field for further study.

2.2.4. Chemical and Physical Manipulations of the Stock Plant. There has been a vast amount of literature published on the subject of plant growth regulators ("hormones") in the culture medium. However, there are not as many examples of literature discussing plant growth regulator treatments of the stock plant. DeLanghe and deBruijne (1976) successfully increased shoot proliferation of tomato leaf segments cultured in vitro by spraying the stock plant with chlormequat chloride (CCC, Cycocel). Read, et al (1979) confirmed this effect and also demonstrated increased callus production by dahlia leaf segments taken from stock plants sprayed with daminozide (B-Nine). Also, more shoots were produced by petunia leaf segments taken from stock plants receiving a 2000 ppm daminozide spray when compared with explants from water-sprayed stock plants (Economou and Read, 1980).

Pre-culture treatments of the explants have been demonstrated to profoundly influence organ proliferation in vitro. For example, dipping petunia leaf segments, leaves or other explants in cytokinin solutions such as benzyladenine prior to culture (Read et al, 1978; Economou and Read, 1980, 1982) has resulted in production of shoot numbers equivalent to, or higher than numbers expected if the cytokinin had been incorporated into the medium in the customary fashion. While treating excised shoots with ABA or NAA was only slightly beneficial to embryo production by alfalfa protoplasts (Dijak and Brown, 1987), von Arnold and Tilberg (1987) found that pulsing buds of Picea abies with 250 μM BA for three hours led to the highest adventitious bud yield of all treatments tested.

The use of forcing solutions has been shown to be a novel and effective method of incorporating plant growth regulating chemicals into potential explants of woody plant material (Read et al, 1984, 1986). The forcing solution technique involves immersing the bases of dormant cut woody stems in solutions similar to those used to extend keeping quality and longevity of cut flowers such as roses, carnations and chrysanthemums. The typical forcing solution contains 55-60μM sucrose and approximately 600μM 8-hydroxyquinoline citrate. To this solution chemicals intended to influence bud break and in vitro response may be added, e.g., GA_3 for bud break and cytokinins and/or auxins for organogenic responses. Responses have been reported for a large number of genera (Yang and Zhang, 1987, Yang and Read, 1990). Table 1 illustrates an example of bud break effects caused by placing GA_3 in the forcing solution. For best results with the forcing solution technique, heavy disinfestation of the stems prior to forcing was employed. This practice led to the discovery that such treatments could also enhance bud break beyond that normally expected (Read and Yang, 1989).

TABLE 1a. Effect of GA3 in the forcing solution on days to bud break of *Vanhoutte spirea* harvested and forced October, 1988.

GA3 (mg/l)	0	1	10	20	30	50	100
Days to Bud Break [z]	9.8a	5.5b	5.8b	5.5b	4.0c	3.8c	4.0c

z Means followed by the same letter are not significantly different at the 0.05 level (Duncan's multiple range test).

TABLE 1b. Effect of BA inclusion for 3 days in the forcing solution on shoot initiation of *Vanhoutte spirea* cultured on PGR-free Linsmaier & Skoog medium.

BA (mg/l)	% of Explants Producing Shoots	Number of Shoots/explant
10	55.7a [z]	3.0a
1	43.3a	1.2b

z Means within column followed by the same letters are not significantly different at the 0.05 level (Duncan's multiple range test). Explants were taken from softwood growth of stems forced in water containing 200 mg 8-HQC per liter plus 2% sucrose.

Physical treatments of stock plants, such as pruning to "rejuvenate" or invigorate the stock plant has been practiced to stimulate new shoot growth that would be suitable for rooting as cuttings. This approach, along with sequential pruning and grafting has enabled researchers to produce more "juvenile" growth conducive to better in vitro response (Franclet et al, 1987). Likewise, wounding has been shown to induce shoots to form from explants of Torenia fournieri that normally would not produce shoots (Takeuchi et al, 1985). A similar effect was observed by Van Aartrijk and Blom-Barnhoorn (1983) when they were able to disrupt the normally basipetal polarity of the bulb regeneration process in Lilium by removing the abaxial epidemis of bulb scale explants prior to in vitro culture.

3. Culture Environment

The environmental conditions most frequently studied for plant tissue cultures are temperature; light intensity, duration or photoperiod, and quality; and atmospheric constituents.

3.1. TEMPERATURE.

Optimum temperatures for plant tissue cultures have been investigated by several researchers and a range of 20-27°C has been employed most often. Murashige (1974) commented on temperature in which the culture is placed. As is true for intact plants, growth and development is strongly influenced by temperature. Its influence on basic, physiological processes such as respiration, and on cell and organ formation is generally accepted.

In plant tissue culture, optimum temperature may range as high as 32-35°C (Hughes, 1981), while reports of temperatures as low as 12°C have been reported as optimum for streptocarpus (Appelgren and Heide, 1972). Using various species of begonia, several workers have reported varying responses to temperature. Optimum temperature ranging from 15 - 24°C have been demonistrated (Heide, 1965, Fonnesbech, 1974). However, at 24°C, cytokinin-induced shoot formation was suppressed. Formation of specialized organs in vitro has also been shown to be temperature-dependent. Tuber formation in potato (Wang and Hu, 1982) and bulblet formation from scales of Lilium (Stimart and Ascher, 1978) have been intensively studied. The latter researchers also reported that alternating temperatures during culture could affect bulblet formation when compared to use of constant temperatures (Stimart and Ascher, 1981).

Heat treatment of intact stock plants prior to in vitro culture is a now-standard treatment to produce plants free of specific pathogens following the pioneering work of Morel (1952, 1955). A thorough review of this approach has been published by TenHouten, et al (1968). In work with dahlia, use of in vitro heat treatment to obtain potentially virus-free meristems has also been investigated (Read, 1989). The meristem thus obtained is very small, but can be successfully cultured. Such in vitro heat treatment may even be used following conventional heat treatments, thus gaining even greater leverage in creating disease-free plants.

3.2. LIGHT

The effects of light on various responses in vitro has been of interest to many research workers in the field of plant tissue culture. Light intensity and duration, including photoperiod, and to a lesser extent, light quality have been studied.

3.2.1. Light Intensity. Since light intensity is a critical factor in photosynthesis, it might be expected that light intensity would be important for in vitro culture of leafy and green explants. However, rather than a primary effect on photosynthesis, early work focussed on physiological effects. Werckmeister (1971) suggested that "full" light intensities were important for the shoots, but that darkness was required around the roots for best results with Cymbidium. He accomplished dark treatment for the roots by inclusion of activated charcoal in the medium, a technique frequently noted in the literature. Pennazio and Redolfi (1973) demonstrated that light intensity was important in the culture of potato, with 4000 lux considered optimum. Hasegawa, et al (1973) obtained best spear production with asparagus cultures given high light intensities, while Pierik, et al (1979) showed that light is detrimental to organ formation in Anthurium andraeanum. The influence of light intensity seems to be related to species, with some benefitting from high intensities, others responding to intermediate levels, while still others are best cultured under low light or darkness (Papachatzi et al, 1981; Miller and Murashige, 1976; Thorpe and Murashige, 1970). Reports for a wide range of species generally place the optimum light intensities for tissue culture of woody plants at about 70 to 80μM/(m^2·sec) (Economou, 1982; Hutchinson, 1984; Villalobos et al, 1984; Read, 1989). However, both high and low light levels have been reported to depress shoot proliferation for species such as black currant and azaleas (Wainwright and Flegman, 1984; Economou and Read, 1987). Furthermore, Hammerschlag (1982) demonstrated an advantage for a dark treatment for rooting of in vitro derived shoots of Prunus spp. (Hammerschlag et al, 1987).

3.2.2. Photoperiod and Light Quality. As is true for intact plants, photoperiod and light duration can produce important physiological responses. In a typical short-day plant, i.e., one that flowers in response to short days (or long nights), it is logical to expect explants that have been taken from long day plants would be preferred. Just as in rooting of cuttings, vegetative tissue should respond better, since there is no floral development to divert metabolites from the vegetative tissue (Read et al, 1978). For this reason and because of possible benefits from photosynthetic activity, a period of 12 to 15 hours of illumination seems to be preferred for most species. These day lengths are the photoperiods expected to be optimum for intact plants. With dahlia cultures, Gavinlertvatana (1979) reported an interaction between short day culture periods and treatment of the stock plant with daminozide prior to culture. However, few direct physiological relationships between photoperiod and culture response have been reported, in spite of the aforementioned reports of Paterson and Rost (1979) and Dunwell and Perry (1973).

Light quality would be expected to have an impact on in vitro cultures, since red/far red light and blue light have been implicated in a number of physiological responses in intact plants, such as flowering, tropisms and other phenomena. As mentioned earlier, enhanced shoot proliferation in vitro was found when explants were taken from stock plants treated with red light (Read et al, 1978). Red light and far red light applied as part of the culture regimen led to better rooting of azalea microcuttings; those taken from cultures grown under two weeks of red light following two weeks of far red light rooted more readily than those cultured under conventional cool-white fluorescent lights (Economou, 1982; Read and Economou, 1982). Blue light has been implicated in both stimulation (Ward and Vance, 1967) and inhibition (Seibert et al, 1975). Furthermore there have been numerous other conflicting reports regarding the relative value and role of light of various wave lengths on in vitro responses (Kadkade and Jopson, 1978; Economou, 1982) and their interaction with cytokinins in the medium (Villalobos et al, 1984; Husemann and Reinert, 1976; Badilla et al, 1985). Although these and most reports relate to a response by the tissue to specific light qualities, a recent report by Hangarter and Stasinopoulos (1991) suggests that evidence exists that light of differing qualities can cause changes in the constituents of the medium. It seems obvious therefore, that much more research is yet to be done to elucidate the effects of various qualities of light on in vitro cultures.

3.3. GASES

The atmospheric constituents adjacent to the culture vessel can profoundly affect the response of plant tissues cultured in vitro. Oxygen, carbon dioxide and ethylene have been the most frequently investigated gaseous constituents in the culture environment. Humidity has also been a subject of concern, since it is imperative that the succulent tissues found in culture are protected from desiccation. For example Debergh et al (1981) have pointed out that a relatively stiff medium resulting from an agar level of around 1.0% will have a lower relative humidity in the atmosphere than found for a medium made with less agar.

Kozai (1988, 1991) has pioneered efforts to employ micropropagation under photoautotrophic conditions. He has clearly demonstrated that CO_2 can be limiting in the culture vessel and that use of gas-permeable closures, larger vessels with a CO_2 supply system and CO_2 enrichment in the culture room can all contribute to greater success in micropropagation and at reduced cost. Critical to the success of Kozai's scheme is use of little or no sucrose (a major cost item in micropropagation) and use of high light intensities. Both high light intensities and increased CO_2 contribute to more rapid photosynthesis and thus more productive micropropagation systems. Furthermore, besides reducing cost, elimination of sucrose helps to minimize the likelihood of contamination by microorganisms. Assuming that larger vessels could be used with liquid culture instead of a gelled medium, flow-through medium and gaseous systems are envisioned. Such systems would almost certainly facilitate automation or robotization of all or parts of the micropropagation system. It seems likely that much more research in manipulating plant micropropagation systems by pursuit of photoautotrophic approaches will be forthcoming.

In general, oxygen is rarely limiting for plant tissue cultures, although it is critical for respiration and numerous other metabolic activities. However, Van der Plas and Wagner (1986) reported improved growth and oxygen uptake by potato tuber callus when cultured in an atmosphere containing 70% oxygen. In contrast, Hegerty et al (1986) reported a detrimental effect on Catharanthus roseus suspension cultures when high rates of aeration were employed.

Ethylene has been implicated as an agent that can reduce growth of plant tissue cultures and it has been shown that auxin contributes greatly to the evolution of ethylene.

The work of Gavinlertvatana et al (1979) has clearly demonstrated that elevated levels of ethylene reduced growth for both dahlia and petunia callus cultures. However, even when the excess ethylene levels were removed by use of potassium permanganate crystals suspended in the culture vessel, the ethylene-induced growth suppression could not be alleviated. Furthermore, when high rates of CO_2 were used to reduce the deleterious effects of ethylene, only minimal success was achieved (Van der Plas and Wagner, 1986).

It is apparent that much work remains to be done in determining optima for various gaseous components of the culture atmosphere and to minimize negative effects of toxic gases such as ethylene. Such research should likely be combined with the approach suggested by Kozai (1991), Aitken-Christic (1988), and others to aid in development of efficient and economically feasible micropropagation systems.

4. Hormones/Plant Growth Regulators in the Medium

Murashige's excellent review (1974) which was mentioned earlier, details many aspects of the culture medium, and much of the other material presented in this paper alludes to the practical applications of the use of auxins and cytokinins in successful micropropagation systems. Many tissue culturists have commented on the need for both substances in tissue culture media and the critical nature of auxin:cytokinin balance or ratio was beautifully elucidated by Skoog and Miller (1957). Various compilations of rates and choices of auxin and/or cytokinin have been presented, including that of Murashige (1974), George and Sherrington (1984) and deFossard (1981). Because of these references and the availability of a wealth of additional literature on the subject, no further reference to the role of conventional auxius and cytokinins will be made.

The use of combinations of several auxins and/or cytokinins in media for preliminary studies is espoused by deFossard (1981). A good example of the effectiveness of this approach is presented by Tabei and Kanno (1989) where they showed that the best rates of embryogenesis in cucumber cotyledon cultures were induced by a combination of 2,4-D and NAA. Such combinations were superior to either auxin used alone. Similarly for cytokinins, Goussard (1982) demonstrated that a combination of zeatin riboside and 6-benzylaminopurine was best for the culture of shoot apices of grapevine.

Other chemicals possessing cytokinin-like attributes have been shown to be potent agents for stimulating responses expected for conventional cytokinins. Substituted phenylureas such as thidiazuron and N-(2-chloro-4-pyridyl)-N'-phenylurea (CPPU) are examples of this active class of compounds. Both thidiazuran and CPPU have been used in our laboratory (Fellman et al, 1987) and thidiazuron was used successfully by Swartz et al (1990) for Rubus and Malus leaf cultures and by Ellis et al (1991) in adventitious bud formation in Picea glauca. Similarly, other auxin-like plant growth regulators have been employed as in the case of the use of pichloram (4-amino-3,5,6-trichloropicolinic acid) and dicamba (3,6-dichloro-2-methoxybenzoic acid) (both potent herbicides) for callus growth and embryogenesis in date palm (Omvar and Novak, 1990).

GA_3 has been used in the medium to stimulate shoot elongation, especially for cases where high cytokinin levels have led to initiation of numerous compact buds and shoots (Economou, 1982). Also, the use of GA_3 has been already mentioned for inclusion in the forcing solution when this technique is employed for bud break, although the GA_3 often leads to a depression of desired in vitro responses. Abscisic acid (ABA) has been included in the medium to help synchronize embryogenic events in Carum carvi (Ammirato, 1974) and to improve desiccation tolerance in celery somatic embryos (Kim and Janick, 1991). As we have seen for most chemicals, however, negative effects are also often observed, as in the case with its depression of growth and photosynthesis in Aronia (Colon et al, 1990).

Numerous other compounds have been used for inclusion in plant tissue culture media, including proline (Kim and Janick, 1991), colchicine (Swartz et al, 1990), jasmonic acid (Ravnikar and Gogala, 1990), $AgNO_3$ and rhizobitoxine (Gavintertvatana et al, 1980), putrescine and cefotaxime (Volabra et al, 1990) and growth retardants such as ancymidol and paclobutrazol (Zin and Ariel, 1991). Much is yet to be learned about these and other compounds purported to have hormone-like activity. It is therefore anticipated that much future research will be concentrated on the exciting field of plant growth regulating chemicals and their potential uses in plant micropropagation.

5. Acknowledgements

The author would like to thank Mr. Guochen Yang for his assistance in preparation of the manuscript and the dozens of former graduate students, technicians and scientists visiting in our laboratory, without whose efforts much of the work reported here would not have been possible.

6. References

Aitken-Christie, J. (1988) 'Development of a semi-automated micropropagation system', Acta. Hort. 230, 81-86.

Ammirato, P.V. (1974) 'The effects of abscisic acid on the development of somatic embryos from cells of caraway (Caruem carvi L.), Bot. Gazette 135, 328-337.

Appelgren, M. and Heide, O. M. (1972) 'Regeneration in streptocarpus leaf discs and its regulation by temperature and growth substances', Physiol. Plant 27, 417-423.

Bachelard, E.P. and Stowe, B.B. (1963). 'Rooting of cuttings of Acer rubrum and Eucalyptus calmaldulensis Dehnh; Aust. J. Biol. Sci. 16, 751-767.

Biran, I. and Halevy, A.H. (1973). 'Stock plant shading and rooting of dahlia cuttings', Scientia Hort. 1, 125-131.

Brandon, D. (1939) 'Seasonal variations of starch content in the genus Rosa and their relation to propagation by stem cuttings', J. Pom. Hort. Sci. 14, 233-253.

Bush, S. R., Earle, E.D. and Langhans, R.W. (1976) 'Plantlets from petal segments, petal epidermis, and shoot tips of the periclinal chimera, Chrysanthemum morifolium 'Indianapolis', Amer. J. Bot. 63, 729-737.

Cade, R. M., Wehner, T.C. and Blazich, F.A. (1990) 'Somatic embryos derived from cotyledons of cucumber', J. Amer. Soc. Hort. Sci. 115, 69-696.

Cheng, T-Y. and Smith, H. (1973) 'The influence of genomes on autonomous growth of pith cultures of Nicotiana glauca-langsdorfii hybrids', Planta 113, 29-34.

Colon, W., Kane, M.E. and Ingram, D.L. (1990) 'Effects of abscisic acid on photosynthesis, growth and development of stage III Aronia arbutifolia (Rosaceae)', Proc. Fla. Hort. Soc. 103, 178-182.

Dijak, M. and Brown, D.C.W. (1987) 'Donor tissue and culture condition effects on mesophyll protoplasts of Medicago sativa', Plant Cell Tissue and Organ Cult. 9, 217-228.

Dhillon, D.S. (1963) 'Influence of stock plants treated with various levels of nitrogen in combination with day length on rooting of black currant stem cuttings', Indian J. Hort. 20, 34-37.

Doerschug, M.R. and Miller, C.O. (1967) 'Chemical control of adventitious organ formation in Lactuca sativa explants', Amer. J. Botany 54, 410-413.

deFossard, R.A. (1981) 'Tissue Culture For Plant Propagators, The Univ. of New England, Armidale.

Duysuis, J.M., Pean, M. and Chagvardieff, P. (1990) 'Plant donor tissue and isolation procedure effect on early formation of embryoids from protoplasts of Helianthus annuus L.', Plant Cell, Tissue and Organ Culture 22, 183-189.

Dunwell, J.M. and Perry, E. (1973) 'The influence of in vivo growth conditions of N. tabacum plants on the in vitro embryogenic potential of their anthers', 64th Ann. Rept. John Innes Inst. pp. 69-79.

Economou, A.S. (1982) 'Chemical and physical factors influencing in vitro propagation of hardy deciduous azaleas (Rhododendron spp.) Ph.D. dissertation, Univ. Minnesota.

---------- and Read, P.E. (1980) 'Effect of benzyladenine pretreatments on shoot proliferation from petunia leaf segments cultured in vitro', Proc. Plant Growth Reg. Working Group 7, 96-103.

----------- and Read, P.E. (1982) 'Effect of NAA on shoot production in vitro from BA-pretreated petunia leaf explants', J. Am. Soc. Hortic. Sci. 107, 504-506.

___________ and Read, P.E. (1986a) 'Influence of light duration and irradiance on micropropagation of a hardy deciduous azalea', J. Am. Soc. Hortic. Sci. 111, 146-149.

----------- and Read, P.E. (1987) 'Light treatments to improve efficiency of in vitro propagation systems', HortScience 22(5), 751-754.

Eizenga, G.C. and Dahleen, L.S. (1990) 'Callus production, regeneration and evaluation of plants from cultured inflorescences of tall fescue (Festuca arundinacea Schreb.)', Plant Cell, Tissue and Organ Culture 22, 7-15.

Ellis, D.D., Barczynska, H., McCown, B.H. and Nelson, N. (1991) 'A comparison of BA, zeatin and thidiazuron for adventitious bud formation from Picea glauca embryos and epicotyl explants', Plant Cell, Tissue and Organ Culture 27, 281-287.

Fellman, C.D., Read, P.E. and Hosier, M.A. (1987) 'Effects of thidiazuron and CPPU on meristem formation and shoot proliferation', HortScience 22, 1197-1200.

Fonnesbech, M. (1974) 'Temperature effects on shoot and root development from Begonia x cheimantha petiole segments grown in vitro', Physiol. Plant. 32, 282-286.

Franclet, A., Bourlay, J., Bekkaoui, F., Fouret, Y., Verschoore-Martouget, B., and Walker, N. (1987) Rejuvenation'. In: Bonga, J.M. and Durzan, D.J. (eds), Cell and Tissue Culture in Forestry 1, 232-248.

Fridborg, G. and Eriksson, T. (1975) 'Partial reversal by cytokinin and (2-chloroethyl) trimethylammonium chloride, of near-ultraviolet inhibited growth and morphogenesis in callus cultures', Physiol. Plant. 35, 162-166.

Gardner, F.E. (1937) 'Etiolation as a method of rooting apple variety stem cuttings', Proc. Am. Soc. Hortic. Sci. 34, 323-327.

Garton, S., Read, P.E. and Farnham, R.S. (1983) 'Effect of stock plant nutrition on macro-and macro-propagability of Salix', Acta Hort. 131, 141-151, Martinus Nijhoff, Dordrecht.

Gavinlertvatana, P., Read, P.E. and Wilkins, H.F.(1980) 'Control of ethylene synthesis and action by silver nitrate and rhizobitoxime in petunia leaf sections cultured in vitro', J. Amer. Soc. Hort. Sci. 105, 304-307.

Gavinlertvatana, P., Read, P.E., Heins, R.D., and Wilkins, H.F. (1979) 'Influence of photoperiod and daminozide stock plant treatments on ethylene and CO_2 levels and callus formation from dahlia leaf segment culture', J. Am. Soc. Hortic. Sci. 104, 849-852.

Gawel, N.J. and Robacker, C.D. (1990) 'Genetic control of somatic embryogenesis in cotton petiole callus cultures', Euphytica 49, 249-253.

Geier, T., Kohlenback, H.W., and Reuther, G. (1979) 'Klonale vermehrung von Cyclamen persicum durch gewebekultur. Gartenbauwiss', 44, 226-237.

George, E.F. and Sherrington, P.D. (1984) 'Plant Propagation by Tissue Culture', Exegetics Ltd, Eversley, Basingstoke.

Grout, J.M. and Read, P.E. (1986) 'Influence of stock plant propagation method on tissue culture and leaf-bud propagation of 'Northblue' blueberry', J. Amer. Soc. Hort. Sci. III, 368-371.

Goussard, P.G. (1982) 'Morphological responses of shoot aspices of grapevine cultured in vitro, effects of cytokinins in routine subculturing', Vitis 21, 293-298.

Haberlandt, G. (1902) 'Culturversuche mit isolierton Pflanzenzellen', Sitzber Kaiser Akad. Wiss. Berlin, Math. Naturw. Kl., Abt. I. iii, 69-92.

Hackett, W.P. (1969) 'Aseptic multiplication of lily bulblets from bulb scales', Comb. Proc. Intl. Plant Prop. Soc. 19, 105-108.

___________ (1983) 'Phase change and intra-clonal variability', HortScience 18, 840-844

Hammerschlag, F. (1982) 'Factors influencing in vitro multiplication and rooting of the plum rootstock Myrobalan (Prunus cerasifera Ehrh.)', J. Am. Soc. Hortic. Sci. 107, 44-47.

----------- Bauchan, G.R., and Scorza, R. (1987) 'factors influencing in vitro multiplication and rooting of peach cultivars', Plant Cell Tissue Organ Cult. 8, 235-242.

Hangarter, R.P. and Stasinopoulos, T.C. (1991) 'Repression of plant tissue culture growth by light is caused by photochemical change in the culture medium', Plant Science 79, 253-257.

Hanson, J. (1976) 'Adventitious root formation induced by gibberellic acid and regulated by the irradiance to the stock plants', Physiol. Plant. 36, 77-81.

Hasegawa, P.M., Murashige, T., and Takatori, F.H. (1973) 'Propagation of Asparagus through shoot apex culture II. Light and temperature requirements, transplantability of plants and cytological characteristics', J. Am. Soc. Hortic. Sci. 98, 143-148.

Haun, R. and Cornell, W. (1951) 'Rooting response of geranium (Pelargonium hortorum) cuttings as influenced by nitrogen, phosphorus and potassium Van dey nutrition of the stock plants, Proc. Am. Soc. Hortic. Sci. 58, 317-323.

Heide, O.M. (1965) 'Interaction of temperature, auxins and kinins in the regeneration ability of begonia leaf cuttings', Physiol. Plant. 18, 891-920.

----------- (1968) 'Auxin level and regeneration of Begonia leaves', Planta 81, 153-159.

----------- (1969) 'Interaction of growth retardants and temperature in growth, flowering, regeneration and auxin activity of Begonia x cheimantha Everett, Physiol. Plant. 22, 1001-1012.

Heins, R.D. and Wilkins, H.F. (1979) 'The influence of node number, light source and time of irradiation during darkness on lateral branching and cutting production in "Bright Golden Anne" chrysanthemum', J. Am. Soc. Hortic. Sci. 104, 265-270.

Henny, P. and Read, P.E. (1971) 'Propagating the new University of Minnesota hardy deciduous azaleas', Proc. Intl. Plant Propagators Soc. 21, 331-339.

Hentig, W. (1977) 'Influence of day length on propagation', Gartenwelt 5, 102-106.

Hughes, B., White, F., and Smith, M. (1978) 'Effect of plant growth, isolation, and purification conditions on barley protoplast yield', Biochem. Physiol. Pflanzen, 172, 67-77.

Hughes, K.W. (1981) 'In vitro ecology: Exogenous factors affecting growth and morphogenesis in plant culture systems', Environ. Exp. Bot. 21, 281-288.

Hutchinson, J.F. (1984) 'Micropropagation of "Northern Spy" apply rootstock', Comb. Proc. Intl. Plant Propagators Soc. 34, 38-48.

Jaramillo, J. and Summers, W.L. (1990) 'Tomato anther callus production: solidifying agent and concentration influence induction of callus', J. Amer. Soc. Hort. Sci. 115, 1047-1050.

Kane, M.E. and Albert, L.S. (1989) 'Comparative short and root regeneration from juvenile and adult aerial leaf explants of variable-leaf milfoil', J. Aquat. Plant Manage. 27, 1-10.

Kawase, M. (1965) 'Etiolation and rooting in cuttings', Physiol. Plant. 18, 1066-1076.

Ko, C.J. (1986) 'The effect of temperature on in vitro bulblet formation of Hyacinthus orientalis L', M.S. thesis, Univ. of Minnesota.

Kozai, T. (1991) 'Micropropagation under photoautotrophic conditions', in P. Debergh and R. Zimmerman (eds.) Micropropagation Technology and Application, Fluwer Academic Publishers, Dordrecht, pp. 447-469.

Kozai, T. (1988) 'Autotrophic (sugar-free) tissue culture for promoting the growth of plantlets in vitro and for reducing biological contamination', Intl. Symp. Application of Biotechnology for Small Industries (pp. 1-21), Bangkok, Thailand.

Kraus, E.J. and Kraybill, H.R. (1918) 'Vegetation and reproduction with special reference to the tomato, Oregon Agric. Exp. Sta. Bull. p. 149.

Liu, Z.R. and Sandford, J.C. (1988) 'Plant regeneration by organogenesis from strawberry leaf and runner tissue', HortScience 23, 1057-1059.

Lu, C.S., Sharma, H.C. and Ahm, H.W. (1991) 'Wheat anther culture:effect of genotype and environmental conditions' Plant Cell, Tissue and Organ Culture 24, 233-236.

Matsubara, S., Chen, D., Masuda, M. and Murakowi, K. (1990) 'Production and proliferation of bulbs from receptacles of virus free garlic plants', Scientific Reports, Faculty of Agriculture, Okayama University, No. 75, 9-13.

Mes, M.G. (1951) 'Plant Hormones', Prog. Pot. Plant Phys. Res. Inst. 1950-51. Univ. of Pretoria, South Africa.

Miller, L.R. and Murashige, T. (1976) 'Tissue culture propagation of tropical foliage plants', in Vitro 12, 797-813.

Moe, R. (1976) 'Environmental effects of stock plants on rooting and further growth of the cuttings in Campanula isophylla Moretti, Acta Hort. 64, 71-81.

Morel, G., and Martin, C. (1955) 'Guerison des plantes atteintes de maladies a virus; par culture de meristemes apicaux', Report of 14th Intern. Hort. Congr. Metherlands, pp. 303-310.

Morel, G., and Martin, C. (1952) 'Guerison de dahlias atteints d'une maladie a virus', Compt. Rend. 235, 1324-1325. 1955.

Murashige, T. (1974) 'Plant propagation through tissue cultures', Ann. Rev. Plant Physiol. 25, 135-166.

Niskio, T., Sakata, Y. Yamagishi, H., Tabei, Y., Sato, T. and Takayanagi, K. (1989) 'Protoplast culture in vegetable crops: Development and improvement of culture procedure', Bull. Natl. Res. Inst. of Vegetables, Ornam. Plants and Tea, Series A, No. 3, 67-96.

Omar, M.S. and Novak, F.J. (1990) 'In vitro plant regeneration and ethylovethanesulphonate uptake in somatic embryos of date palm (Phoenix dactylifera L.)', Plant Cell, Tissue and Organ Culture 20, 185-190.

Papchatzi, M., Hammer, P.A., and Hasegawa, P.M. (1981) 'In vitro propagation of Hosta decorata Thomas Hogg using cultured shoot tips', J. Am. Soc. Hortic. Sci. 106, 232-236.

Paterson, K. and Rost, T. (1979) 'Effects of light and hormones on regeneration of Crassula argentea from leaves', Am. J. Bot. 66, 463-470.

Pennazio, S. and Radolfi, P. (1973) 'Factors affecting the culture in vitro of potato meristem tips', Potato Res. 16, 20-29.

Pierik, R.L.M. (1967) 'Regeneration, Vernalization and Flowering in Lumaria', Wageningen, Veenman and Zonen, 71 pp.

Pierik, R.L.M. and Ruibing, M.A. (1973) 'Regeneration of bulblets on bulb scale segments of hyacinth in vitro', Neth. J. Agric. Sci. 21, 129-138.

Pierik, R.L.M., van Leeuwen, P., and Rigter, G.C.C.M. (1979) 'Regeneration of leaf explants of Anthurium Andreanum Lind. in vitro', Neth. J. Agric. Sci. 27, 221-226.

Read, P.E. (1990) 'Environmental effects in micropropagation',in P. Ammirato, D.A. Evans, W.R. Sharp and Y.P.S. Bajaj (eds.), Handbook of Plant Cell Culture, McGraw Hill, New York, pp. 95-125.

Read, P.E. and Economou, A. (1982) 'Supplemental lighting in the propagation of deciduous azaleas', Proc. Intl. Plant Propagators Soc. 32, 639-645.

----------- Gavinlertvatana, P., Suriyajantratong, P., Garton, S., and Brenner, M. (1978) 'Stock plants affect tissue culture success. In: Propagation of Higher Plants Through Tissue Culture', (K. Hughes, R. Henke, M. Constantin, eds.), Univ. of Tennessee, Knoxville, p. 249.

----------- Garton, S., Carlson, C., Conviser, L., and Preece, J. (1979) 'Programming stock plants for tissue culture success, Proc. Plant Growth Regulator Working Group 6, 197-204.

----------- Economou, A.S., and Fellman, C.D. (1984) 'Manipulating stock plants for improved in vitro mass propagation', Proc. Intl. Symp. Plant Tissue Cell Cult.: Application to Crop Improvement, (F.J. Novak, L. Havel, and J. Dolezel, eds.), Czech, Acad. Sci., Prague, pp. 467-473.

----------- Fellman, C.D., Economou, A.S., and Yang, Q-G. (1986) 'Programming stock plants for propagation success', Proc. Intl. Plant Propagators Soc. 35, 84-91.

Read, P.E. and Yang, G. (1989) Response in vitro of explants chemically treated via forcing solutions', Proc. Intem. Plant Prop. Soc. 38, 406-408.

Samish, R.M. and Spiegel, P. (1957) 'The influence of nutrition of the mother vine on the rooting of cuttings', Ktavim 8, 93-100.

Sato, T., Nishio, T. and Hirai, M. (1989) 'Culture conditions for the initiation of embryogenesis from isolated microspores in Chinese cabbage (Brassica carupestris L.)', Bull. Natl. Res. Inst. of Vegetables, Ornamental Plants and Tea, Series A, No. 3, 55-65.

Skoog, F. and Miller, C.O. (1957) 'Chemical regulation of growth and organ formation in plant tissue cultured in vitro', Symp. Soc. for Exp. Biol. 11, 118-131.

Stamp, J.A., Colby, S.M. and Meredith, C.P. (1990) 'Direct shoot organogenesis from leaves of grape (Vitis spp.) 'Plant Cell, Tissue and Organ Culture 22, 127-133.

Starring, C.C. (1924) 'Influence of carbohydrate-nitrate content of cuttings upon the production of roots', Proc. Am. Soc. Hort. Sci. 20, 288-292.

Stimart, D.P. and Ascher, P.D. (1978) 'Foliar emergence and growth of in vitro generated bulblets from bulb scale explants of Lilium Longiflorum', HortScience 13, 356 (abstract).

----------- and Ascher, P.D. (1981) 'Developmental response of Lilium longiflorum bulblets to constant or alternating temperatures in vitro', J. Am. Soc. Hortic. Sci. 106, 450-454.

Stoltz, L. and Hess, C. (1966) 'The effect of girdling upon root initiation: carbohydrates and amino acids', Proc. Am. Soc. Hortic. Sci. 89, 734-743.

Swartz, H.J., Bors, R., Mohamed, F. and Naess, S.K. (1990) 'The effect of in vitro pretreatments on subsequent shoot organogenesis from excised Rubus and Malus leaves', Plant Cell, Tissue and Organ Culture 21, 179-184.

Takeuchi, N., Tanimoto, S., and Harada, H. (1985) 'Effects of wounding on adventitious bud formation in Torenia stem segments cultured in vitro', J. Exp. Bot. 36, 841-847.

tenHouten, J.G., Quak, F., and van der Meer, F.A. (1968) 'Heat treatment and meristem culture for the production of virus-free plant material', Neth. J. Plant Path. 74, 17-24.

Thorpe, T.A. and Murashige, T. (1970) 'Some histochemical changes underlying shoot initiation in tobacco callus cultures', Can. J. Botany 48, 277-285.

Tran Thanh Van, K. (1973a) 'Direct flower neoformation from superficial tissue of small explants of Nicotiana tabacum L.', Planta 115, 87-92.

Tran Than Van, M. (1973b) 'In vitro control of de novo flower, bud, root and callus differentiation from excised epidermal tissues', Nature 246, 44-45.

Van Aartrijk, J. and Blom-Barnhoorn, G.J. (1983) 'Adventitious bud formation from bulb-scale explants of Lilium Speciosum Thumb, in vitro. Effects of wounding, TIBA, and temperature, A. Pfanzenphysiol. 110, 355-363.

Villalobos, V.M., Leung, D.W.M., and Thorpe, T.A. (1984) 'Light-cytokinin interaction in shoot formation in cultured cotyledon explants of radiata pine', Physiol. Plant. 61, 497-504.

Von Arnold, S., and Tillberg, E. (1987) 'The influence of cytokinin pulse treatments on adventitious bud formation on vegetative buds of Picea abies', Plant Cell Tissue Organ Cult. 9, 253-261.

Volabra, C.P. and James, D.J. (1990) 'In vitro shoot regeneration from leaf discs of Betula pendula 'Dalecarlica' EM 85', Plant Cell, Tissue and Organ Culture 21, 51-54.

Wainwright, H. and Flegmann, A.W. (1984) 'The influence of light on the micropropagation of black currant', J. Hort. Sci. 59, 387-393.

Wang, P.J. and Hu, C.Y. (1982) 'In vitro mass tuberization and virus-free seed-potato production in Taiwan', Am. Potato J. 59, 33-37.

Weiser, C. and Blaney, L. (1960) 'The effect of boron on the rooting of English holly cuttings', Proc. Am. Soc. Hortic. Sci. 75, 704-710.

Werckmeister, P. (1971) 'Light induction of geotropism and the control of proliferation and growth of Cymbidium in tissue culture', Bot. Gaz. 132, 346-350.

White, P.R. (1963) 'The cultivation of animal and plant cells (2nd ed.)', New York: Ronald Press.

Williams, J., Pink, D.A.C. and Biddington, N.L. (1990) 'Effect of silver nitrate on long-term culture of callus from Brassica oleracea var. gemmifera' Plant Cell, Tissue and Organ Culture 21, 61-66.

Yang, G. and Read, P.E. (1990) 'Use of forcing solutions to study plant growth regulator effects on Vanhoutte's spiraea cultured in vitro', HortScience 25, 1124.

Yang, Q-G, and Zhang, L. (1987) 'Studies on the sprouting of the dormant shoots and their shoot proliferation and root initiation of the Japanese apple "coloured Fiji No. 2" in vitro', J. Anhui Agric. Coll. (PRC) 14, 31-34.

Yu, Dan-hua and Meredith, C.P. (1986) 'The influence of explant origin on tissue browning and shoot production in shoot tip cultures of grapevine', J. Am. Soc. Hortic. Sci. 111, 972-975.

Ziv, M. and Ariel, T. (1991) 'Bud proliferation and plant regeneration in liquid-cultured philodendron treated with ancymidol and paclobutrazol', J. Plant Growth Regul. 10, 53-57.

THE IN VITRO ENVIRONMENT AND ITS CONTROL IN MICROPROPAGATION

T. KOZAI and K. FUJIWARA
Faculty of Horticulture
Chiba University
Matsudo, Chiba 271
Japan

M. HAYASHI
Dept. of Bio. Sci.
and Technology
Tokai University
Nishino, Numazu
Shizuoka 410-03
Japan

J. AITKEN-CHRISTIE
Forest Research
Institute
Rotorua
New Zealand

ABSTRACT. The reasons for high production costs of micropropagated plantlets are analyzed. Problems requiring solutions by proper control of the *in vitro* environment for improving shoot/plantlet quality and cost effectiveness are revealed. Features of the *in vitro* environment and responses of shoots/plantlets cultured *in vitro* to the environment in conventional micropropagation are discussed in detail. The features and responses are interpreted from an environmental control point of view. Similarity of growth patterns between plantlets and seedlings under different *in vitro* environmental conditions is shown. Strategies for environmental control in micropropagation are discussed. The environmental effect on growth and development of shoots/plantlets cultured *in vitro* under controlled environments is shown. Advantages and disadvantages of heterotrophic, photomixotrophic and photoautotrophic micropropagation are discussed. Some novel environmental control systems are introduced.

Abbreviation: PPFD – photosynthetic photon flux density, MS – Murashige and Skoog (1962)

1. Introduction

The demand for high quality transplants has recently been increasing rapidly worldwide for reforestation, food/forage production, urban/indoor horticulture and global environmental protection. In many cases, micropropagation gives some superior transplant qualities to seedling production and conventional vegetative propagation (Debergh and Zimmerman, 1990).

In fact, its commercial use in horticulture is currently expanding worldwide. However, widespread use of micropropagation for major crops in agriculture and forestry is still restricted due to relatively high production costs. The high production costs of micropropagated shoot are mainly due to high labor costs, limited rates of plantlet growth during multiplication, and a low percentage of survival of plantlets (plantlet=shoot with roots) during acclimatization.

Labor costs in commercial micropropagation account for approximately half of the total production cost. Therefore, introduction of labor-saving methods into the production process is essential for achieving the reduction in production cost because labor costs keep increasing all over the world.

Labor-saving methods do not necessarily mean the introduction of full automation or robotization. Partial introduction of automation or robotization into some labor-intensive procedures in micropropagation can reduce the labor costs to some extent. However, production costs will not be largely decreased if conventional micropropagation procedures are automated without any procedural improvement. Automation must be achieved based on novel concepts with thorough understanding of environmental effects on the growth and development of shoots/plantlets cultured *in vitro* (hereafter, shoots/plantlets *in vitro*). Because, as will be shown

K. Kurata and T. Kozai (eds.), Transplant Production Systems, 247–282.

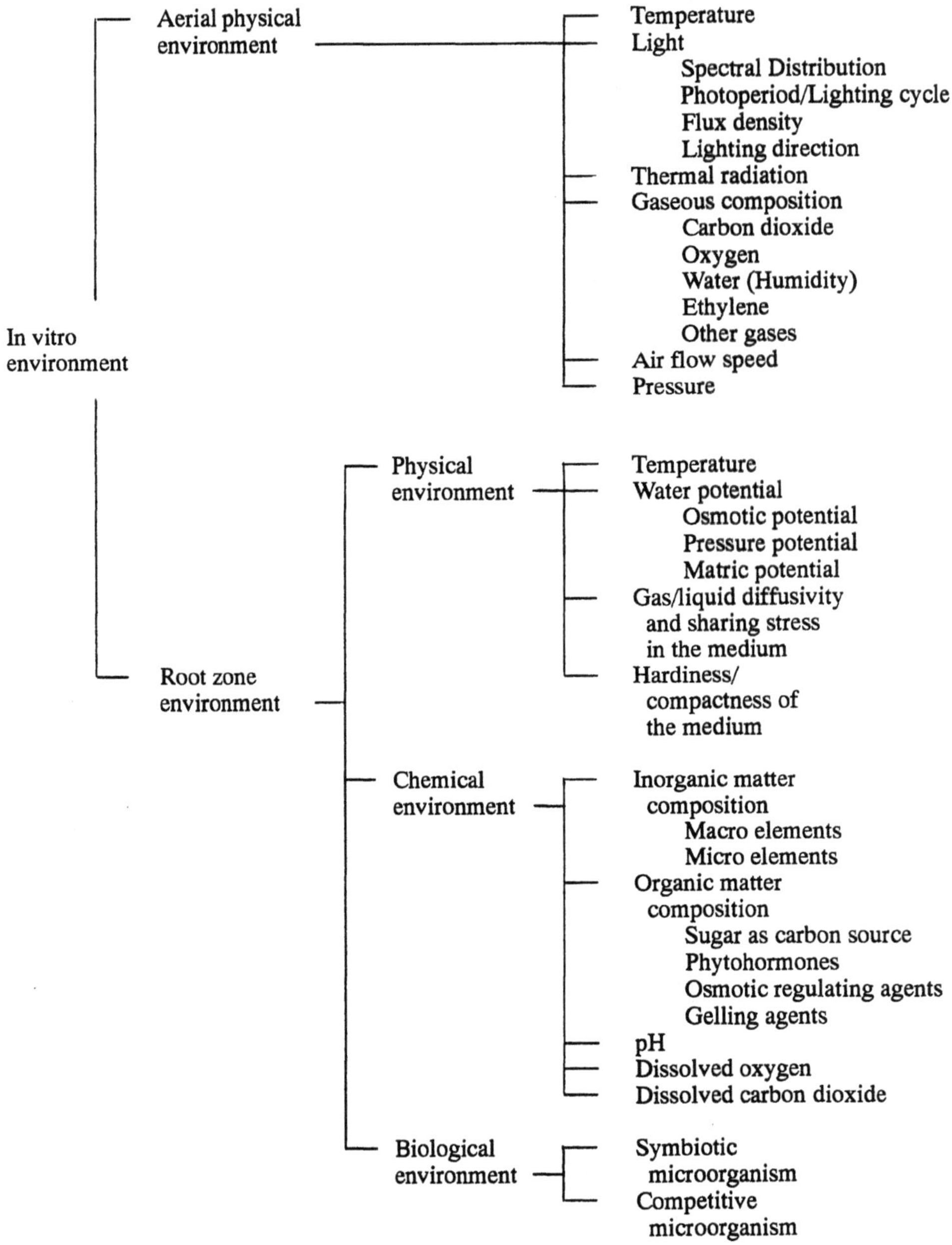

Figure 1. Classification of *in vitro* environmental factors which have an effect on growth, development and morphogenesis of shoots/plantlets *in vitro*.

in this chapter, rates of growth and development and many of the physiological/morphological characteristics of plantlets are largely influenced by the environment (physical, chemical, gaseous). Systematic comprehension of the environmental effect and its control in micropropagation is therefore required for realizing the full potential of micropropagation.

Many advances have been achieved recently in research on bioreactor systems and automatic liquid medium feeding systems (e.g., Weathers and Giles, 1988; Aitken–Christie, 1988; Takayama et al., 1990; Kurata et al., 1991; Hale and Young, 1991). The bioreactor systems approach and the automatic feeding systems approach seems to be promising in many cases. Description of the systems is, however, beyond the scope of this article.

2. Features of the In Vitro Environment in Conventional Micropropagation

2.1. ENVIRONMENTAL FACTORS AND THE MECHANISM OF ENVIRONMENTAL CHANGES

Major environmental factors in plant tissue culture (i.e., *in vitro*) can be classified as shown in Figure 1. The aerial and root zone environments are the environments above and below the level of the medium, respectively. The mechanism of environmental changes in a culture vessel is somewhat similar to that of the environmental changes in a greenhouse. In fact, a culture vessel is, in a sense, a miniature greenhouse or a small growth chamber maintained under aseptic conditions. Moreover, an explant to be cultured *in vitro* may be thought of as a miniature cutting in conventional vegetative propagation (Read, 1990).

Hence, ecological, ecophysiological, environmental physiological, and environmental engineering studies on the micropropagation process should be important (Hughes, 1981; Smith and McClelland, 1991). In reality, however, micropropagation research has ignored physical environmental factors for too long and has relied on pharmaceutical levels of exogenous phytohormones application based on huge factorial experiments (Wilkins, 1988).Some important environmental factors and their effect on growth and development of shoots/plantlets *in vitro* are discussed from an environmental control engineering view point. Problems requiring solutions by environmental control or reasons for high production costs of micropropagated plantlets are summarized in Table 1.

2.2. GENERAL FEATURES OF THE IN VITRO ENVIRONMENT

Table 2 summarizes general features of the environment in plant tissue culture, i.e., '*in vitro* environment', in conventional micropropagation. Noticeable features of the *in vitro* environment are low flow rates of material and energy due to its semi-closed system, constant temperature and high humidity conditions, and relatively large diurnal changes of CO_2 concentrations in the vessel. Other unique features of the *in vitro* environment are the presence of sugar in the medium and the absence of microorganisms (asepsis) in the vessel. These features are seldom observed in nature, horticulture, agriculture and forestry.

2.3. CO_2 CONCENTRATION

It has been found that the CO_2 concentration in a vessel containing chlorophyllous shoots/plantlets often reduces to just above the CO_2 compensation point (50–100 umol mol^{-1} or ppm) during most of the photoperiod (Ando, 1978; Fujiwara et al., 1987; Pospisilova et al., 1988; Desjardins

Table 1. Problems requiring solutions by environmental control

1) A low growth rate of shoots/plantlets at each culture stage
2) A significant percentage of dead shoots/plantlets due to bacterial or fungal contamination and/or physiological/ morphological disorders such as vitrification during the multiplication stage
3) A significant percentage of dead plantlets due to environmental stress during the acclimatization stage
4) A large variation in shoot/plantlet size, shape and quality
5) Significant energy costs for lighting, air conditioning (cooling), autoclaving, washing, drying, etc.
6) Significant costs for phytohormones, culture vessels, sugar, etc.
7) An overproduction or shortage of plantlets at various times due to unpredictable circumstances
8) A significant space and its related costs required for multiplication and acclimatization

Table 2. General features of the *in vitro* environment in conventional micropropagation with respect to state and rate variables

State variables	Rate variables
AERIAL ENVIRONMENT	FLOW OF MATERIAL AND ENERGY
1) High relative humidity	1) Low transpiration rate
2) Constant air temperature	2) Low photosynthetic rate
3) Low CO_2 concentration in light and high CO_2 concentration in dark	3) High dark respiration rate
	4) Low photosynthetic photon flux
	5) Low net thermal radiation flux
4) High C_2H_4 concentration	6) Negative CO_2 balance
ROOT ZONE ENVIRONMENT	
5) High sugar concentration	7) Low water uptake rate
6) High salt concentration	8) Low ion uptake rate
7) Low dissolved oxygen concentration high concentration of phenolic or medium other toxic substances	9) Low sugar uptake rate
	10) Low transport rates of medium components
8) Low density of microorganisms	
9) High concentration of phytohormones	

Notes) State variables quantify conserved properties of the system. Rate variables quantify the rate of change with time of the state variables.

et al., 1988; Infante et al., 1989). It increases with time during the dark period up to 5 to 10 mmol mol^{-1} but decreases sharply over time to less than 100 umol mol^{-1}) within a few hours after the onset of the photoperiod and remains the same until the dark period begins (Figure 2). Even in loosely capped vessels with gas permeable film, the concentration deceased to 100–200 umol mol^{-1} during the photoperiod (Kozai and Sekimoto, 1988) as shown in Figure 3. (The atmospheric CO_2 concentration is 350 umol mol^{-1}.).

It should be noted, however, that the reduction in CO_2 concentration during the photoperiod is observed only when PPFD in the vessel is above the light compensation points of chlorophyllous plantlets *in vitro* (roughly 5 umol m^{-2} s^{-1}).

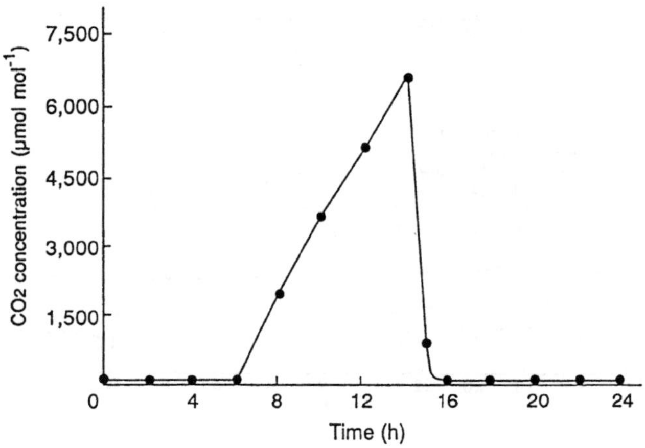

Figure 2. Diurnal change in CO2 concentration in a culture vessel containing *Ficus lyrata* plantlets. Photoperiod was from 0 to 6 h and from 14 to 24 h. PPFD at culture level and air temperature were approximately 65 μmol m^{-2} s^{-1} and 25±1 °C, respectively. (Fujiwara et al., 1987)

2.4. GAS EXCHANGE CHARACTERISTICS OF CULTURE VESSELS

It has been shown that the type of vessel closure affects the gaseous composition, and hence occurrence of vitrification (a physiological disorder) and growth of plantlets *in vitro*; the looser types being, in general, better than the tighter ones for reducing vitrification of plantlets *in vitro* (Hakkaart and Versluijs, 1983; Dillen and Buysens, 1989; Blazkova et al., 1989) and for promoting growth of strawberry plantlets *in vitro* (Kozai and Sekimoto, 1988).

The natural air exchange characteristics of a vessel with its closure are best expressed by the number of air changes (infiltration) per hour of the vessel, E, which is defined as the hourly air exchange rate of the vessel divided by the air volume of the vessel. The E can be estimated using CO_2 gas as a tracer gas by Seidel's equation (Kozai et al., 1986b):

$$E = -\ln((K-Kou)/(Ko-Kou))/T$$

where T is the time interval from time 0 to t, K the gas concentration in the vessel at time t, Ko the gas concentration in the vessel at time 0, Kou the constant gas concentration outside the

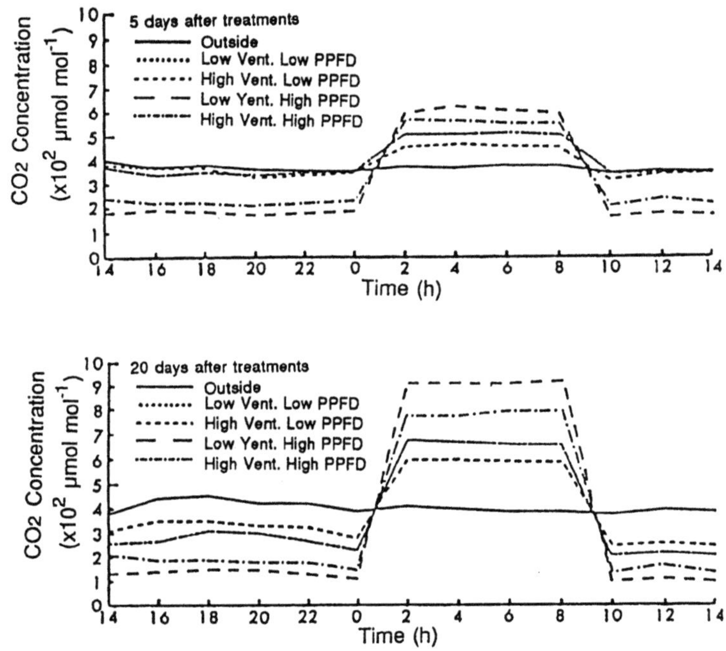

Figure 3. Diurnal changes in CO_2 concentrations inside (C_{in}) and outside (C_{out}) the tissue culture vessels containing strawberry plantlets at the multiplication stage. (Kozai & Sekimoto, 1988)

Low vent.: Number of air changes per hour of the vessel=1.5 h^{-1}

High vent.: Number of air changes per hour of the vessel=2.7 h^{-1}

Low PPFD: Photosynthetic photon flux density=34.3 µmol m^{-2} s^{-1}

High PPFD: Photosynthetic photon flux density=133 µmol m^{-2} s^{-1}

vessel. Ko is usually taken in a range between 0.5 and 1.5%(v/v) when Kou is 0.035%(v/v). T is taken in a range between 1 and 10 hours depending on the air tightness of vessel.

The E values for a flat bottom glass test tube (air volume: 45 ml) closed with an aluminum foil cap, plastic formed cap and silicon foam rubber plug are, respectively, 0.18, 1.5 and 0.6 per hour (Kozai et al., 1986b). The E value can be increased up to several times greater by incorporating a gas-permeable microporous polypropylene film into the vessel closure (Kozai and Sekimoto, 1988). It is difficult to have the E value less than 0.1 per hour even for the vessels capped with aluminum foil and sealed with plastic tape.

The E is a type of physical property of the vessel and is basically constant with time at a constant air flow speed. It increases with increasing air flow speed around the vessel. The gas concentration in the vessel containing shoots or plantlets and medium varies with the E value (Figure 3), the gas concentration outside the vessel, and gas evolution/absorption rates of the plantlets/medium in the vessel. The mathematical relationship among the above factors is given by Fujiwara et al. (1987). The E can be similarly defined for forced air changes of the vessel.

2.5. WATER POTENTIALS OF THE AIR AND THE CULTURE MEDIUM

The interchange of water in the gaseous and/or liquid phase among plantlets, aerial and root zone environments in the vessel, and the air outside the vessel plays an important role in plant growth and development. Directions of water flow are determined by the spatial distribution of water potentials inside and outside the vessel. Water potential is a measure of the free energy status of water. Generally, the water potential gradients and diffusion coefficients of air and medium in the vessel are much smaller, compared with those outside. Hence, the water flow rates among the medium, plant tissue and air in the vessel are low (Sallanon and Goudret, 1990).

Water potential of the air, Pr (unit: Pa (Pascal); 1 bar= 100 kPa = 0.1 MPa), can be calculated by the following equation:

$$Pr = (R\ T/v)\ \ln(RH/100)$$

where R is the perfect gas constant (8.31 J mol^{-1} K^{-1}), T is the absolute temperature (K), v is the partial molar volume of water (18 10^{-6} m^3 mol^{-1}), and RH is the relative humidity (%). The water potential is zero for air saturated with water vapor (H = 100%). The water potentials of the air for H being 99, 98, 97, 96, 95, 90, and 80% are, respectively, −1.38, −2.78, −4.19, −5.62, −7.06, −14.5, and −30.7 MPa at a temperature of 25°C.

The transpiration rate of shoots/plantlets *in vitro* is almost proportional to the absolute value of water potential for the air when RH is high (RH>90%). Then, the transpiration rates at RH=98, 97, 96 and 90% are, respectively, 2, 3, 4 and 10 times that at RH=99%. That is, a small difference in RH results in a large difference in transpiration rate, although the absolute values are all low.

Relative humidity is generally higher than 95% in an air−tight vessel containing plantlets on gelled or liquid medium. Although condensation on the inner surface of the vessel is often observed, it does not necessarily mean that the relative humidity is always 100%. The humidity varies in space and time in accordance with air temperature fluctuations and gas exchange characteristics within the vessel.

The water potential of a liquid culture media is equal to the osmotic potential which is a negative osmotic pressure of the liquid culture medium. (The water potential of pure water is zero.) The water potential of the medium has lower values with increasing total molar concentration of the basic inorganic components, disaccharides, and monosaccharides. Examples of water potentials based on the basic components of the widely used culture media in liquid form are given in Table 3.

The following equation has been obtained to estimate approximately the water potential of a liquid medium, Pw (unit: kPa)(Kozai, et al., 1986a):

$$Pw = Pb - 78\ Cd - 146\ Cm$$

where Pb is the water potential due to the basic component (kPa), Cd the concentration of disaccharide such as sucrose (w/v %), Cm the concentration of monosaccharide such as glucose, fructose, mannitol and sorbitol (w/v %). Water availability to plantlets is, in general, higher with increasing the water potential of the medium.

The water potential of gelled medium or plant tissue is the sum of several component potentials: osmotic, matric, pressure and gravitational. Gravitational and pressure potentials of the medium are minimal in a small vessel under atmospheric pressure conditions. Matric potential measurements of gelled media are difficult to make and there is some doubt on the contribution of matric potential to water potential of the medium (Fujiwara, 1991). The pressure potential in plant tissue is a negative turgor pressure. The water potential of plant tissue has a value between the water potential of the air and the medium, when water flows from the medium to the air through the plant tissue.

Table 3. Water potential values caused by the basic components of some widely used culture media in liquid form (Kozai et al., 1986a and Fujiwara et al., unpublished).

Basic component	Water potential	Basic component	Water potential
Heller (1953)	-89 kPa	Nitsh (1969)	-106 kPa
Knop (1865)	-49	B5 (1968)	-139
Knudson C (1946)	-69	R2 (1973)	-219
MS (1962)	-218	N6 (1975)	-177
White (1963)	-111	Modified LP (1984)	-171
SH (1972)	-153	Enshi-Shoho (1966)	-76

The values are converted for use at 25°C. The basic components consist of macro and micro inorganic salts, vitamins and the other micro organic substances. Measurement was made after adjustment of pH and autoclaving.

2.6. DEPENDENCE OF CHANGES WITH TIME IN ION CONCENTRATIONS IN THE MEDIUM ON THE MEDIUM VOLUME

The concentration of inorganic ions in the culture medium decreases with time as they are taken up by shoots/plantlets *in vitro*. The rate of decrease is greater when smaller amounts of medium volume are used because it is detérmined by the amount of absorbed ions divided by the medium volume. In other words, the ion uptake rate or amount of absorbed ion per unit time, is equal to the rate multiplied by the volume. Thus, the medium volume is an important state variable to characterize the rate. In turn, the uptake rate is dependent on the concentration in the medium. Therefore, optimum (initial) nutrient composition should be dependent on the medium volume when the medium is not renewed during culture.

Figure 4 shows the dependence of changes in ion concentration with time on the medium volume. Figure 5 shows the dependence of dry weight increase of potato plantlets cultured *in vitro* under photoautotrophic culture conditions on the absorbed amount of ions. The dry weight increased with increasing the absorbed amounts of ions during culture regardless of the strength of MS medium and the medium volume. The dry weight increase was more related to the integral of ion uptake rate or the amount of absorbed ion during culture than to the initial strength of the medium component and the change of concentrations with time.

In principle, the same applies for carbon source and for the air in the vessel. The volume of the medium and air per plantlet are, respectively, in a range between 2 and 20 ml and 20 and 100 ml. Figure 6 shows the dependence of dry weight increase of *Cymbidium* PLB (Protocorm like body) *in vitro* under heterotrophic conditions on the total amount of absorbed sugar (Kubota and Kozai, unpublished).

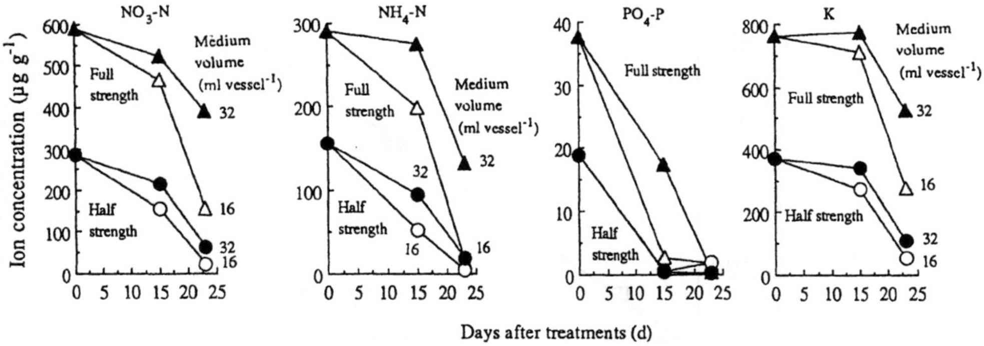

Figure 4. Dependence of time courses of the ion concentration of full and half strength MS sugar-free medium on the medium volume (ml vessel^{-1}). Numbers in the figure indicate the medium volume per vessel. Leafy single node cuttings of potato plantlets were used as explants. (Murai et al., unpublished)

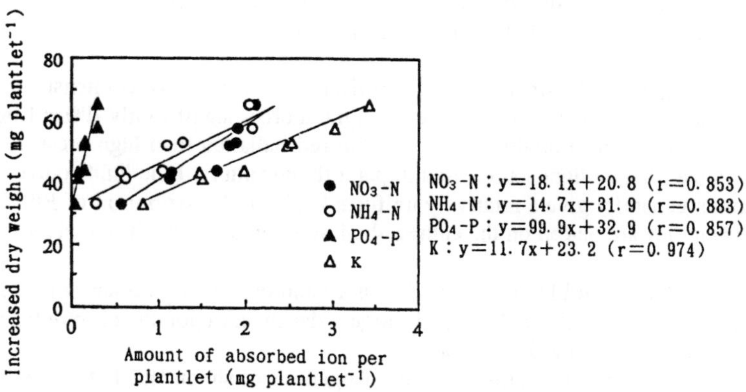

Figure 5. Relationship between amount of absorbed ion per plantlet and increased dry weight for 23 days. (Murai et al., unpublished)

2.7. LIGHT

Illuminance (lx) is not an appropriate indicator to describe the light environment for plants because it deals with the measurement of light seen by the 'standard human eye'. Photon (quantum) flux density (mol m^{-2} s^{-1}) and irradiance or radiation flux density (W m^{-2}) are, respectively, appropriate indicators to describe the light environment in terms of photosynthesis, photomorphogenesis and energy budgets.

The photosynthetic reaction of plants is driven by light in the spectral range between 400 and 700 nm. Photon flux density and irradiance in this wave band are, respectively, called photosynthetic photon flux density (PPFD) or simply PPF and photosynthetically active radiation (PAR).

The photomorphogenic reaction of plants is controlled by wavelengths in the blue (400–460 nm), red (620–680 nm), near-red (700–800 nm), and near-violet (300–400 nm) regions. It re-

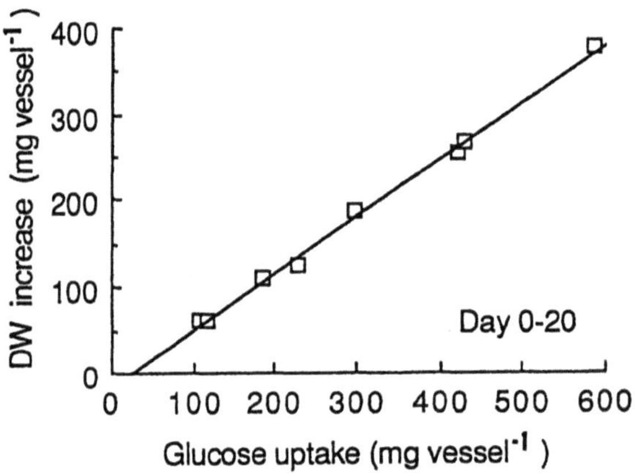

Figure 6. Relationship between glucose uptake by *Cymbidium* PLB (protocome like body) from the medium and dry weight increase of the PLB for 20 days. (Kubota et al., unpublished)

sults from a low level of flux density and is not significantly influenced by its intensity.

Spectral distributions of radiation from different light sources significantly differ from each other. Fluorescent lamps which produce much less far-red radiation than high pressure sodium lamps have been the primary light source used in plant tissue culture. Line light sources such as fluorescent lamps generally give a relatively uniform horizontal distribution of PPFD on the culture shelf when they are placed on the culture shelf at a height of 40-50 cm above the cultures.

There is a large difference in PPFD between the inside and outside the vessels, and the distribution of PPFD in culture vessels on the shelf is largely dependent upon the closure type, vessel type and their arrangement on the shelf (Figure 7).

It is also known that light quality affects the morphogenesis of shoots/plantlets grown *in vitro* (Read, 1990; Seabrook, 1987; Dooley, 1991). Photon ratios of red to far red regions and of blue to red regions are, in general, important parameters for photomorphogenesis.

Spectral transmissivities of vessel materials differ from each other (Dooley, 1991). For example, polycarbonate, glass and polystyrene vessels do not transmit the light which wavelength is shorter than 390, 290 and 300 nm, respectively. Thus, spectral distribution of light received by the plantlets *in vitro* is determined by both the light source and vessel material. Flux densities in the specified wavelengths and photo-/dark periods (lighting cycle) are all important environmental variables in plant tissue culture.

2.8. TEMPERATURES IN THE VESSEL AND HEAT BALANCE OF THE VESSEL

The light source produces thermal radiation and convective heat as well as PAR and photomorphologically active radiation. In fact, only about 25% of electric energy consumed by the light source is converted to PAR.

Due to radiative heating of the vessel by the light source, the air temperature inside the vessel is almost always about 0.5-1°C higher than that outside the vessel during the photoperiod. The difference in air temperature between inside and outside the vessel is nearly proportional to the

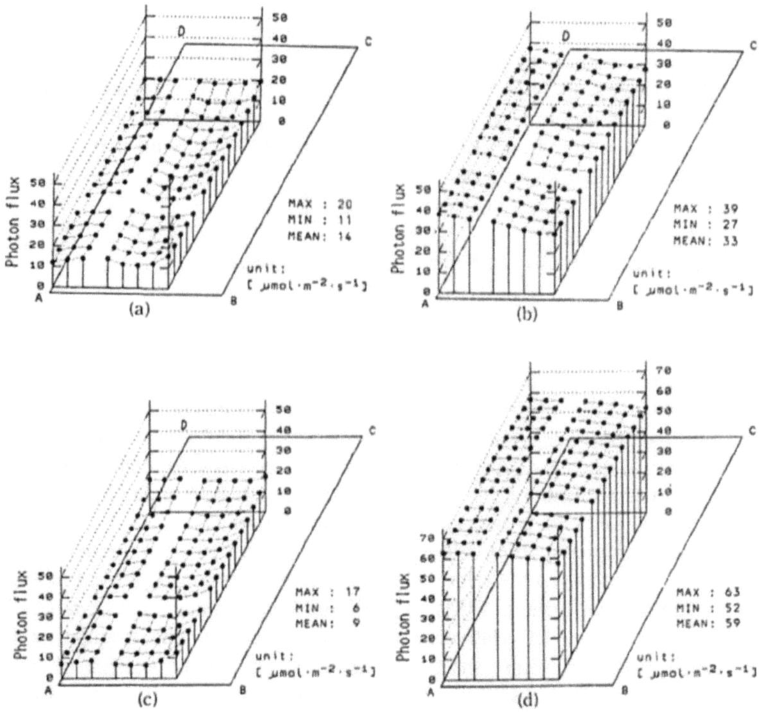

Figure 7. Effects of closures on Photosynthetic photon flux density (PPFD) in empty glass test tubes closed with either aluminum foil caps (a), polypropylene formed caps (b) or silicon foam rubber plugs (c) and PPFD on the shelf surface (d). (Fujiwara et al., 1989)

net radiation flux density (or approximately to PAR) just above the vessel. Most radiative energy transmitted into the vessel is released as convective heat from the outer surface of the vessel and only a very small fraction of the transmitted radiative energy is released by air exchange between inside and outside the vessel. The temperatures of shoots/plantlets exposed to the light should be 0.1–1°C higher than the inside air temperature due to the low air flow speed and low transpiration rate in the vessel.

Under a given net radiation flux density, the temperature difference increases with decreasing air flow speed around the vessels and with decreasing the area ratio of sides and cap to bottom surface of the vessel. Uraban and Jaffrin (1990) developed a model for heat and mass transfers in the vessels and applied it to various physical conditions to predict thermal phenomena and air temperature in the vessel.

2.9. SPATIAL VARIATIONS OF CONCENTRATIONS AND FLUXES

There must be spatial variations of CO_2 concentration, temperature, humidity, etc. in the vessel, due to little air movement in the vessel, gas and energy exchanges between inside and outside the vessel and between the shoots/plantlets and the inside air. There must also be a spatial variation of light flux density in the vessel due to interception of light by the plantlets. Furthermore, there must be spatial variations of medium component concentrations in the vessel.

The variations of concentrations and fluxes are affected by the presence of shoots/plantlets *in vitro* and the culture medium, the physical characteristics of the vessels, and by the environment outside the culture vessels and interactions between all of these. The variations in size and shape of shoots/plantlets often observed *in vitro* may be attributed, in part, to the spatial variation of the microenvironment in the vessel. Honjo and Takakura (1991) found that the humidity profiles in the vessel could be measured using a dew sensor.

3. Responses of Shoots/Plantlets In Vitro to the Environment

3.1. GENERAL

Table 4 summarizes some of the typical responses of shoots/plantlets *in vitro* at the tissue and whole plantlet levels to the *in vitro* environment in conventional micropropagation. These responses are results of, or interaction with, features of the *in vitro* environment given in Table 2.

Plantlets cultured *in vitro* are sensitive to environmental stresses, especially to water stress (Grout, 1977). At the acclimatization stage, excess transpiration of the plantlets often causes growth retardation (or negative growth rate) and death of plantlets.

In conventional micropropagation, therefore, the main effort of environmental control at the acclimatization stage is to keep the relative humidity high in order to prevent excessive plantlet transpiration (Dunstan and Turner, 1984). High humidity is generally maintained by covering plantlets with plastic film under relatively dense shade, together with frequent misting (Preece and Sutter, 1990). However, dense shading, in turn, causes the suppression of photosynthesis, rooting, leaf expansion, etc., and subsequently growth and a vicious circle is created (Table 5).

Kozai et al. (1987a) showed that micropropagated plantlets could be well acclimatized under

Table 4 Responses of shoots/plantlets *in vitro* in conventional micropropagation

Tissue level	Whole plantlet level
1) Little or wrong type of epicuticular wax	1) Low growth rate
2) Stomatal malfunction	2) Succulent growth (thin, tall and succulent shoot)
3) Low chlorophyll content	3) Physiological/morphological disorders such as vitrification
4) Low percent dry matter	4) Incomplete rooting/few secondary roots
5) Restricted leaf area expansion	5) High variation in size, shape and developmental stage
6) Less number of stomata on leaves than normal	6) Occurrence of mutation
7) Poorly-structured spongy and palisade tissues	7) Death of plants due to contamination

TABLE 5 The vicious circle of environmental control at the acclimatization stage in conventional micropropagation

1. Plantlets are sensitive to water stress and have not developed pho toautotrophy at an early stage of acclimatization.
2. Relative humidity has to be kept high to reduce damage or death of the plantlets due to excess plantlet transpiration.
3. Dense shading with frequent misting is necessary during the daytime to keep the humidity high and temperature low under natural solar light.
4. Photosynthesis of plantlets is suppressed by wilting and/or dense shading.
5. Suppression of photosynthesis suppresses the development of pho toautotrophy of the plantlets.
6. Suppression of photoautotrophy may suppress rooting and leaf emergence of the plantlets.
7. The water and nutrient uptake is suppressed due to an insufficient amount of secondary roots and/or high humidity with dense shading.
8. A small amount of water uptake and relatively high transpiration compared to water uptake, leads to damage or death of the plantlets.
9. Part of the leaves should then be removed from plantlets when acclimatized in order to reduce the excess transpiration.
10. Photosynthesis is suppressed by reduced leaf area.
11. Back to No.5.

relatively high PPF conditions if the relative humidity was accurately controlled at a certain high level using a microcomputer–controlled, ultrasonic humidifier. Under such conditions, CO_2 enrichment at the acclimatization stage was effective for promoting growth of plantlets (Hayashi and Kozai, 1987; Hayashi et al., 1990; Kozai, 1991d).

In conventional micropropagation, many observations are required at the beginning of the multiplication stage to distinguish normal plantlets from abnormal ones and to excise the normal ones for obtaining normal explants. Not all normal explants, however, produce normal plantlets. A significant part of a normal plantlet may also be discarded because it is not the highest quality. Much labor is also required to select and transplant normal plantlets at the beginning of the rooting and acclimatization stages. Loss of transplanted plantlets at the end of acclimatization can not be disregarded.

3.2. PHOTOSYNTHETIC CHARACTERISTICS

Photosynthetic response curve parameters of 12 different ornamental plants were estimated based on the measured data on diurnal changes in CO_2 concentration *in vitro* at a PPFD of 65 umol m^{-2} s^{-1} and a temperature of 25 °C (Fujiwara et al., 1987). Estimated net photosynthetic rates at saturated CO_2 concentrations ranged between 1.5 and 4.5 cm^3CO_2 gDW^{-1} h^{-1}. These were roughly 1000 times the actual net photosynthetic rates obtained for the plantlets *in vitro* and in situ under steady state conditions during the photoperiod. Such low actual rates are, of course, mainly due to the CO_2 concentrations being close to CO_2 compensation points.

Photosynthetic response curves for intact *Cymbidium* plantlets *in vitro* cultured on Hyponex-agar medium with 20 g l^{-1} sucrose were obtained as a function of CO_2 concentration, PPFD and temperature (Kozai et al., 1990b). CO_2 saturated net photosynthetic rates were about 3, 6 and 7 mgCO_2 gDW^{-1} h^{-1}, respectively, for a PPFD of 35, 102 and 226 umol m^{-2} s^{-1} at a temperature of 25°C (Figure 8). The response curves did not differ greatly from those of *Cymbidium* plants grown under shade in the greenhouse. Actual net photosynthetic rates of the plantlets in loosely capped vessels were, respectively, about 1, 2 and 2 mgCO_2 gDW^{-1} h^{-1} for these PPFD values, when the CO_2 concentration outside the vessel was kept at approximately 350 umol mol^{-1}.

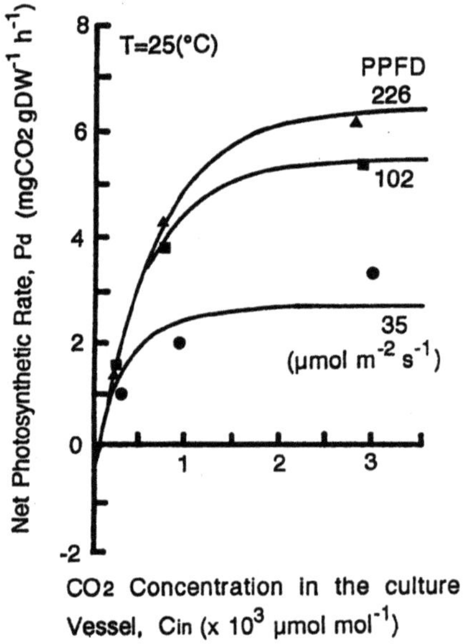

Figure 8. CO2-net photosynthetic curves in situ at different PPFD levels for Cymbidium plantlets *in vitro*. Each plot is an average of 10 measurements. (Kozai et al., 1990)

Leafy, single–node stem cuttings of potato (*Solanum tuberosum* L.) were cultured for 3 days on MS agar medium with 3% sucrose and without sugar. Net photosynthetic rates of the cuttings per leaf dry weight were 8–10 times greater in the sugar–free medium treatment than in the sugar–containing treatment when CO_2 concentration in the vessel was in a range between 500 and 1000 umol mol^{-1} (Figure 9). A decrease in net photosynthetic rate with increasing sucrose concentration in the medium has also been shown for rose (Capellades, 1989 and 1991), *Spathiphyllum* (Watanabe et al., 1990), *Chrysanthemum morifolium* (Tanaka et al., 1991a), Tani et al.,1991 etc.

Figure 10 shows the net photosynthetic rates of potato plantlets in situ and *in vitro* as influenced by CO_2 concentration and relative humidity (Tanaka et al., unpublished). The rates are higher with increasing relative humidity and CO_2 concentration.

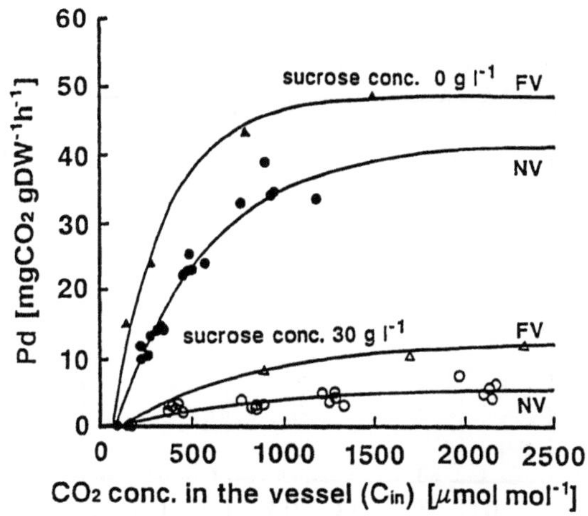

Figure 9. Net photosynthetic rates per leaf dry weight (Pd) of potato explants *in vitro* on day 3 as influenced by CO_2 concentration in the vessel and the presence/absence of sucrose in the medium under natural (NV) and forced (FV) ventilation conditions. (Photosynthetic photon flux density: 200 μmol m^{-2} s^{-1}, Air temperature: 25 °C) (Nakayama et al., 1991)

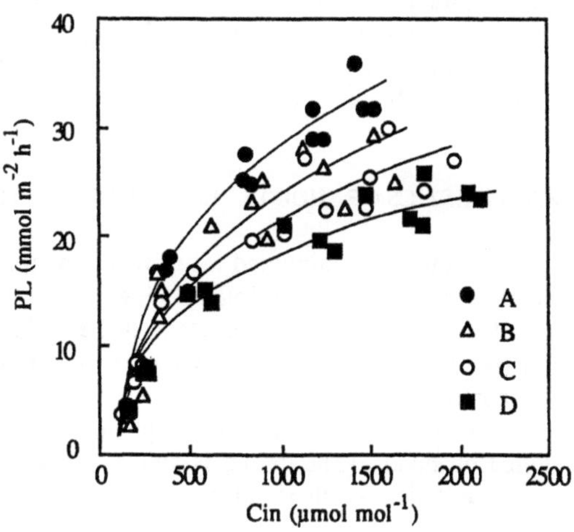

Figure 10. Effect of CO_2 concentration in the culture vessel (Cin) on net photosynthetic rates per leaf area (PL) of potato plantlets *in vitro*, cultured for 21 days at different relative humidities in the culture vessel: 94-95% (A), 92-95% (B), 86-92% (C) and 81-88% (D). (Tanaka et al., unpublished)

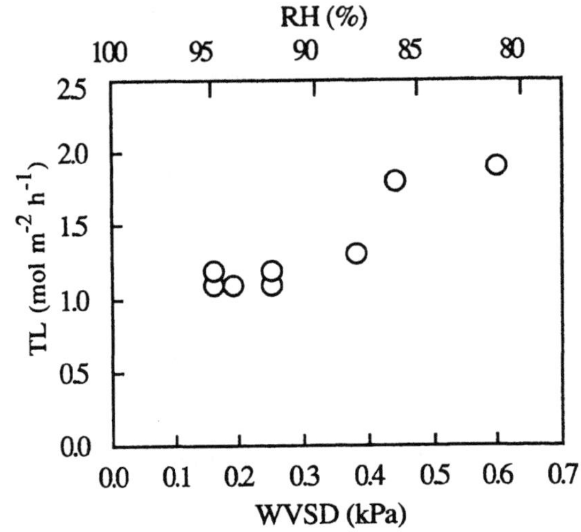

Figure 11. Effect of water vapor saturation deficit of the air (WVSD) on transpiration rate per leaf area (TL) of potato plantlets *in vitro*. (Tanaka et al., unpublished)

Rates of net photosynthesis, transpiration and dark respiration of *in vitro* plantlets under different environmental conditions remain unknown for almost all crops. These rates should be studied in detail at the leaf and whole plant levels over different periods of time during different stages of development in order to find better control strategies for the *in vitro* environment (Reuther, 1988).

3.3. TRANSPIRATION CHARACTERISTICS OF SHOOTS/PLANTLETS IN VITRO

There are many reports on stomatal malfunctioning and insufficient epicuticular wax formation on shoots/plantlets *in vitro* from anatomical and biological view points. Plantlets cultured *in vitro* under saturated humidity and low PPFD conditions tend to have thin epicuticular layers (Sutter and Langhans, 1979). However, there are only a few reports on the transpiration characteristics from an environmental control engineering view point.

Figure 11 shows the transpiration rate per leaf area of potato plantlets *in vitro* as influenced by the water vapor saturation deficit of the air. (The deficit is almost proportional to the negative water potential at high relative humidities but their values are quite different each other.) The rate increases with increasing the saturation deficit. The absolute value is lower for shoots/plantlets *in vitro* (less than 2 molH_2O m^{-2} h^{-1}) than for field–grown potato plants (2–12 molH_2O m^{-2} h^{-1}).

Leaf (stomatal plus cuticular) resistance to transpiration of potato plantlets was 0.5 s cm^{-1} when cultured *in vitro* at 94–95% relative humidity (RH) or the saturation deficit of 0.16–0.19

kPa, 1.0 s cm^{-1} at 92–95% RH (0.16–0.25 kPa), 1.7 s cm^{-1} at 86–92% RH (0.25–0.44 kPa) and 5.0 s cm^{-1} at 81–88% RH (0.38–0.60 kPa) on day 25 after transplanting on the medium. The resistance on day 0 was considered to be 0.5 s cm^{-1} for all the treatments. The resistance increased with time under lower relative humidity conditions. Thus, the transpiration rate per plantlet on day 25 was between 840 and 930 umolH_2O h^{-1} for all treatments regardless of differences in leaf resistance and relative humidity between treatments. Thus, control of relative humidity in the vessel seems to be important in order to control the transpiration characteristics, transpiration rates, and also photosynthetic characteristics of *in vitro* plantlets.

The transpiration rate through cuticular layers on leaves of greenhouse–grown crops is often less than 10% of the total transpiration rate during the photoperiod. However, transpiration through cuticular layers would be significant if the layers are poorly structured and thin, such as in shoots/plantlets cultured *in vitro*. There are few reports on the quantitative analysis of cuticular and stomatal transpiration of shoots/plantlets *in vitro*. This analysis is important to find a rational way to prevent excess transpiration.

3.4. SIMILARITY OF GROWTH PATTERNS BETWEEN PLANTLETS AND SEEDLINGS UNDER DIFFERENT IN VITRO ENVIRONMENTAL CONDITIONS

It has been demonstrated that the growth pattern of plantlets in a vessel is similar to that of seedlings in a vessel when cultured under the same *in vitro* environmental conditions (Kozai et al., 1991b). Explants and seeds of *Brassica campestris* L. were cultured for 3 weeks on MS medium with and without sucrose in a vessel with different numbers of air changes per hour under different PPFD conditions. The growth of plantlets/seedlings was greater for treatments with a higher number of air changes per hour and a higher PPFD regardless of the sucrose concentration in the medium (Table 6 and Figure 12). Similar results have been obtained for tobacco plantlets/seedlings *in vitro* (Kozai et al., 1990a).

It is expected that for a cultivar or plant species, that the growth pattern of seedlings grown *in vitro* would only be different from that of seedlings grown outside when the environment is different each other. It suggested that the growth pattern and physiological characteristics of plantlets *in vitro* should be similar to those of seedlings *in vitro* when the environments are the same as each other.

4. Heterotrophy, photomixotrophy and photoautotrophy

4.1. INTERPRETATION OF THE MEASURED RESULTS OF CO_2 CONCENTRATION AND PHOTOSYNTHETIC CHARACTERISTICS OF PLANTLETS IN VITRO

Based on the results of the CO_2 measurement given in Section 2.3 and photosynthetic characteristics given in Section 3.2, the following interpretation or hypotheses can be made: 1) chlorophyllous leafy explants/shoots/plantlets have photosynthetic ability; 2) their photosynthesis is restricted by low CO_2 concentration during the photoperiod; 3) they are forced to develop heterotrophy or photomixotrophy to grow; 4) a higher PPFD will not increase net photosynthetic rates under such low CO_2 concentration conditions; 5) plantlets grow faster under photoautotrophic, high CO_2 and high PPFD conditions than under hetero– or photomixotrophic conditions; 6) the initial growth rate under photoautotrophic conditions is greater for an explant with a larger leaf area and high photosynthetic ability; 7) the presence of sugar in the medium itself inhibits the photosynthetic activity of plantlets *in vitro*, 8) the medium composition widely used for hetero- or photomixotrohic micropropagation is not optimal for photoautotrophic micropropagation, 9) not only low CO_2 concentration but also low air movement (stagnant air) in the vessel are limiting photosynthesis, 10) air mixing or forced ventilation promotes photosynthesis and subsequently growth (Kozai, 1991f).

To confirm the above interpretation a series of experiments on growth and development responses of *in vitro* plantlets have to be conducted under different but controlled environment conditions.

4.2. STRATEGIES FOR ENVIRONMENTAL CONTROL

The growth of tissues/shoots/plantlets *in vitro* is basically heterotrophic in conventional micropropagation. Under heterotrophic conditions, small (air volume: 50–300 ml) and relatively airtight vessels must be used to minimize the loss of shoots/plantlets *in vitro* due to biological contamination.

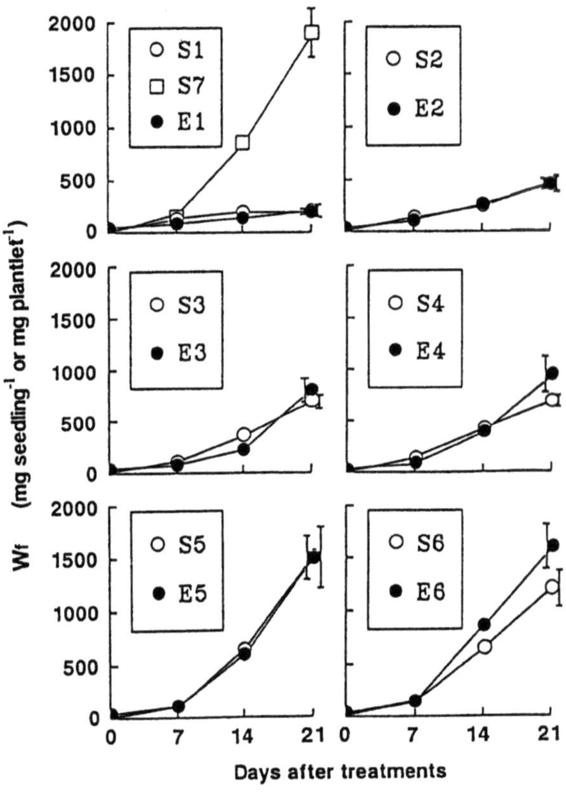

Figure 12. Time courses of fresh weight per plantlet/seedling (Wf) of *Brassica campestris* L. under different environmental conditions (vertical bars indicate confidence interval at 5% level; for details of the treatments, see Table 6). (Kozai et al., 1991)

Table 6. Description of the treatments.

Treatment	Plant material	Sucrose ($g\,l^{-1}$)	PPFD[a] ($\mu mol\,m^{-2}s^{-1}$)	N[b] (h^{-1})
S1	Seed	0	80	0.3
S2	Seed	20	80	0.3
S3	Seed	0	80	4.7
S4	Seed	20	80	4.7
S5	Seed	0	170	4.7
S6	Seed	20	170	4.7
S7	Seed	0	170	—[c]
E1	Explant	0	80	0.3
E2	Explant	20	80	0.3
E3	Explant	0	80	4.7
E4	Explant	20	80	4.7
E5	Explant	0	170	4.7
E6	Explant	20	170	4.7

[a] PPFD: Photosynthetic photon flux density on the empty culture shelf. PPFD of 80 and 170 $\mu mol\ m^{-2}\ s^{-1}$ corresponded, respectively, to an illuminance of 6.1 klx and 13 klx.

[b] N: Number of air changes per hour of vessel (see text).

[c] An accurate measurement of N was unsuccessful for the vessel without a cap. The N is considered however to be well over 50 h^{-1}

It is difficult to install environmental sensors into small vessels for controlling the *in vitro* environment aseptically. The physical environmental factors controlled in the culture room, if at all, are the air temperature and photoperiod. All other physical environmental factors, as well as chemical ones, are uncontrolled during the culture process in conventional micropropagation. Chemical components and pH of the culture medium are controlled only at the beginning of culture period.

At the acclimatization stage, i.e., in the 'ex vitro environment', growth is photoautotrophic and the environment varies a lot with time. This shift in the trophic phase and the drastic environmental changes may cause serious damage or loss of plantlets at the acclimatization stage, as shown in Table 5.

Taking the trophic and environmental changes into consideration, there are several strategies for environmental control in future micropropagation systems for obtaining high quality transplants at a lower production cost.

One is to control the environment for growing shoots/plantlets heterotrophically and aseptically at the multiplication and rooting stages. The environment would then be carefully and gradually modified with time at the acclimatization stage to induce shoots/plantlets to grow and develop photoautotrophically in the greenhouse or open field. This way is widely accepted in conventional micropropagation.

The second is to control the environment for growing shoots/plantlets photoautotrophically (but aseptically) at the multiplication and rooting stages. In this case, there would be no drastic environmental and trophic phase changes when plantlets are transferred to the greenhouse or open field conditions. Because, acclimatization would have been basically completed at earlier stages (*in vitro* acclimatization). The latter approach brought about the concept of photoautotrophic (sugar–free) micropropagation under a controlled environment (Kozai, 1991f), in which environmental control procedures are essential for maintaining growth, development and morphogenesis of shoots/plantlets *in vitro*.

There is another approach which may be called photomixotrophic micropropagation. It is a mixture of heterotrophic and photoautotrophic micropropagation in varying degrees. The best approach depends, perhaps, on plant species, number of transplants required, etc. For example, heterotrophic micropropagation may be a better choice to produce microtubers and bulblets. To make such a choice correctly, features of the *in vitro* environment and physiological responses of shoots/plantlets *in vitro* should carefully examined.

In the sense mentioned above, measurement, analysis and control of total carbon, CO_2 metabolism and carbohydrate balance are important areas for future study (Dogane et al., 1989; Fujiwara et al.,1990; Falque et al., 1991).

4.3. ADVANTAGES AND DISADVANTAGES OF PHOTOAUTOTROPHIC MICROPROPAGATION

Photoautotrophic micropropagation seems to have many advantages for increasing productivity and plantlet quality and for reducing production costs (Kozai, 1991a). Some of the advantages over heterotrophic or conventional micropropagation have been proven in recent experiments but many have not been and additional research is required. Potential advantages of photoautotrophic micropropagation are summarized in Table 7.

Table 7. Advantages of photoautotrophic micropropagation over heterotrophic micropropagation.

1. Growth and development of shoots/plantlets *in vitro* are promoted.
2. Physiological/morphological disorders and mutation are reduced and plantlet quality is improved.
3. Relatively uniform growth and development are expected.
4. Leafy single or multi node cuttings can be used as explants.
5. Procedures for rooting and acclimatization are simplified.
6. Application of growth regulators and other organic matter can be minimized.
7. Loss of shoots/plantlets due to contamination is reduced.
8. A larger vessel can be used with minimum contamination.
9. Then, environmental control of the vessel becomes easier.
10. The control of growth and development by means of environmental control becomes easier.
11. Asepsis in the vessel is not required if no pathogen is guaranteed.
12. Automation, robotization and computerization become easier.

In photoautotrophic micropropagation, leafy single node or multi node cuttings are used as explants at the multiplication stage. With this multiplication method, explants are relatively equal in size and quality, resulting in relatively uniform growth of plantlets. Occurrence of mutation or other abnormalities can be minimized by reduced use of phytohormones, by not using callus or cells, and by maintaining a favorable environment. In photoautotrophic micropropagation, asepsis is not a necessity if the plantlets are kept free from pathogens because no fungi and bacteria develop quickly in the sugar-free medium and loss of plantlets due to biological contamination is reduced.

Above all, the whole micropropagation system can be simplified. Furthermore, research and development on environmental effects on the micropropagation process could be simplified and more knowledge on growth and development of seedlings and cuttings could be applied in micropropagation research.

However, photoautotrophic micropropagation can be applied only when chlorophyllous explants are available. The main disadvantages of photoautotrophic micropropagation are the increase in cost for CO_2 enrichment (high CO_2 concentration), lighting (high PPFD) for better photoautotrophic growth of shoots/plantlets, and cooling (high PPFD produces more heat in the culture room).

However, in reality, the cost of CO_2 enrichment will not be significant (Kozai, 1991c). CO_2 enrichment has been widely commercialized for major horticultural crops in large (air leaky) greenhouses at minimum cost. There are several methods of increasing CO_2 in the vessel at a reasonable cost, as will be shown later.

There are also several methods of reducing the costs for lighting and cooling, as will be shown later. Development of lighting system for micropropagation is still at a primitive stage and there is much room for improvement. Other disadvantages of photoautotrophic micropropagation over heterotrophic micropropagation are discussed elsewhere (Kozai, 1991c).

5. Control of growth and development of shoots/plantlets *in vitro* by environmental control

The purpose of environmental control in micropropagation is to solve the problems shown in Table 1 and to modify shoot/plantlet responses shown in Table 4. In the following sections, recent research on the effect of environmental factors on the growth and development of *in vitro* plantlets under controlled *in vitro* environments are reviewed, with an emphasis on control of shoot length.

5.1. IMPORTANCE OF SHOOT LENGTH CONTROL

Among the responses given in Table 4, shoot length (or internode length) of shoots/plantlets *in vitro* is becoming an important variable to be controlled. The control of shoot length has often been achieved by application of plant growth regulating agents. However, its widespread application is not generally recommended from the view point of reduction in possibilities of environmental pollution. Thus, the control of shoot length by environmental control is becoming an essential technique. There are two reasons for the importance of controlling shoot length (Kozai et al., 1991c).

One reason is that when micropropagated plantlets are transferred for acclimatization to the greenhouse or the open field, the morphology of plantlets is as important as their physiological characteristics. Generally required morphological characteristics of transplants are: (1) short and thick main stem or short plantlet height, (2) low shoot/root weight ratio, and (3) relatively large leaf area with normal stomata and thick cuticular layers.

Among them, the short and thick main stem or short plantlet height is often one of the important morphological characteristics of high quality transplants. In reality, however, some micropropagated transplants tend to be tall and thin due to characteristics of the *in vitro* environment. Tall and thin transplants tend to result in lodging as they grow in the greenhouse or in the open field.

The other reason is that nodal sections and axillary buds are the most commonly used explants in commercial micropropagation at the multiplication stage. When nodal sections are used as explants at the multiplication stage, excision can be more easily completed if the internode is longer than, for example, 5–10 mm. This is also true for automatic or robotic excision (Kozai, et al., 1991d; Kozai, 1991e). In addition, the initial growth rate of explants at the multiplication stage is often improved with increasing their internode length or dry matter weight. Therefore, the control of shoot length, plantlet height or internode length is as important as the control of other parameters of growth (weight, leaf area, etc.) and development (the number of unfolded leaves, rooting, etc.) both at the multiplication and acclimatization stages.

5.2. TEMPERATURE

Air temperature in the culture room is kept more or less constant throughout the day at around 25°C in most cases. This is reasonable for heterotrophic micropropagation, because a high dark respiration rate can be considered, in many cases, as an indicator of active growth. In photoautotrophic or photomixotrophic micropropagation, however, growth may be inhibited if the dark respiration rate is too high. A high, dark respiration rate may indicate a loss of carbon which was fixed by photosynthesis during the preceding photoperiod (Kozai, 1991b). Optimum temperatures for photo- and dark periods in photoauto- and photomixotrophic growth of shoots/plantlets *in vitro* are an area for further study.

Heins et al.(1988) showed that the plant height (or internode length) of many ornamental plants such as lily and chrysanthemum increased with increasing the difference in air temperature between day and night, named as 'DIF'. The plants became taller with greater positive DIF values and became shorter with greater negative DIF values. This DIF procedure is now commercially applied for controlling plant height by many American greenhouse crop growers. Erwin et al. (1989) showed that the DIF procedure could also be applied to control the height of plug seedlings of some ornamental bedding plants.

An experiment was then conducted to see whether the DIF procedure could be applied for the plantlets *in vitro* under photoautotrophic and CO_2 enriched conditions (Kozai et al., 1992). Single node leafy cuttings (fresh weight: 30 mg each) excised from potato (*Solanum tuberosum* L. 'Benimaru') plantlets *in vitro* were cultured for 28 days on MS agar (8 g l^{-1}) medium in glass test tubes (one explant per vessel, 100 explants per treatment) with a number of air changes per hour of 1.5. The photoperiod was 12 h d^{-1} with a PPFD (photosynthetic photon flux density) of 74 (low PPFD) or 147 (high PPFD) umol m^{-2} s^{-1}. The CO_2 concentration in the culture room was kept at 1300–1500 umol mol^{-1} during the photoperiod and was uncontrolled during the dark period. The daily mean air temperature in the culture room was kept at 20 °C with photo-/dark period temperatures of 25°C/15°C, 20°C/20°C, and 15°C/25°C, which provided DIF values of 10, 0, and –10 °C, respectively.

Figure 13 shows the effect of the DIF and PPFD on the shoot length of potato plantlets *in vitro* by day 28. The length was greater with increasing DIF under both low and high PPFD conditions. There were no statistically significant differences in dry and fresh weight among the different DIF treatments under high PPFD conditions (data not shown). The shoot length was slightly greater for low PPFD than for high PPFD in each DIF treatment, but the dry and fresh weight, leaf area and number of unfolded leaves were greater in high PPFD than in low PPFD in each DIF treatment (data not shown).

Since the culture room is, in general, thermally well insulated, the cost for cooling is largely dependent upon the amount of heat energy generated from the lamps, and is much less dependent upon the air temperatures inside and outside the culture room. The cost for heating during the dark period is, on the other hand, minimum even during the winter if the room is thermally well-insulated. Therefore, the DIF procedure should be an efficient way of controlling the height of plantlets *in vitro* with minimum heating/cooling costs.

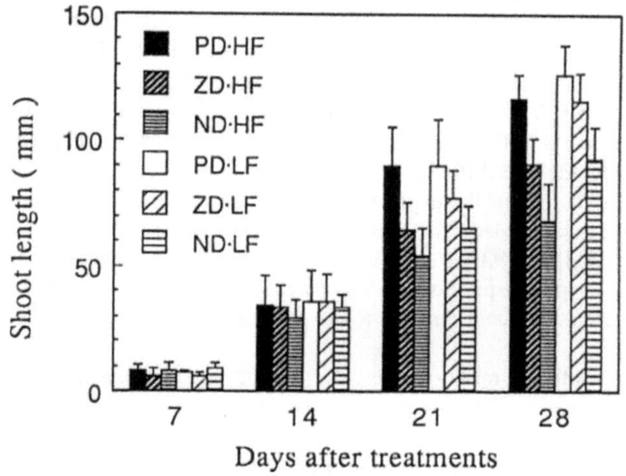

Figure 13. Effect of DIF on shoot length of potato plantlets *in vitro* 7, 14, 21 and 28 days after treatments. DIF is defined as the difference between photoperiod and dark period temperatures (Heins et al., 1988). Vertical bars represent S.D. from the mean of duplicate. PD: Positive DIF (Photoperiod: 25 °C, Dark period: 15 °C), ZD: Zero DIF (Photoperiod: 20 °C, Dark period: 20 °C), ND: Negative DIF (Photoperiod: 15 °C, Dark period: 25 °C), HF: High PPFD (147 µmol $m^{-2} s^{-1}$), LF: Low PPFD (74 µmol $m^{-2} s^{-1}$). Daily mean temperature is 20 °C in all the treatments. (Kozai et al., unpublished)

5.3. HUMIDITY AND WATER POTENTIAL

It is well known for plants grown outside that high humidity, in general, elongates their shoot length. A similar effect for shoots/plantlets *in vitro* is widely recognized by plant tissue culture researchers but there are few reports clearly indicating this effect under controlled relative humidity conditions.

An experiment was then conducted to evaluate the effect of relative humidity on shoot length of potato (*Solanum tuberosum* L. 'Benimaru') plantlets *in vitro*, using single node leafy cuttings (fresh weight: 45 mg each) as explants (Tanaka et al., unpublished). One hundred and ninety two explants were cultured in 96 cylindrical plastic vessels (air volume: 750 ml, 9.9 cm in diameter,

10.7 cm high, 2 explants per vessel) with a number of air changes per hour of 0.95. Two small plastic cups (caps for closing test tubes) were placed in the vessel and each cup was filled with MS agar medium containing 20 g l^{-1} sucrose (medium volume: 15 ml cup^{-1}). Then, the explant was placed on the medium (one explant per cup).

Distilled water was then poured into the vessel (80 ml $vessel^{-1}$) to keep the relative humidity at nearly 100%. (The measurement showed that the average humidity was a few percent lower than 100%.) The explants were cultured for the first 8 days under the high humidity conditions.

On day 8, the distilled water of 24 vessels out of the 96 vessels was replaced with saturated solution of NaCl to maintain the humidity at 75% or little higher. Similar treatments for 2 other groups of 24 vessels were done using KCl and KNO_3 to maintain the humidity at 84% and 93% or higher, respectively. (Equilibrium relative humidities of saturated solutions of NaCl, KCl and KNO_3 at 25°C are, respectively, 75, 84 and 93%.) The remaining 24 vessels were unchanged to maintain the humidity at nearly 100%. The explants were cultured for an additional 14 days (from day 8 to day 22) under the four different humidity conditions.

The humidity increased with time as the explants grew (the transpiration increased) in each of the treatments, even with the salt added. The humidities in the vessels on day 10 were, respectively, 81% (NaCl), 86% (KCl) and 93% (KNO_3), and those on day 22 were, respectively, 88%, 92% and 95%. Part of the evapotranspired water was absorbed by the saturated solution and the rest was transported to the culture room through a gas permeable membrane filter incorporated into the vessel.

The air temperature and relative humidity in the culture room were kept, respectively, at 25 °C and 75% throughout the experiment. The photoperiod was 16 hours followed by 8 hours of dark period with a PPFD of 100 umol m^{-2} s^{-1}. The CO_2 concentration in the culture room was increased with time so as to keep the average concentration in the vessels no lower than 350 umol mol^{-1}.

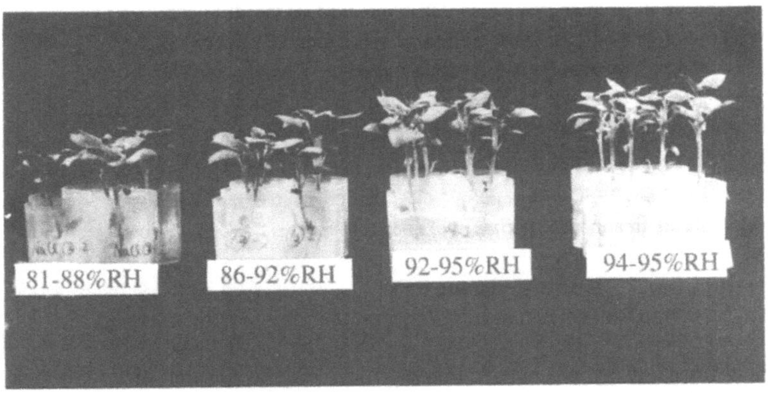

Figure 14. Potato plantlets cultured *in vitro* on MS agar medium with 20 g l^{-1} sucrose for 22 days under different relative humidity (RH) conditions. (Tanaka et al., unpublished)

Fig. 14 shows the potato plantlets cultured *in vitro* for 14 days (on day 22) under the four different humidity conditions. The shoot length increased with increasing the initial (day 8) relative humidity. If the humidity had been kept at the same level as the humidity on day 8 throughout the experiment, the differences in shoot length among the treatments might have been much greater than those shown in the figure. The fresh weight in RH 100 and 93% treatments was greater than in RH 84 and 75% treatment. There were, however, no significant differences in dry weight among the treatments. The specific leaf area (m^2 gDW^{-1}) decreased with decreasing the relative humidity.

There are many reports on the effect of humidity on physiological disorders such as vitrification and microscopic characteristics of plant tissues cultured *in vitro* (Ziv, 1990). Responses of plantlets to gel type and concentration has been reported previously for a variety of shoots/plantlets *in vitro* (e.g., Von Arnold and Eriksson, 1984). These physiological responses to gel type and concentration are often attributed to water potential (matric and osmotic) of the medium or water availability of shoots/plantlets *in vitro* (Debergh et al., 1981). Owens and Wozniak (1991) found recently that water availability is also affected by the ease with which liquid is expressed from gels in response to localized pressure caused by explant/root expansion and contortion, called gel expressibility.

5.4. LIGHT

5.4.1. *Lighting direction* A sideward (lateral) lighting system has recently been developed to promote the growth of plantlets *in vitro* and to control plantlet height or shoot length and over all shoot quality (Kozai et al., 1991a). Figure 15 shows the schematic diagram of sideward lighting system using fluorescent lamps as a light source. This lighting system has been developed for a large scale micropropagation facility and automated handling of culture vessels.

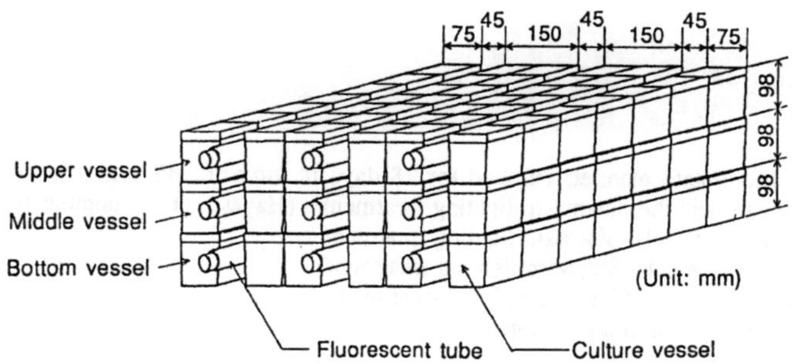

Figure 15. Schematic diagram of sideward lighting system using fluorescent tubes as light source. -Geometry of the tubes and culture vessels- (Hayashi et al., unpublished)

The vessels and the lamps are stacked vertically to save space, without reducing the amount of light energy received by each of the shoots/plantlets. With this system, the electricity costs per shoot/plantlet for lighting and cooling are reduced because the ratio of light energy received by the shoots/plantlets to the light energy emitted from the light source is increased. Furthermore, the light energy emitted is more evenly received by the upper, middle and lower leaves of shoots/plantlets, resulting in relatively large leaves at the bottom and relatively small leaves at the top when compared with those of the shoots/plantlets cultured under downward (normal) lighting conditions (Kozai et al, 1991a).

Figure 16 shows the potato plantlets cultured with the sideward lighting system shown in Figure 15 and those cultured with conventional, downward lighting system. Leafy single node cuttings were used as explants and they were cultured *in vitro* for 28 days on MS sugar–free medium in the vessel with the number of air changes per hour being 3.9. The shoot length on day 28 was approximately 3.5 cm shorter in the sideward lighting treatments than in the downward lighting treatments. However, the dry weight and leaf area per plantlet were 1.8 times greater in the sideward lighting treatment than in the downward lighting treatments. In the sideward lighting treatment, there were no significant differences in growth and development between the plantlets cultured in top, middle and bottom vessels. The quality of the plantlets when used as transplants was much higher in sideward lighting treatment than in downward lighting treatment.

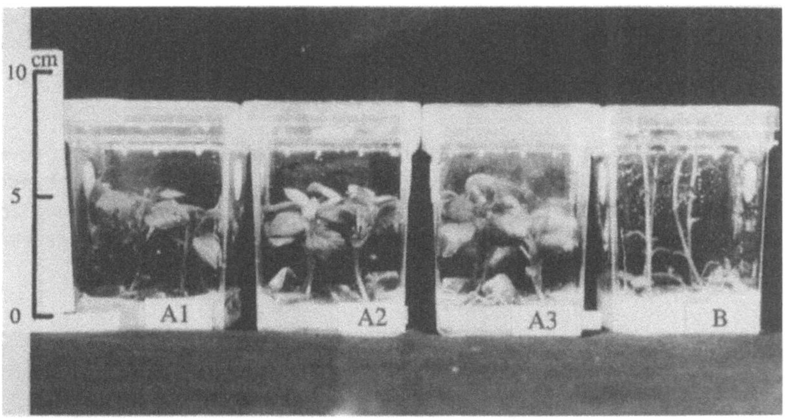

Figure 16. Potato plantlets cultured for 28 days *in vitro* on MS sugar-free medium in the sideward and the downward lighting treatments. (Hayashi et al., unpublished)
A1: Bottom vessels, Sideward lighting treatment
A2: Middle vessels, Sideward lighting treatment
A3: Upper vessels, Sideward lighting treatment
B : Downward lighting treatment

The required air space between the light source and the shoots/plantlets is reduced with the sideward lighting system, especially when diffusive optical fibers or light emitting diodes are used as a light source (Figure 17). They produce less conductive heat and less longwave radiant energy than fluorescent lamps. The diffusive fibers are string light sources and the diodes are point light sources, and thus they can be placed closer to the shoots/plantlets and thereby improving space utilization of the culture area (Kozai, 1991b). In this case, however, an automatic vessel transportation system should be incorporated.

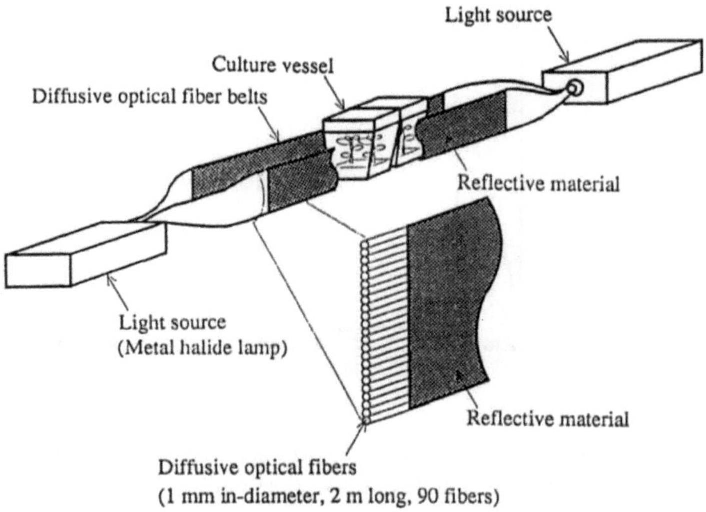

Figure 17. Schematic diagram of the sideward light distribution system using diffusive optical fibers. (Kozai et al., unpublished)

5.4.2. *Lighting Cycle* Morini et al. (1990 and 1991) found that the growth of fruit tree shoots *in vitro* was affected by the lighting cycle (photoperiod plus dark period). Ratios of the photoperiod to the dark period (16 h/8 h, 8 h /4 h, 4 h/2 h and 2 h/1 h) over a 24 hour period, provided a total daily photoperiod of 16 hours for each of the treatments. In their experiment, the lighting cycle of 6 hours (4 h/2 h) gave the best shoot growth. The authors explained these results by possible roles of the phytochrome and of a higher CO_2 availability in the vessel. Doi et al. (1989) have reported that day-lengths (ratios of the photoperiod to dark period) affect the time course of CO_2 concentration in culture vessels.

The shoot length as well as growth of potato plantlets *in vitro* are also affected by the lighting cycle (Hayashi, et al., unpublished). Leafy single node cuttings were cultured *in vitro* for 28 days on MS agar medium with 20 g l^{-1} sucrose in a vessel with the number of air changes per hour being 0.8 at a temperature of 23°C, a PPFD of 50 umol m^{-2} s^{-1} and a CO_2 concentration in the culture room of 400–500 umol mol^{-1}. The shoot length was significantly greater when cultured at a lighting cycle of 1.5 hours (1 h photoperiod/0.5 h dark period) than when cultured at lighting cycles of 24 h, 6 h and 0.375 h; the ratio of photoperiod to dark period was 2 for all the treatments. The fresh and dry weight was greatest in the shortest lighting cycle treatment than in the other treatments.

Figure 18 shows the difference in CO_2 concentrations between inside and outside the vessel on days 14 and 28 for different lighting cycle treatments. Daily CO_2 uptake by the plantlets is defined as the negative of the daily integral of the difference. The daily net CO_2 uptake increased with decreasing the lighting cycle on both days 14 and 28.

5.4.3. *Light Quality* Appelgren (1991) showed that exposure of *Pelargonium in vitro* to red light significantly increased stem elongation compared with white (fluorescent) light, while blue light strongly inhibited elongation. Moe et al.(1991) showed that stem elongation and plantlet height

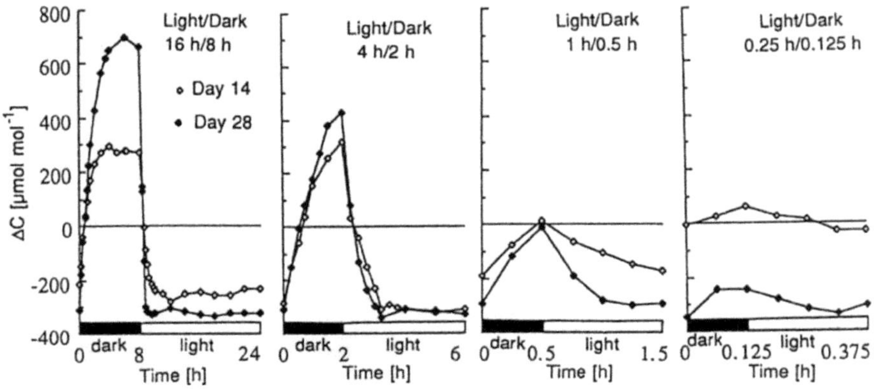

Figure 18. The difference in CO2 concentrations between inside and outside the vessel on days 14 and 28 for different lighting cycle treatments. ($\Delta C = C_{in} - C_{out}$, C_{in}: CO2 concentration inside the vessel, C_{out}: CO2 concentration outside the vessel. Number of air changes of the vessels = 0.8 h^{-1}.) (Hayashi et al., unpublished)

at flowering in *Campanulla isophylla* were greater when plants were exposed to far red light or light from incandescent lamps which had a low red/far–red ratio. Control of plantlet height or morphology by light quality is an important area for further study.

5.5. CARBON DIOXIDE CONCENTRATION

Based on the data given in Section 3, one can expect an increase in net photosynthetic rate and hence growth of shoots/plantlets *in vitro* by increasing CO_2 concentration in the vessel under high PPFD conditions. Practical ways of increasing the concentration include:

a) Use of gas–permeable film as closures.
b) Use of gas–permeable film as closures in a CO_2 enriched culture room.
c) Use of a larger vessel with a CO_2 supply system or a forced ventilation system.

5.5.1.*Gas–permeable film as closures* There are many reports indicating positive effects of use of gas–permeable film as closures under high or middle PPFD on increases in net photosynthetic rate and growth of shoots/plantlets *in vitro*. When leafy or chlorophyllous explants are used, the growth of carnation (Kozai and Iwanami, 1988), strawberry (Kozai and Sekimoto, 1988), radiata pine (Aitken–Christie et al., 1990) and tobacco (Kozai et al, 1990a) shoots/plantlets *in vitro* is greater when cultured on a medium without sugar in vessels closed with the film than when cultured on a medium with sugar in relatively air–tight vessels. Even with the film, however, the increase in net photosynthetic rate per leaf area is reduced as the shoots/plantlets grow mainly because the CO_2 concentration in the vessel decreases with time to 150–200 umol mol^{-1}. Thus, promotion of photosynthesis of shoots/plantlets *in vitro* is limited. The number of air changes per hour of the vessel should be in the range between 3–5 to keep the relative humidity higher than 90–95% and CO_2 concentration higher than 150–200 umol mol^{-1}.

With this approach, the effect of the film under high or middle PPFD on the growth of shoots/plantlets *in vitro* can be verified with minimal changes in an existing micropropagation facility. Only the film and a higher PPFD are necessary to conduct these kind of experiments and micropropagation.

The percentage of vitrification is also decreased with the use of the film possibly due to lower humidity and increased gas exchange of the vessel and drying out of the medium (Aitken–Christie et al., 1990). The film enhances stomatal function, photosynthesis and subsequently acclimatization of tobacco plantlets *in vitro* (Safadi and Hughes, 1991; Safadi et l., 1991)

5.5.2. *CO_2 enrichment in the culture room* CO_2 enrichment under high PPFD (100–200 umol $m^{-2}s^{-1}$) was effective for promoting shoot/plantlet growth of tobacco (Mosseau, 1986), *Cymbidium* (Kozai et al., 1987b), carnation (Kozai and Iwanami, 1988), potato (Kozai et al., 1988a; Cournac et al., 1991), etc., when cultured on media both with and without sugar.

For some plant species such as radiata pine, a PPFD higher than 100 umol mol^{-1} seemed not to be beneficial (Davies et. al., 1992) when conventional nutrient media were used. There are extensive reviews on CO_2 enrichment for shoots/plantlets *in vitro* (Kozai, 1991b and 1991c).

5.5.3. *A larger vessel with a CO_2 supply system or a forced ventilation system* The dry weight and net photosynthetic rate of strawberry plantlets cultured with sugar–free liquid medium increased several times when cultured in a large vessel with a forced ventilation system under a PPFD of 96 umol m^{-2} s^{-1}, compared with those cultured using conventional method (Fujiwara et al., 1988). There are several papers indicating the positive effect of a forced ventilation system (e.g. Kubota and Kozai, 1992).

Forced ventilation with atmospheric air or nitrogen–oxygen–carbon dioxide mixtures when applied at the multiplication stage, however, resulted in reduced propagule weights and shoot numbers of *Rhododendron* in vessels with a 400 ml air volume when cultured under a PPFD of 39 umol m^{-2} s^{-1} (Walker et al., 1988). A possible reason for the detrimental effect in the experiment could be that the experiment was carried out under low PPFD conditions.

In many of the experiments on CO_2 enrichment, not only CO_2 concentration but also relative humidity, ethylene, air flow speed, etc. in the vessel were affected and changed by the use of gas permeable films. Therefore, the changes in growth of plantlets *in vitro* resulting from the experiments cannot be attributed only to a CO_2 enrichment effect. However, the changes are considered to be brought primarily by CO_2 enrichment in most cases. Further detailed studies should clarify the effect of a gaseous environment on photosynthetic growth and development.

5.5.4. *CO_2 enrichment for Cymbidium* Honjo and Takakura (1987) examined the possibility of photoautotrophic growth of *Cymbidium* PLB (Protocorm like body) in liquid medium the multiplication stage at different CO_2 concentrations and PPFD conditions. The PLBs increased in dry weight when grown photoautotrophically. However, The increase in dry weight was 2–5 times greater under heterotrophic conditions than under photoautotrophic conditions. This may be be cause the PLBs are submerged in the liquid medium.

5.6. INORGANIC NUTRIENT COMPOSITION

The most commonly used nutrient compositions in conventional micropropagation have been optimized for heterotrophic cultures. Therefore, the nutrient compositions may not be optimal for photoautotrophic and photomixotrophic micropropagation. Furthermore, some organic substances such as vitamins and phytohormones may be eliminated from the medium under photoautotrophic and photomixotrophic culture conditions. Because, photosynthetically active plant–

lets cultured *in vitro* have the capability to produce some of these substances endogenously.

The growth of carnation plantlets cultured in loosely capped vessels under high PPFD was greater on a sugar–free medium containing the inorganic nutrient components, widely used for hydroponic culture in greenhouses, than on a sugar–containing medium with either half–strength MS medium or with the hydroponic components (Kozai et al., 1988b). In half–strength MS, the ion concentrations of NH_4 and Mn^{2+} are, respectively, 7.6 and 5.5 times those in the hydroponic medium, and for PO_4^{3-}, Ca^{2+} and Mg^{2+} are, respectively, 0.46, 0.37 and 0.37 times.

Explants of strawberry plantlets were cultured for 21 days on half–strength MS medium with 0 and 20 g l^{-1} sucrose under CO_2 enriched and non–enriched conditions (Kozai et al., 1991c). The best growth was obtained without sucrose and with CO_2 enrichment, for which the percent residual inorganic ions in the medium on day 21 was 3% for PO_4^{3+}, 41% for Mg^{2+} and 53% for NO^{3-}. For each ion, this amount was much less than in the other treatments. Growth might have been promoted further if some PO_4^{3-} had been added to the medium in the latter part of the culture period.

In conventional micropropagation, rates of water uptake or transpiration are very low due to low saturation water vapor deficit of the air, low PPFD and low air flow rate in the vessel. Under such environmental conditions, uptake of some ions or other elements in the medium by the plantlets might be restricted even if sufficient quantities of them are contained in the medium. Nutrient uptake rate is influenced not only by the nutrient concentrations in the medium and shoots/plantlets but also by the aerial and root zone environments. Control and efficiency of nutrient uptake under *in vitro* environmental conditions remains to be investigated.

5.7. OXYGEN

Photoautotrophically grown C_3 plants lose up to 50% of photosynthetically fixed carbon due to photorespiration in a normal atmospheric concentration of O_2 (21%) and CO_2 (350 umol mol^{-1}). However, photo respiration is repressed with decreasing O_2 concentration and is almost completely repressed at 2% O_2 (Zelitch, 1975).

The net photosynthetic rate was measured after short term exposure to O_2 concentrations of 1, 10 and 21% for *Primula malacoides* (C_3 plants) cultured *in vitro* for 20 days on half strength MS agar medium with 30% g l^{-1} sucrose at a PPFD of 110 umol m^{-2} s^{-1} (Shimada et al., 1988; Tanaka et al., 1990). The rate at 1% and 10% O_2 were, respectively, nearly 3 times and 1.5 times greater than that at 21% O_2 when the CO_2 concentration was approximately 200 umol mol^{-1}.

The effect of long term exposure to low O_2 concentrations on the growth of *Chrysanthemum morifolium* (C_3 plants) was also studied by Tanaka et al.(1991b). The dry weight increase of the plantlets for 30 days was significantly greater in 10% O_2 treatment than in 5%, 15% and 21% O_2 treatments.

Increasing the net photosynthetic rate and the growth of shoots/plantlets *in vitro* by decreasing O_2 concentration during the photoperiod has a practical application, because it is relatively easy to control O_2 concentration in such a small and airtight vessel for the entire culture period. Another possible advantage of decreasing O_2 concentration is the prevention of anaerobic bacteria and fungus growth *in vitro*.

5.8. OTHER ENVIRONMENTAL FACTORS

There are many other environmental factors as listed in Figure 1 that are not discussed in this section. They include ethylene in the air (e.g., Jackson et al., 1991), pH of the medium (e.g.,

Song et al., 1989), carbon source in the medium, etc.

6. Concluding Remarks

Growth and development of shoots/plantlets *in vitro* is strongly influenced by their environment. Control of growth and development by environmental control is possible and has many important practical applications. Moreover, environmental control is becoming essential to improve the quality of shoots/plantlets *in vitro* and to reduce the production cost of the plantlets. The above statements also apply to other transplant production methods using seeds, cuttings, tubers, bulblets, etc.

The methodology and techniques for transplant production via conventional micropropagation are currently different in many respects from those in transplant production by using seeds, cut tings, etc. However, in photoautotrophic micropropagation using leafy explants, there are some similarities in growth patterns and physiological characteristics between seedlings and plantlets when they are grown under the same *in vitro* environmental conditions. This suggests that there are many common concepts, methodology, techniques, etc. among different transplant production systems. Furthermore, many of the environmental control systems, and equipment for production, transportation and transplanting can be applied to any type of transplant production with minimum modifications. A unified approach towards future transplant production systems for more efficient research, development and production of transplants is therefore necessary. Exchange of information and ideas among people working on transplant production in different fields is essential for this goal.

Acknowledgement

The authors would like to thank their students and colleagues who have contributed to the experimental work behind this paper. Thanks are specially extended to Mr. K. Tanaka and S. Kino and Ms. Y. Murai and C. Kubota, for their permitting us to use part of their unpublished work in the text. Special thanks are also extended to Ms. K. Watanabe and Mr. S. Kira for their preparing the text.

7. References

Aitken-Christie, J. (1988) Development of a semi-automated micropropagation systems, *Acta Hort*. 230, 81-86.

Aitken-Christie, J., Davies, H., Kubota, C. and Kozai, T. (1990)Benefits of using a microporous membrane in the tissue culture vessel lid for the micropropagation of *Pinus radiata*. *Abst. VIIth Intr. Congr. on Plant Tissue and Cell Culture*, Amsterdam, pp. 86 (Code No. A3-15).

Ando, T. (1978) Gaseous environment in the airtight culture vessel containing orchids. *Abst. Annual Autumn Meet. Jap. Soc. Hort. Sci.*, pp. 368-369. (in Japanese).

Appelgren, M. (1991) Effects of light on stem elongation of *Pelargonium in vitro*. *Scientia Horticulturae* 45, 345-351.

Arnold, S. von and Eriksson, T. (1984) Effect of agar concentration on growth and anatomy of adventitious shoots of *Picea abies* (L.) Karst. *Plant Cell Tissue Organ Culture* 3, 257-264.

Blazkova, A., Ullmann, J., Josefusova, Z., Machackova, I. and Krekule, J. (1989) The influence of gaseous phase on growth of plants *in vitro* – The effect of different types of stoppers. *Acta Hort.* 251, 209–214.

Capellades, M.Q. (1989) Histological and ecophysical study of the changes occurring during the acclimatization of *in vitro* cultures. Dissertation. Gent University, Belgium, 98 pp.

Capellades, M., Lemeur, R. and Debergh, P. (1991) Effects of sucrose on starch accumulation and rate of photosynthesis in Rosa cultured *in vitro*. *Plant Cell Tissue Organ Culture*. 25, 21–26.

Cournac, L., Dimon, B., Carrier, P., Lohou, A. and Chagvardieff, P. (1991) Growth and photosynthetic characteristics of *Solanum Tuberosum* plantlets cultured *in vitro* in different conditions of aearation, sucrose supply and CO_2 enrichment. *Plant Physiol.* 97, 112–117.

Cuello, J.L., Walker, P.N. and Heuser, C.W. (1989) Effects of ventilated CO_2, light and sucrose levels on stage II micropropagation of *Buddleia alternifolla* (Butterfly Bush). ASAE Paper No.89–6091. (ASAE/CSAE presentation paper).

Davies, H.E., Kubota, C., Kozai, T. and Aitken–Christie, J. (1992) Sugar–free growth of *Pinus radiata in vitro* for conventional and automated micropropagation. *Australian J. of Botany* (in press).

Debergh, P., Harbaoui, Y. and Lemeur, R. (1981) Mass propagation of globe artichoke (*Cynara scolymus*): Evaluation of different hypotheses to overcome vitrification with special reference to water potential. *Physiol. Plant.* 53, 181–187.

Debergh, P. and Zimmerman, R. (eds.)(1990) Micropropagation: Technology and Application, Kluwer Academic Pub., Dordrecht, 484pp.

Desjardins Y., Laforge, F., Lussier, C. and Gosselin, A. (1988) Effect of CO_2 enrichment and high photosynthetic photon flux on the development of autotrophy and growth of tissue–cultured strawberry, raspberry and asparagus plants. *Acta Hort.* 230, 45–53.

Dillen, W. and Buysens, S. (1989) A simple technique to overcome vitrification in *Gypsophila paniculata* L. *Plant Cell Tissue Organ Culture* 19, 181–188.

Dogane, Y., Ando, T., Nakagawa, M., and Mii, M. (1989) Trophic phase: A new concept for *in vitro* culture and acclimatization. *J. Japan. Soc. Hort. Sci.* 58 (suppl. 2), pp. 510–511.

Doi, M., Oda, H. and Asahira, T. (1989) *In vitro* atmosphere of cultured C_3 and CAM plants in relation to day–lengths. *Environ. Control in Biol.* 27, 9–13.

Dooly, J.H. (1991) Influence of lighting spectra on plant tissue culture, ASAE Paper No. 917530, Amer. Soc. Agr. Eng., St. Joseph, MI, USA.

Dunstan, D.I. and Turner, K.E. (1984) The acclimatization of micropropagated plants. In: Cell Culture and Somatic Cell Genetics of Plants. Vol.1. Ed. Vasil, I.. pp. 123–129. Academic Press.

Erwin, J., Heins, R., Carlson, W. and Biernbaum, J. (1989) Do cool days/warm nights work with plugs ? *GrowerTalks* 3, 46–49.

Falque, Compan, M.D. and Galzy, R. (1991) A method for the study of CO_2 exchanges *in vitro* cultured *Vitis rupestris* plantlets. *Plant Cell Tissue Organ Culture* 27, 175–181.

Fujiwara, K., Kozai, T. and Watanabe, I. (1987) Fundamental studies on environment in plant tissue culture vessels (3). *J. Agric. Meteorol.* 43, 21–30.

Fujiwara, K., Kozai, T. and Watanabe, I. (1988) Development of a photoautotrophic tissue culture system for shoots and/or plantlets at rooting and acclimatization stages.

Acta Hort. 230, 153–158.
Fujiwara, K., Kozai, T., Nakajo, Y. and Watanabe, I. (1989) Effects of closures and vessels on light intensities in plant tissue culture vessels. *J. Agric. Meteorol.* 45 (3), 143–149 (in Japanese with English summary).
Fujiwara, K., Dogane, Y., Ando, T. Mii, M. and Kozai, T. (1990) A classification system of trophic phases of plant cultures in vitro. *Abst. VIIth intr. Congr. on Plant Tissue and Cell Culture*, Amsterdam, pp. 100 (Code No. A3–70).
Fujiwara, K. (1991) Water environment in plant tissue culture vessels. *J. Agric. Meteorol.* 47, 101–107. (written in Japanese)
Grout, B.W.W. and Ashton, M.J. (1977) Transplanting of cauliflower plants regenerated from meristem culture. I. Water loss and water transfer related to changes in leaf wax and to xylem regeneration. *Hort. Res.* 17, 1–7.
Hakkaart, F.A. and Versluijs, J.A. (1983) Some factors affecting glassiness in carnation meristem tip cultures. *Neth. J. Pl. Path.* 89, 47–53.
Hale, A. and Young, R. (1991) Plant micropropagation bioreactor development, ASAE Paper No. 917542, Amer. Soc. Agr. Eng., St. Joseph, MI, USA.
Hayashi, M. and Kozai, T. (1987) Development of a facility for accelerating the acclimatization of tissue–cultured plantlets and the performance of test cultivations. *Proc. Symposium Florizel 87* Arlon, Belgium. Plant micropropagation in Horticultural Industries. pp. 123–134.
Hayashi, M., Kozai, T., Watanabe, K. and Watanabe, I. (1990) Effects of CO_2 enrichment and high solar radiation on the growth of potato plantlets in direct ex–vitro rooting method. Environ. *Control in Biology* 28, 147–154. (E with J)
Heins, R., Erwin, J., Berghage, R., Karlsson, M., Biernbaum, J. and Carlson, W. (1988) Use temperature to control plant height. *Greenhouse Grower* 6, 32–34.
Honjo, T. and Takakura, T. (1987) Effects of CO_2 concentration, light intensity and liquid medium composition for the growth of *Cymbidium* PLB *in vitro*. *J. Agric. Meteorol.* 43, 223–227. (J with E)
Honjo, T. and Takakura, T. (1991) Measurement of humidity profiles in micropropagation vessels. ASAE Paper No. 911512, Amer. Soc. Agr. Eng., St. Joseph, MI, USA.
Hughes, K. W.(1981) *In vitro* ecology: Exogenous factors affecting growth and morphogenesis in plant culture systems. *Environ. and Experimental Botany*. 21, 281–288.
Infante, R., Magnanini, E. and Righetti, B. (1989) The role of light and CO_2 in optimizing the conditions for shoot proliferation of *Actinidia deliciosa in vitro*. *Physiol. Plant.* 77,191–195.
Jackson, M.B., Abbot, A.J., Belcher, A.R., Hall, K.C., Butler, R. and Cameron, J. (1991) Ventilation in plant tissue culture and effects of poor aeration on ethylene and carbon dioxide accumulation, oxygen depletion and explant development. *Annals of Botany* 67, 229–237.
Kozai, T., Fujiwara, K. and Watanabe, I. (1986a) Relation between the culture medium composition and water potential of liquid culture media. *J. Agric. Meteorol.* 42, 1–6. (J with E)
Kozai, T., Fujiwara, K. and Watanabe, I. (1986b) Effects of stoppers and vessels on gas change rates between inside and outside of vessels closed with stoppers. *J. Agric. Meteorol.* 42, 119–127. (J with E)
Kozai, T., Hayashi, M., Hirosawa, Y., Kodama, T. and Watanabe, I. (1987a) Environmental control for acclimatization of *in vitro* cultured plantlets (1). *J. Agric. Meteorol.* 42,

349–358. (J with E)
Kozai, T., Oki, H. and Fujiwara, K. (1987b) Effects of CO_2 enrichment and sucrose concentration under high photosynthetic photon fluxes on growth of tissue–cultured *Cymbidium* plantlets during the preparation stage. *Proc. Symposium Floriezel 87* Arlon, Belgium. Plant micropropagation in Horticultural Industries. pp. 135–141.
Kozai, T. and Sekimoto, K. (1988) Effects of the number of air changes per hour of the closed vessel and the photosynthetic photon flux on the carbon dioxide concentration inside the vessel and the growth of strawberry plantlets *in vitro*. *Environ. Control in Biol.* 26, 21–29. (J with E)
Kozai, T. and Iwanami, Y. (1988) Effects of CO_2 enrichment and sucrose concentration under high photon fluxes on plantlet growth of carnation (*Dianthus caryophyllus* L.) in tissue culture during the preparation stage. *J. Jap. Soc. Hort. Sci.* 57(2), 279–288.
Kozai, T., Koyama, Y. and Watanabe, I. (1988a) Multiplication of potato plantlets *in vitro* with sugar free medium under high photosynthetic photon flux. *Acta Hort.* 230, 121–127.
Kozai, T., Kubota, C. and Watanabe, C. (1988b) Effects of basal medium composition on the growth of carnation plantlets in auto– and mixotrophic tissue culture. *Acta Hort.* 230, 159–166.
Kozai, T., Takazawa, A., Watanabe, I. and Sugi, J. (1990a) Growth of tobacco seedlings and plantlets *in vitro* as affected by in vitro environment. *Environ. Control in Biol.* 28, 31–39. (J with E)
Kozai, T., Oki, H. and Fujiwara, K. (1990b) Photosynthetic characteristics of *Cymbidium* plantlet *in vitro*. *Plant Cell Tissue Organ Culture* 22, 205–211.
Kozai, T. (1991a) Photoautotrophic micropropagation. *In Vitro Cell Dev. Biol.* 27P, 47–51.
Kozai, T. (1991b) Controlled environments in conventional and automated micropropagation. In: Cell culture and somatic cell genetics of plants, Vol.8 (ed. Vasil, I.), Academic Press, Inc. pp.213–230.
Kozai, T. (1991c) Autotrophic micropropagation, In: Biotechnology in Agriculture and Forestry, Vol.17, High–Tech and Micropropagation I (ed. Bajaj, Y.P.S.), Springer–Verlag, Berlin, pp.313–343.
Kozai, T. (1991d) Acclimatization of micropropagated plantlets, In: Biotechnology in Agriculture and Forestry, Vol.17, High–Tech and Micropropagation I (ed. Bajaj, Y.P.S.), Springer–Verlag, Berlin, pp. 127–139.
Kozai, T. (1991e) Environmental control and automation in micropropagation. In: "Automation in Biotechnology (Proceedings of the 4th Toyota Conference)" (ed. Karube, M.), Elsevier Sci. Pub., Amsterdam, pp. 279–304.
Kozai, T. (1991f) Micropropagation under photoautotrophic conditions. (In: Micropropagation: Technology and Application, eds. Debergh, P. and R. Zimmerman), Kluwer Academic Pub., Dordrecht, pp. 447–469.
Kozai, T., Hayashi, M. and Ochiai, M. (1991a) Effect of the sidewards lighting on the growth and morphogenesis of potato plantlets *in vitro*. *J. Jap. Soc. for Hort. Sci.* 60. (Suppl. 1), pp. 228–229.
Kozai, T., Ohde, N. and Kubota, C. (1991b) Similarity of growth patterns between plantlets and seedlings of *Brassica campestris* L. under different *in vitro* environmental conditions. *Plant Cell Tissue Organ Culture* 24, 181–186.
Kozai, T., Iwabuchi, K., Watanabe, K. and Watanabe, I. (1991c) Photoautotrophic and photomixotrophic growth of strawberry plantlets *in vitro* and changes in nutrient

composition of the medium. *Plant Cell Tissue Organ Culture* 25, 107–115.

Kozai, T., Ting, K.C. and Aitken–Christie, J. (1991d) Considerations for automation of micropropagation systems, *Automated Agriculture for the 21st Century* (Proceedings of the ASAE 1991 Symposium), Amer. Soc. Agr. Eng., St.Joseph, MI, USA. pp. 503–517.

Kozai, T., Fujiwara, K. and Giacomelli, G. (1991c) Environmental control in micropropagation, ASAE Paper No. 911511, Amer. Soc. Agr. Eng., St. Joseph, MI, USA.

Kozai, T., S. Kushihashi, C. Kubota. and K. Fujiwara (1992) Effect of the difference in air temperature between photoperiod and dark period on morphogenesis and growth of potato plantlets *in vitro* under photoautotrophic and CO_2 enriched conditions. *J. Jap. Soc. Hort. Sci.* 61 (in press). (J with E)

Kubota, C. and Kozai, T. (1992) Application of high forced/natural ventilation for photoautotrophic micropropagation of potato (*Solanum tuberosum* L.). *Hortscience* (ac cepted for publication).

Kurata, K. Ibaraki, Y. and Goto I. (1991) Systems for micropropagation by nutrient mist supply. *Transactions of the ASAE* 34, 621–624.

Moe, R., Heins, R. and Erwin, J. (1991) Stem elongation and flowering of the long–day plant *Campanula isophylla* Moretti in response to day and night temperature alternations and light quality. *Scientia Horticulturae* 48, 141–151.

Mousseau, M. (1986) CO_2 enrichment *in vitro*. Effect on autotrophic and heterotrophic cultures of *Nicotiana tabacum*. *Photosynthesis Research* 8, 187–191.

Morini, S., Fortuna, R., Sciutti, R. and Muelo, R. (1990) Effect of different light–dark cycles on growth of fruit tree shoots cultured *in vitro*. *Adv. Hort. Sci.* 4, 163–166.

Morini, S., Trinci, M. and Zacchini, M. (1991) Effect of different photoperiods on *in vitro* growth of Mr.S.2/5 plum rootstock. *Plant Cell Tissue Organ Culture* 25, 141–145.

Murashige, T. and Skoog, F. (1962) A revised medium for rapid growth and bio assay with tobacco tissue cultures. *Physiol. Plant.* 15, 473–497.

Nakayama, M., Kozai, T. and Watanabe, K. (1991) Effect of the presence/absence of the sugar in the medium and natural/forced ventilation on the net photosynthetic rates of potato explants *in vitro*. *Plant Tissue Culture Letters* 8, 105–109. (J with E)

Owens, L.D. and Wozniak, C.D. (1991) Measurement and effects of gel matric potential and expressibility on production of morphogenic callus by cultured sugarbeet leaf discs. *Plant Cell Tissue Organ Culture* 26, 127–133.

Pospisilova, J., Catsky, J., Solarova, J. and Ticha, I. (1987) Photosynthesis of plant regenerants. Specificity of *in vitro* conditions and plantlets response. *Biologia Plantarum* (Praha). 29, 415–421.

Preece, J.E. and Sutter, E.(1990) Acclimatization of micropropagated plants to the greenhouse and field. In: Micropropagation: Technology and Application. Kluwer Academic Publishers. Dordrecht. pp.71–94.

Read, P.E., (1990) Environmental effects in micropropagation. In: Handbook of Plant Cell Culture, Vol.5, eds. Ammirato, P.V. et al., McGraw–Hill Pub., New York, pp. 95–125.

Reuther, G. (1988) Comparative anatomical and physiological studies with ornamental plants under *in vitro* and greenhouse conditions. *Acta Hort.* 226, 91–98.

Safadi, F., Morgan, J. and Hughes, H. (1991) Semipermeable covers enhance stomatal function and photosynthesis of *in vitro* cultured tobacco. *HortScience* 26, 726.

Safadi, F. and Hughes, H. (1991) Semipermeable covers enhance acclimatization of micropropagated tobacco plantlets. *Hort Science* 26, 757.

Sallanon, H. and Coudret, A. (1990) Water fluxes between *in vitro* plants and atmosphere in micropropagation. *C. R. Acad. Sci. Paris* 310, 607–613.

Seabrook, J.E.A. (1987) Changing the growth and morphology of potato plantlets *in vitro* by varying the illumination source. *Acta. Hort.* 212, 401–410.

Shimada, N., Tanaka, F. and Kozai, T. (1988) Effects of low O_2 concentration on net photosynthesis of C_3 plantlets *in vitro*. *Acta Hort.* 230, 171–175.

Smith, M.A.L. and McClelland, M.T. (1991) Gauging the influence of *in vitro* conditions on in vivo quality and performance of woody plants. *In Vitro Cell. Dev. Biol.* 27P, 52–56.

Song, J.E., Kim, M.K., Honjo, T. and Takakura, T. (1989) Influence of CO_2 enrichment in Jar fermentor culture on the pH of culture media. *J. Agric. Meteorol.* 45, 1–5.

Sutter, E. and Langhans,R.W. (1979) 'Epicuticular wax formation on carnation plantlets regenerated from shoot tip culture. *J. Amer. Soc. Hort. Sci.* 104, 493–496.

Takayama, S., Swedlund, B. and Miwa Y. (1990) Automated propagation of microbulbs of lilies, In: Scale–up and automation in plant propagation (ed. Vasil, I.), Academic Press, N.Y., pp.112–131.

Tanaka, F., Watanabe, Y. and Shimada, N. (1990) Effect of O_2 concentrations on photorespiration in *Chrysanthemum morifolium* plantlets in plant tissue culture. *Plant Tissue Culture Letters* 7, 85–91. (J with E)

Tanaka, F., Watanabe, Y. and Shimada, N. (1991a) Effect of O_2 concentrations on photorespiration in *Chrysanthemum morifolium* plantlets *in vitro* cultured photoautotrophically and photomixotrophically. *Plant Tissue Culture Letters* 8, 87–91. (J with E)

Tanaka, F., Watanabe, Y. and Shimada, N. (1991b) Effect of O_2 concentrations on growth of *Chrysanthemum morifolium* plantlets *in vitro* cultured photoautotrophically and mixotrophically. *Environ. Control in Biol* 29, 107–116. (J with E)

Tani, A., Kiyota, M., Taira, T. and Aiga, I. (1991) Growth and photosynthesis of alfalfa plantlets cultured *in vitro* under photoautotrophic and photomixotrophic conditions. *Environ. Control in Biol.* 29, 127–132.

Urban, L. and Jaffrin, A. (1990) Steady state thermal conditions inside plant tissue culture vessels submitted to a constant level of irradiation. *Biotronics*. 19,71–81.

Walker, P.N., Heuser, C.W. and Heinemann, P.H. (1988) Micropropagation: Studies of gaseous environments. *Acta Horticulturae* 230, 145–151.

Watanabe, K., Watanabe, Y. and Shimada. N. (1990) Effect of sucrose concentration in the medium on growth, apparent photosynthesis and ribulose–1,5–bisphosphate carboxylase of Spathyphyllum plantlets in aeration culture. *Plant Tissue Culture Letters* 7, 74–79.

Weathers, P.J. and Giles, K.L. (1988) Regeneration of plants using nutrient mist culture. *In Vitro Cellular & Development Biology* 24, 727–732.

Wilkins, H.F. (1988) Techniques to maximize cutting production. *Acta Hort.* 226, 137–143.

Ziv, M., (1990) Vitrification: morphological and physiological disorders of *in vitro* plants. In: Micropropagation: Technology and Application, eds. Debergh, P and R. Zimmerman, Kluwer Academic Pub., Dordrecht, pp. 45–69.

Note: (J with E) in the list means that the article is written in Japanese with English summary and captions

ROBOTICS AND IMAGE ANALYSIS APPLIED TO MICROPROPAGATION

F. R. BROWN
Mechanical Engineering Division
Silsoe Research Institute
Wrest Park
Silsoe
Bedfordshire
MK45 4HS
Great Britain

ABSTRACT. This paper describes the underlying research leading to the development of a fully automated experimental robotic work cell for the dissection and transfer of microplants. All operations within the work cell, including handling of the plant containers, the dissection and transfer of the microplants and sterilisation of the robot tools are controlled by an IBM compatible PC. Flexible robot control software provides appropriate levels of access for both the casual user and software development engineer. As described, this experimental work cell is particularly suited to the evaluation of the effectiveness of multi-task robot tools and the corresponding image analysis algorithms. However, if an appropriate image analysis algorithm is not available, the plant specific data for guidance of the robot can be provided by an operator working from a video image overlaid on the computer screen.

1. **Introduction**

Micropropagation is a tissue culture technique capable of producing large numbers of genetically identical plants. The process starts with the selection, sterilisation and establishment of a part of the selected mother plant and its acclimatization to growing under artificial conditions. Subsequently, during multiplication, the microplants are routinely divided into viable portions and grown on in a sterile medium. This medium supports the growth of the plant material and consists of a mixture of salts, a carbon source and the necessary vitamins and hormones. Because the media will also support the growth of common bacteria and fungi, contamination by micro-organisms must be excluded. Therefore, all operations must be carried out under strictly aseptic conditions. Once the required number of plants has been produced they are rooted and weaned so that they may continue to flourish in a more natural environment.

Throughout the multiplication process all tasks directly involved in the manipulation of plant material are carried out manually. Thus, despite the many advantages of micropropagation, compared with the more conventional propagation techniques, its use as an industrial process is limited by the cost of labour. This cost may be as high as 70% of the total plantlet cost.

When considering how best to automate the process, most researchers

K. Kurata and T. Kozai (eds.), Transplant Production Systems, 283–296.

Figure 1. Open structured plants - Chrysanthemum

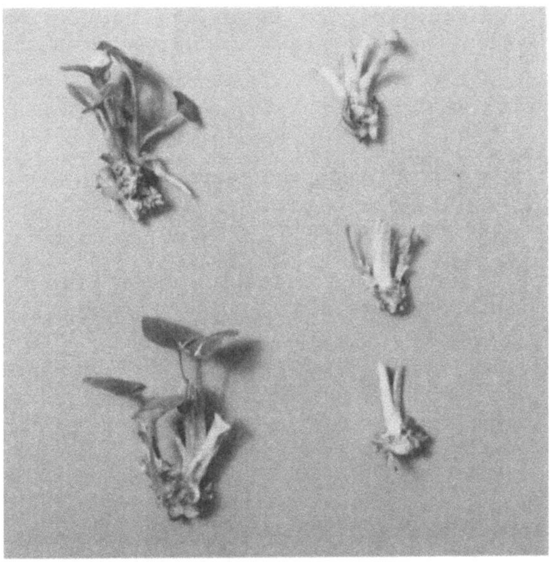

Figure 2. Shoot clusters - Syngonium

agree that the greatest challenge lies in automating the dissection and transfer of individual plantlets. Because these tasks account for the greater part of the labour input their automation represents a desirable goal. Also, it appears likely that the route to success will combine the following features:-

i) advanced robotic techniques for the identification, dissection and handling of the individual plants.

ii) more conventional automation to regulate the flow of material to the robot(s)

iii) automated sterilisation techniques to maintain a high level of hygiene.

Of the above features ii) and iii) are largely independent of the plants growing within the containers. Even so, the approach to materials handling is likely to include detail appropriate to a specific design of container. Within the robotic system however, the state of the art is such as to suggest that overall capability will initially be limited to specific plant forms.

Two contrasting forms of plant have been studied (Fig. 1 and 2). Within the UK plants of a form similar to Chrysanthemum (Fig.1) are commercially less important than those having a growth habit comparable with Syngonium (Fig.2). However, selection of Chrysanthemum for the initial studies was based upon the perceived benefits of starting with a simple, clearly defined structure. For example, interpretation of the dissection strategy was unambiguous. As reported elsewhere[1], these initial studies were limited to those parts of the system calling for advanced robotic techniques rather than conventional automation. This paper extends the earlier work. In particular, it describes the current experimental facility and introduces the idea of multi-task robot end effectors.

2. Robot work cell

The robot work cell was designed as an experimental facility and is housed within a laminar flow cabinet. Shown schematically in Fig.3, the work cell comprises:-

1) a cylindrical robot to carry the multi-task end effector(s) used for dissection and planting of the microplants.

2) a plant container handling unit to dispense, open and deliver individual containers from a small stack to a position within the field of view of the camera used for vision guidance

3) a variable power CO_2 laser for destructive opening of the plant containers and optional use for plant dissection

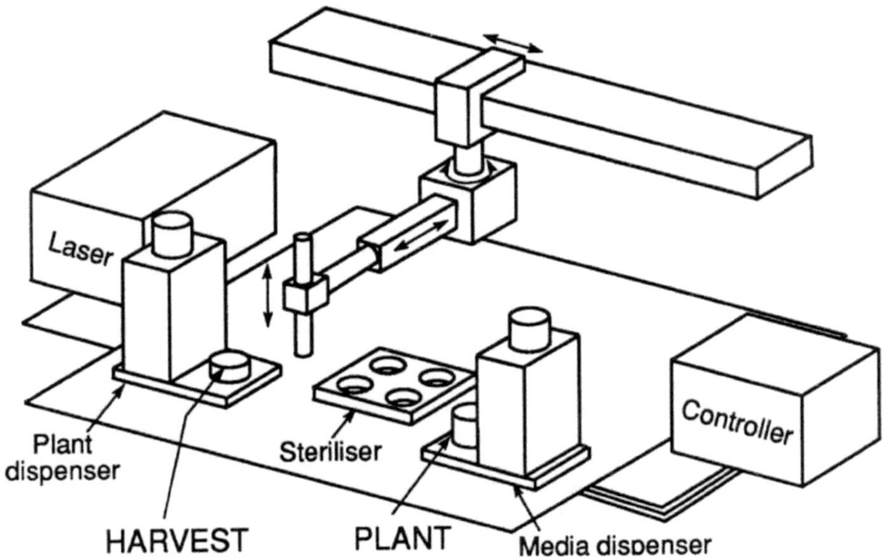

Figure 3. Schematic of robot work cell

Figure 4. Dispenser for containers of fresh media

4) a fresh media container handling unit incorporating means for non-destructive removal and replacement of the lids

5) a sterilisation station, centrally located, for sterilisation of the robot end-effector

6) provision for manual guidance

7) an automatic vision guidance capability

8) an IBM compatible 386/33 PC for image interpretation and control.

As a research facility the features provided differ from those required in a production unit in two important respects. Firstly, the container handling units have limited holding capacity and require the intervention of an operator to remove spent and newly filled plant containers. Secondly, the supervisory control software is readily accessible for ease of modification during system development.

2.1 Robot description

The robot is installed with the track uppermost to give unobstructed access to the work space beneath the robot arm. It is a cylindrical robot, mounted on a linear track to give five degrees of freedom. All five axes motions are driven by stepper motors, under open loop control. The robot end-effectors are described in section 5.

The robot controller is based around an industrial computer with a Z80 processor. The controller can be user programmed in ROBOFORTH 2, a derivative of FORTH for use in robotics. It has facility for on-line interaction with an external computer and thus allows predefined movements to be intermixed with plant specific motion.

2.2 Robot guidance

Methods of guiding the robot tools automatically have centred on the use of computer vision. The system used in the work cell consists of a colour CCD TV camera and proprietary frame grabber. The digitised images are processed within the supervisory PC. An additional facility permits the use of the work cell to test a robot end-effector in isolation from the corresponding image analysis algorithm. In this mode, the controlling intelligence is supplied by the operator. Points on the video image, overlaid on the computer screen, are picked out by the operator using a tracker ball to guide the graphics cursor. This information is then processed by the computer to provide positional commands to the robot controller. Such a facility may also be useful in an industrial environment.

2.3 System control software

To meet the special requirements of a research facility, the software is structured to allow users to edit, debug and execute their systems control procedures as well as programmers to modify and debug all the modules. In operation it behaves like an interpreter, explaining the user message sentence by sentence. It organises all the necessary modules within a user-friendly software operating environment. At the simplest level of access single characters can be defined to actuate pneumatic devices or execute a predetermined sequence of movements. In this way individual mechanisms can be set up and evaluated by a designer lacking detailed knowledge of the computer software.

3. Container handling

In the absence of an industry standard container, it is perhaps inevitable that any handling system will be customised for a particular type. The system described below is most directly suited to the circular disposable 'margarine tub' widely used in the UK. These are formed from thermoplastic sheet and are approximately 90mm in diameter at the aperture and 55mm deep. The containers are pre-filled with media and closed ready for use. In the USA greater use is made of reusable 'magenta' square section tubs requiring a different approach to lid removal and replacement. Elsewhere, custom designed rectangular containers are used and filled on line with both media and plant material.

3.1 Media containers

The container dispenser (Fig.4) is of a conventional single column design, in which pneumatically actuated levers are used to release the lowermost container. The column is pre-loaded by hand. On release, the freed container is held on a suction cup and carried on a linear carriage to a lid removal station. A second, freely rotating suction cup, situated above the container is then lowered to hold the container lid during and after removal. To break the seal between the lid and container, a pair of diametrically opposed rollers are raised to apply an upward pressure to the underside of the lip of the lid during rotation of the container. The lid is then raised clear of the container before it is moved to the innermost end of its track. After the plant segments have been inserted into the container, the lid is replaced through the action of a second pair of diametrically opposed rollers engaging on the upper surface of the rim.

3.2 Plant containers

The procedure and equipment described in 3.1 may also be used for the plant containers. However, several advantages were foreseen in cutting through the side wall of the container, approximately in line with the surface of the growing medium. These advantages include:-

1) an unobstructed view of the plants inside the container

2) greater freedom of design and therefore effectiveness of the robot tools

3) the opportunity, where required, to remove plant leaves before removing the plants from the container.

To open a container destructively, it is released onto the suction cup and carried to the innermost end of the track. A low powered CO_2 laser focused on the sidewall of the container is then used to cut through the sidewall as the container is rotated about the axis of the suction cup. It is then only necessary to lift away the container lid complete with sidewall to gain free access to the plants.

Whilst manual intervention was considered acceptable when removing spent material, the task of disposing of the lid was automated in the interest of operator safety. This was simply achieved by the application of a pulse of compressed air to the underside of the rim of the container.

4. Laser system

The laser unit comprises a 40 watt Carbon Dioxide (CO_2) laser and a 5 watt Helium-Neon (HeNe) sighting laser. The combined unit is mounted outside but adjacent to the left hand end of the laminar flow cabinet. A series of mirrors within the laser cabinet collimate the two beams. Laser power is brought within the working envelope of the robot by an articulated beam guide. After alignment of the two beams with the longitudinal axes of the beam guide the latter is locked in position. The beam guide is terminated with a variable focus end piece. The focal characteristics of the emergent laser light can be varied by adjustment of the distance between the Zinc Selenide focusing lenses within the end piece. The range of adjustment and corresponding focal characteristics of the CO_2 laser are presented in table I.

Table I. Focal characteristics for the CO_2 laser beam.

Focal length (mm)	Spot diameter (mm)	Depth of field (mm)
24	0.050	0.372
140	0.710	75

Power output may be adjusted by varying the current within the HT circuits.

In addition to use of the laser to destructively open the plant containers, it has also been used for dissection of the plant material (section 5.2.4). In both applications the use of the laser simplifies the sterilisation procedure because, as a cutting device, it is inherently sterile.

5. Multi-task robot end-effectors

In a commercial laboratory reliant upon manual labour, the operators are trained to carry out their duties in accordance with a carefully devised sequence of tasks. Thus, for example, a portion of the contents of each container would be removed and placed upon a sterile surface ready for dissection. These plants are then dissected to separate the viable segments from the waste. Finally, each wanted segment is picked up and inserted in the growing medium.

During the previously reported work with Chrysanthemum, separate end-effectors and corresponding image analysis algorithms were developed for each task. In expanding the research to include other plant forms a somewhat different approach was adopted. This new approach was designed to reduce the number of tools and intelligent decisions required to effect each transfer. The resultant multi-task tools are described below.

5.1 Spider cutter

This tool was designed for use with plants of a form similar to the Syngonium shown in Fig.2. It requires only a single instruction set to complete all three tasks of harvesting, dissecting and planting. Guided by positional information derived from a plan view of the plant container, the robot centres the tool above the selected plant. The tool then descends so that the flexible wire grippers surround the plant. As the grippers close, the cutting blade descends to dissect the plant before removing it from the container. On transfer of the whole, but pre-cut, plant to the first planting station, the planting element moves downward, passing between adjacent grippers, to slide the first plant segment into the growing medium. As the robot moves to the next planting position, the grippers and the cutting blade are rotated to bring the second plant segment into the line of action of the planting element.

The design is equally suitable for dissecting plants into three or more comparable sized segments, each containing developing buds. The guidance instructions comprise a mix of plant specific and pre-programmed data. The plant specific data can be derived from human or computer interpretation of the video image. The pre-programmed data controls the height of the tool and defines the planting positions within the fresh growing medium. Working at maximum capability and using a multiple toolhead, the yield from each plant can be adjusted in response to plant size. All the plant specific movements are built into the robot tool. Therefore, the robot requires only three degrees of freedom and the rate of work will depend on the time taken by the robot to move back and forth between the two plant containers. Commercial systems should be capable of achieving transfer times significantly better than the 4 seconds recorded with the experimental system.

5.2 Roller feed

The roller feed device (Fig.6) was designed for use with open structured plants, similar in form to the Chrysanthemum shown in Fig.1. It combines a gripper with a transport mechanism and is used in

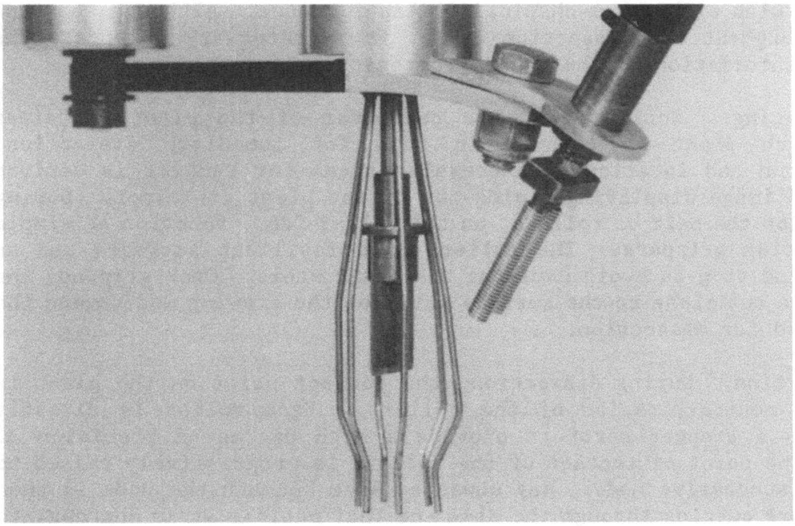

Figure 5. Spider cutter

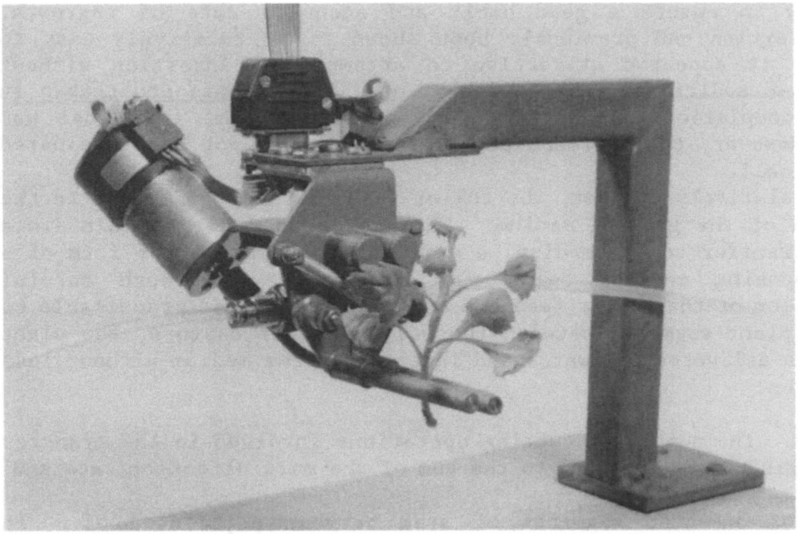

Figure 6. Roller feed device

combination with either a mechanical or laser cutter. Although the tool is used throughout the dissection cycle, it is necessary to update the positional information as the cycle progresses.

5.2.1 *Harvesting.* Working from the perimeter of the plant container inwards, each plant is removed in turn for immediate dissection. Identification and location of successive stems for removal is derived from a video image displaying a sideview of the plant container. During this operation the pair of rollers, on the roller feed, function as simple parallel action grippers. The rollers have resilient surfaces and an adjustable end stop to avoid damaging the plant stems. Once gripped, the plant stem is cut close to the surface level of the growing medium and the plant removed for dissection.

5.2.2 *Dissection.* During dissection, the contact point on the plant is adjusted by contra-rotation of the rollers. Each roller is directly connected to a stepper motor to provide a high degree of precision of movement. The point of contact of the rollers is progressively raised to reveal each successive node. Any unwanted stem beneath the node is then removed before cutting through the stem and leaf petiole at an appropriate height above the node.

5.2.3 *Planting.* For some plant species the growth cycle can be continued simply by dropping the Y-shaped plant segments onto the surface of the growing medium. However, for other plants insertion in the growing medium is necessary to ensure a good habit and adequate rate of regrowth. Although insertion had previously been shown to be relatively easy to automate[1], it appeared attractive to attempt the insertion without introducing an additional robotic operation. Experiments undertaken in advance of completion of the work cell confirmed that this goal was feasible. However, the approach described below has not been evaluated in the work cell.

Although relatively compact, the roller feed has limited access to the surface area of the growing medium. Therefore, in order to maintain a controlled transfer to the medium, a vertical conveyor in the form of a pair of opposing endless belts was introduced. Through careful synchronisation of the roller feed and conveyor, it proved practicable to capture the plant segments between the belts. After capture, the plant segments were delivered downwards and inserted in the medium without loss of orientation.

5.2.4 *Cutter.* The number of cutting operations involved in the transfer of each incoming plant is equal to the sum of the multiplication rate and the number of unwanted segments of stem between adjacent buds. In effecting these cuts the vertical position of the cutting line relative to the point of contact of the gripper changes. Also, the lowest bud may be no more than 2 or 3mm above the surface of the growing medium. Therefore, whilst it is technically feasible to use a shear cutter, the task is ideally suited to a laser. For example, the initial cut to free the plant can be exceedingly close to the media surface without risk of transferring contaminated media. Also, because the diameter of the plant stem and leaf petiole is in the order of 1.5 to 3mm, the power requirement is modest, e.g. <20 watts for a cutting rate of 1 cm/sec.

6. Requirements of the image analysis system

One problem in using image analysis for the identification of salient plant features is their natural physical variability, even though they may be genetically identical. Therefore, the algorithms must have the flexibility to cope with this variability. Other considerations include processing speed and the possible interactions with both the design and operation of the tool. These further considerations led to an approach to image analysis in which embryo robot tools were evaluated under manual guidance, before finalising the requirements of the vision algorithms. Detailed descriptions of the mathematical steps involved in derivation of the algorithms are presented elsewhere[1,2,3,4,5]. To supplement that information, this paper presents the system considerations giving rise to the specific approaches.

6.1 Shoot clusters

During multiplication, the parts of the plant transferred to fresh medium are in the form of developing buds at the base of the cluster. This leads to a requirement to defoliate the more advanced shoots to check their dominance. Prior to defoliation, the position of each cluster is most clearly observed from beneath. However, removal of the container sidewall as described in Section 3.2 allows defoliation to take place before removal of the plants. The resulting overhead view of the plants contains more structural information than that derived from below.

The detail shown in an overhead view is considerably less than the 3D information available to an operator. However, observation of a sample of operators indicated that they frequently use the developed shoot stems, rather than the position of the developing buds, to guide their scalpel. Therefore it was decided to attempt the dissection and transfer relying only on intelligence extracted from a plan view. From this starting point an algorithm[4] was developed to determine the centre and size of each cluster. Subsequently the preferred direction of cut was added to the requirement, to avoid inadvertent fragmentation observed when some clones were randomly dissected. In adopting this approach it was recognised that defoliation was not acceptable prior to transfer to a rooting medium. Therefore, for this additional stage in the process a second algorithm was developed[5]. This algorithm identifies cutting planes between adjacent shoots, after removal of the cluster from the growing medium.

6.2 Open structured plants

With open structured plants, it is necessary to locate, separate and transfer the shoot tip and each of the axillary buds. However, although the dissection strategy is unambiguous, the specific approach to automation can influence the image analysis requirement.

In the previously reported research at the Institute, the approach closely mimicked the action of an operator. This resulted in the need for separate algorithms and robot end effectors for the tasks of harvesting[2], dissection[1][3] and transfer[1]. However, despite the successful development of appropriate algorithms, their implementation in a robotic system required the provision of a camera and robot manipulator for each task. Thus the resultant system was likely to be capital intensive.

When using the multitask roller feed the plant is held substantially vertically throughout the transfer process. During harvesting the tool acts as a conventional parallel action gripper and has no influence on the decision making processes involved in stem location. However, subsequent location of the axillary buds is simplified. Successive buds are presented in a window beneath the contact point of the rollers. The rollers also impart a lifting action to the leaves, thereby reducing the risk of masking of the buds. The algorithm used for planting is redundant.

7. Sterilisation

Current practice within a micropropagation laboratory calls for frequent sterilisation of all tools which are used for cutting and manipulating plant material. Usually this requirement is met by providing each operator with duplicate sets of hand tools and a heat source. Thus one set of tools can be heated and cooled whilst the other set is in use, thereby minimising the down time involved in tool sterilisation. To avoid the duplication of robot tools implicit in adopting the above approach in an automated system, investigations were undertaken into chemical sterilisation procedures. In initiating these investigations it was considered desirable, if possible, to adopt a process already in use within the micropropagation industry. Ethanol is widely used to clean down the working surfaces within a laminar flow cabinet, whilst sodium hyperchlorite finds use in the sterilisation of initiating plant material.

7.1 Experimental procedure

7.1.1 *Working Area.* All working surfaces within the laminar flow cabinet were thoroughly cleaned with a 70% solution of ethanol containing a wetting agent. Petri dishes containing Tryptone soya growing medium were laid at different locations in the laminar flow cabinet. On completion of several hours running of the robotic system, the Petri dishes were removed, for incubation at 25°C in a darkened incubator. Subsequent examination of these samples showed the laminar flow cabinet to be functioning effectively. In particular, there was no detectable deposition of contaminants under the working arc of the robotic arm and tool.

7.1.2 *Tools.* In preparation for the experiment, the robot tool (spider cutter) was soaked overnight in a 10% solution of sodium hyperchlorite. Samples were then taken from the blade and spine of the tool, using sterilised, moist cotton buds, which were then plunged into sterile Tryptone Soya broth prior to incubation. Thereafter, as the experiment progressed, further samples were taken following brief periods of immersion of the tool in a second container of sodium hyperchlorite. Care was taken to ensure that any visible debris (plant or growing medium) was removed from the tool by rotating it during immersion. Immersion times were progressively reduced from 60 secs. to as little as 5 secs. Although the control samples taken from the growing medium show the presence of both fungi and bacteria, all samples following immersion of the tool in 10% sodium hyperchlorite were clean. However when the concentration of the sodium hyperchlorite was reduced to 5% the majority of samples were contaminated.

8. Discussion and conclusions

The components of the experimental robotic work cell, described in this paper, provide all the facilities necessary to evaluate the effectiveness of a robot end effector and the associated vision guidance algorithm. The design of the workplace is, however, not suitable for direct demonstration of the potential output rate of a fully developed industrial system. In particular, components are not optimised for speed, and interruptions occur when changing from one container to the next and during tool sterilisation. Rounded to the nearest multiple of the time taken to transfer each plant segment (4 secs), the respective delays are 20, 32 and 8 secs for plant containers, media containers and sterilisation.
Sterilisation of the robot tools is necessary to safeguard against the transfer of contaminants in one plant container to the next. However, because it takes only 8 secs compared with 20 secs to change plant containers it need not affect output. Similarly, if the multiplication rate is an integer, the requirement to change a plant container will coincide with a change of media containers. Thus, in the experimental system, the only delay having a direct bearing on output is that involved in changing media containers. For example, if there are 15 plant segments in each container and it takes 4 secs to insert each segment the total time to fill each container will be 60 secs. However, because plant transfer and container replacement take place sequentially, the actual output will be one container every 92 secs.

In an industrial system it will be desirable to minimise non-productive delays by arranging, where possible, for ancillary operations to run in parallel with the dissection and transfer. This is feasible for the plant and media containers but some delay is likely to be associated with tool sterilisation, even if automatic tool changing is introduced. With this arrangement the kinematic characteristics of the robot can then be used to set the performance specification for the other components, including the speed of the image processing hardware.

9. Acknowledgements

Advanced robotics research is, by its very nature, a multidisciplinary task calling for a high level of collaboration. In addition to the individuals named in the references, special acknowledgement is made to Peter Billington, Peter Grundon and Wenbin He from our Biomaterials Handling Group. Also acknowledged are Horticulture Research International and Munton & Fison plc for the supply of plants and an essential biological input. Finally, personal thanks are due to the organisers of the symposium and to John Reed for his presentation of this paper.

REFERENCES

1. Tillett, R.D; Brown, F.R; McFarlane, N.J.B; Onyango, C.M; Davis, P.F; Marchant, J.A. (1990). 'Image-guided robotics for the automation of micropropagation'. IEEE. International Workshop on Intelligent Robots and Systems. IROS'90.

2. McFarlane, N.J.B. (1992) 'Image guidance for robotic harvesting of microplants'. Computers and Electronics in Agriculture (in press).

3. Davis, P.F. (1991) 'Orientation independent recognition of Chrysanthemum nodes by an artificial neuron network'. Computers and Electronics in Agriculture. 5, 305-314.

4. Onyango, C.M. (1991). 'Location of defoliated syngonium growing in containers'. Divisional Note DN1617, Silsoe Research Institute, Silsoe, Bedford.

5. Davis, P.F; Tillett, R.D; Brown, F.R; Marchant, J.A. (1990). 'Automatic image analysis of syngoniums to select cutting planes for robotic micropropagation'. Divisional Note DN 1567, AFRC Institute of Engineering Research, Silsoe, Bedford.

AUTOMATED MICROPROPAGATION AND THE APPLICATION OF A LASER BEAM FOR CUTTING

D.P. HOLDGATE AND E.A. ZANDVOORT
PhytoNova B.V.
P.O. Box 98
2230 AB Rijnsburg
The Netherlands

ABSTRACT. The basic micropropagation process has been analysed with a view to improving operational efficiency. Labour content is only one of several aspects which require upgrading to fit into more industrialised production schemes. Reduction in culture variability and losses and accurate progress information for management control is required. An automated cutting system based on image analysis, computer controlled laser cutting and robotised planting of explants has been developed. The system provides for a substantial reduction in labour, more uniformity, data collection and thereby current status reports, forward planning and overall upgraded management control.

1. Introduction

Tissue Culture as a tool for mass propagation has proven to be very successful for many plant types since 1966. The major advantages are the possibilities for a fast, disease-indexed clonal multiplication. These features can lead to a rapid introduction to the market of new valuable plant varieties obtained through selection, mutation, breeding programmes or genetic manipulation. The micropropagation industry, however, has its limitations because of the high costs involved, frequently represented by skilled labour, necessary for high outputs in a short time span. Activity has therefore been limited mainly to the ornamental sector with the majority of production on a direct producer to grower arrangement, or by vertically integrated operations. In 1990 a total of over 210 million plants were produced by micropropagation in Western Europe. The majority of these plants are ornamental crops (82%), followed by fruit trees (10%) and small fruits (5%). Only 3% consisted of agricultural crops, vegetables and forest trees (Pierik 1990).

K. Kurata and T. Kozai (eds.), Transplant Production Systems, 297–311.

The major markets are The Netherlands, Germany, Italy and France.

During the past few years there has been a significant number of micropropagation operations going out of business or modifying their activities away from micropropagation. Evenso there are still well over 200 commercial operations in Europe. A few companies continue to expand their total market volume, and others to specialise, and/or enter a degree of vertical integration. The relatively large European market is increasingly offered products from other regions and particularly from USA, Poland, India and Malaysia, whilst local production has also increased despite increasing costs. The end result is a slippage in price and unfortunately in many crops a decrease in quality. This quality reduction is in part a consequence of growers previously demanding more and more at lower and lower prices, but subsequently requiring reduced numbers still at the lower prices! The downward spiral stops when the micropropagator refuses to produce, or cannot maintain minimum quality standards required by the grower, resulting in the micropropagators going out of business rather than a more realistic pricing structure being agreed for the quality products which can be produced.

The need for mechanisation has been related to the high labour requirement of the micropropagation process, and the assumption that there would be an immediate reduction in the price of plants produced. This in turn would create an opportunity to widen the application to more plant crop species. This may not be the reality. Mechanisation, for low cost production reasons alone may not be a viable proposition.

It is generally considered that expansion to those plants with a lower individual value (vegetables, arable crops) and to plants from which returns can be expected only after long periods (forest trees) will be stimulated by reducing the cost price of the micropropagated plants. It appears that growers and processors are unwilling to recognise the added value which quality micropropagated plants secure via benefits of genetic gain and healthy starting plant material. The real added value of the micropropagated plantlets is referred to only rarely and not in the context of higher prices for the micropropagator, but only in higher benefits for the growers and processors. Irrespective of price arguments, as in other industries, mechanisation and finally automation is considered to be the logical route for reducing labour intensity, costs and to improve quality. However any savings achieved cannot be passed entirely to the grower - the micropropagator must benefit from the investment risks via a reasonable profit in real terms.

2. The current micropropagation process

2.1. WHAT IS IT?

Micropropagation is the multiplication of plants in vitro via the proliferation of axillary buds or adventitious shoots arising from individual cells, callus or other tissue, organs or structures, by manipulation of nutrients, sugars, growth regulators, the growing environment and by mechanical subdivision. The basic process of micropropagation is illustrated in figure 1.

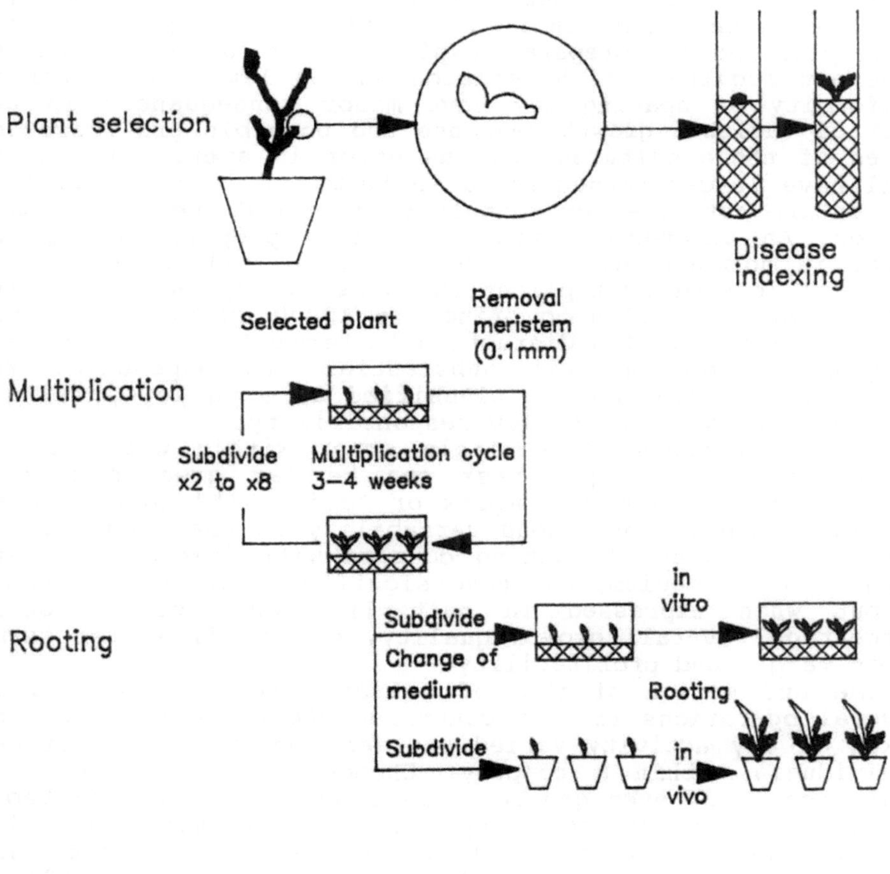

Figure 1. Basic process of micropropagation

Cultures are initiated from meristems derived from buds of selected parent plants. Each developed culture must be tested as free from contaminating fungi and bacteria, and disease-indexed as free from specified pathogens irrespective of nature - fungal, bacterial, viral or viroidal. The propagation is a cyclic procedure. The initiated cultures are multiplied up to a base stock from which shoots are transferred to the glasshouse, either as rooted plantlets or microcuttings used for in vivo rooting or sold.

2.2. ANALYSIS OF THE STATE OF THE ART

A critical path analysis highlights manual activities at each stage of the process, the level of decision making required and the consequences of error and variabilities. Thus the relevance of each activity and the cost liability of a wrong decision can be assessed. For example the uniform planting of prepared explants into a new container does not require a high decision making level. An error in uniformity of spacing may have minor consequences in the form of variable growth response and possible proliferation rates of those cultures, but an error in sterile technique will have larger consequences in terms of lost output from the entire container or containers involved due to contamination. An undetected error in medium preparation eg. an error in hormone level in 100 litres of media, may result in total loss of multiplication, a delivery, and/or require extra transfer cycles to bring the 15000 cultures involved back to sequence. The disruption to production planning and programming may be both substantial and expensive. Yet medium preparation may be classified as of a low intellectual demand and a job of low responsibility.

The consequences of a single error within what may be classified as a simple task can be far greater than a single error in a more complex or "difficult" task such as cutting. However continuous variability in the higher level decision activity of cutting coupled with inadequate planting into the medium, can have significant effects on total output when expressed in multiplication rates, rooting percentage, overall growth quality, or, in financial terms gross margin and profitability.

Time and motion studies of transfer room activities in several operations in four countries demonstrated that time taken on any activity varied between companies and between individuals within a company. Change-over to a new crop could result in even greater variations, with unpredictable performance by individual operatives. The most intensive decision making activity of where and how to cut, and actual cutting, took 16% to 37% of total time. Over 60% of real work time was spent on container handling, examination, de- and re-planting, labelling and sterilisation of instruments and operating surfaces. Such study reveals, in more detail, that micropropagation, as a multiphased process, is complicated by the biological nature of the products, and that the level of variability in growth may be increased by even minor errors or changes in any one of the numerous production cycles. Thus operative and culture performance, against the planned, must be a subject of continuous monitoring if efficient production of forecasted or contracted sales are to be achieved in a cost effective manner.

Various techniques to secure cost savings have been reported. Maene and Debergh (1985) and Vanderschaeghe and Debergh (1988) reported that cost reductions, improved growth rates, and conversion from one growth stage (multiplication) to another (rooting) can be achieved by repeated addition of appropriate liquid medium to cultures growing on a solid medium. No subdivision is required. Although this has been achieved in several crops substantially more labour is required subsequently in vivo to subdivide and grade plants of different sizes pre or post hardening in the glasshouse. This system may not be sufficiently flexible to secure the economic production of plantlets of specified dimensions without a significant input of labour, intensive glasshouse operations, and excess production. However aseptic transfer work can be significantly reduced.

Increasing the speed of transfer can be achieved by the hedging system described by Aitken-Christie and Jones (1987) for forest tree cultures and variations on this theme employed for fruit-trees (Jones 1977). In these cases the periodic removal of shoot tips, in combination with the addition of further medium to the original culture vessel to stimulate new axillary shoot growth from the retained basal tissue, may increase yield per operative. Various problems such as overcrowding, etiolation, vitrification, contamination, shoots of variable strength and rate of rooting have been experienced with this system. It is possible to utilise the hedging procedure for short runs to achieve an apparent cost reduction, but it is not so suitable for mechanisation, or for long run high volume production levels requiring high product uniformity.

Although the reduction of labour costs is frequently given as the main need for mechanisation, other criteria may prove to be even more relevant. The maintenance of product uniformity and quality, reduction in contamination levels, elimination of human errors and the upgrading of production control are essential elements in the advancement of tissue culture procedures to an industrial level.

3. Automation of micropropagation procedures

3.1 LOGISTICS

The study reported on here started from analyses of the micropropagation process and its separate parts, including planning and control systems, the efficient usage of specialised space requirements and facilities, and the limitations and variabilities caused via a large labour force. The conclusion was that if conventional micropropagation was to contribute to major agroindustry developments a significant change in the current operational techniques, which were introduced more than 25 years ago, must be made.

The earlier studies on the operational sequences employed by micropropagation companies revealed that:

1. The time taken for cutting and subdivision of cultures is a minor but significant time consumer.

2. Operatives, irrespective of training level, tended to produce variable explants, a source of overall non-uniformity, and, as such, a major and progressive source of declining efficiency leading to higher rather than reduced costs with increased volumes.

3. With increased pressure to complete order levels the error rate and culture contamination increase, leading to reduced production and quality.

4. Labelling, recording and keeping track of specific cultures and sales orders where a multicrop and multicustomer operation exists, can cause substantial management problems.

5. Increasing output from a single production unit to and beyond the 6 million plant level has been a rare occurrence although many companies have the growth room and other space to achieve such production numbers.

Completion of the preliminary studies suggested that all parts of the micropropagation system could be mechanised on a modular basis. The opportunity to progress to an integrated fully automated system was presented by the rapid advances during recent years in other technologies, particularly in robotics and computer hard- and software. The system requires the following:

- media preparation, sterilisation and dispensing
- development of a recognition system and integration to a controlled cutting system
- development of a cutting system

- collection of required cut pieces - explants
- recognition of the explants
- pick and place system for transfer of freshly cut explants into new containers
- sterilisation of operational parts
- controlled positioning of containers in, and their sequential recovery from the growth rooms
- automated recording system linked to advanced planning control for subsequent activities
- mechanised planting system for transfer of selected cultures to the in vivo condition
- total integration

The study in mechanisation and automation by PhytoNova B.V., within its subsidiary PPS B.V., was targetted to progressively develope a cutting and handling system to achieve 10 million units per annum with a concomitant reduction in labour requirement. Further through modular steps provide a system which would allow a further increase in production to be built up.

The simplest route to achieve the target appeared to be the continuous supply of grown cultures to operatives for subdivision only. The subdivided pieces would be passed by conveyor to a pick and place unit, which in turn has a continuous feed of medium-filled containers. Theoretically a very substantial increase in cutting rates could be secured. The main disadvantage of such a concept, is the monotony of the cutting task, which could lead to substantial loss of concentration and consequently product variability. It is doubtful whether operative job satisfaction could be achieved, and the probability would be an unacceptable and expensive staff turnover rate.

Consideration was therefore centred on the development of a completely integrated automated production line.

3.2. DEVELOPMENT OF BASE EQUIPMENT

3.2.1. <u>Containers</u>. The use of containers made from various materials in a multitude of shapes and sizes was unacceptable. After examination the VEG-box described by Fari (1987), and which can be supplied now with two lid forms to allow for variations in plant height, was chosen to be the working standard.

The boxes, made from a durovinyl plastic allowing excellent light transmission, are supplied as gas sterilised separate bases and lids in easy-to-open sleeves. Convenient handling trays holding eighteen boxes have been made.

3.2.2. <u>Medium preparation</u>. Using a standard autoclave sequence can be time consuming, inefficient and expensive, whilst the dispensing of hot media, which is still a common practice in many companies, can be a safety hazard. A batch medium preparator and dispensing apparatus modified after Fari (1987) has been designed and manufactured. The steriliser consists of a 50 litre reservoir with fast heating and cooling facilities. Cooled sterilised medium is pumped directly to a dispensing unit held within a laminar sterile air flow cabinet. Presterilised VEG-boxes are fed from a vertical stock to a conveyor. Each box stops under the filling unit to allow a preset volume of medium to be dispensed. The box then moves on to receive a lid which is manoevered into its final position by roller pressure after which the box leaves the dispensing area. The preparation, sterilising and dispensing of 50 litres of medium can be completed within two and a half hours. Larger runs of the same medium may be accomodated by increasing the original concentration of the medium constituents and diluting by mixing with sterilised water at the pump dispensing head. Under these conditions 100 litres of medium may be prepared and dispensed in three and a half hours. The filling machine may be linked with the pick and place apparatus (described later) to allow placement of cut tissues into the medium prior to lid placement and closure.

3.3. AUTOMATED CUTTING AND TRANSFER OF PLANTS

In general, manual cutting and transfer of cultures includes the following handlings:

- examination of container for contaminated / abnormal growth
- opening of containers
- removal of individual cultures
- selection of cultures
- presentation of culture for cutting
- dead plant material is removed, cultures are trimmed
- cultures are cut into desired explants and sometimes graded or classified (for instance into nodal cuttings and shoot tips)
- discarding of waste
- explants are planted in the appropriate containers, which are then closed
- sterilisation of used instruments before re-use
- labelling of containers and placement into trays

Nearly all these steps include time consuming decision making - where to hold the culture, where to cut, what to do with the cut explant, all of which must be completed in an aseptic manner. The potential for increasing productivity per operative appears limited.

The potential variability of cultures, but desired uniformity of the end products, and the need for sterility present substantial problems for the mechanisation engineer. Some successes have been previously described.

A completely mechanised separating and transfer system of bulb scales for propagation of lilies is described by Takayama et al. (1991). Lily bulblets produced in vitro are mechanically placed, one by one, onto two rollers that remove the roots. Scales are separated by pressure at the, more or less, natural breaking point at their junction with the basal plate. Separated scales are transferred to a fresh medium by a robotic hand controlled by a vision system.

Mechanised and automated systems have been claimed for explants of the nodal cutting type derived from stems with long internodes, and shoot tips. Schonstein and Johnson (1986) describe a system, developed for Commonwealth Industrial Gases of Australia which can be used to subdivide and transplant shoot material. Individual shoots are automatically removed from containers, placed on sterile conveyor belts, and a two dimensional image is created to detect the nodal regions to be cut. Following cutting by computer controlled double-bladed instruments, the cut explants are transferred to a new medium by robotic handling.

The Toshiba Company (Fujita 1989) have developed a robotic cutting and transfer system which identifies nodes to be cut by a three dimensional vision survey. Identified nodal regions are held by a robotic-hand and cut with scissors within the same robot. Held cut pieces are transferred to a new container, and the robotic-hand returns to repeat the process.

A mechanised cutting and planting system has been developed by Brown (1992; see report by F.R. Brown in this volume).

Since these mechanised systems may operate on a continuous or semi continuous basis total handlings predicted as feasible within any one year reach the multimillion level. However none of the described systems appear to have entered into routine production operations.

In developing the PhytoNova concept various cutting systems have been considered. The selection criteria for the cutting tool were:

- speed
- accuracy and consistency
- longevity of use
- access to cut pieces
- decontamination of the instrument
- maintenance of sterility
- nature of the cut piece, subsequent growth, and eventual plant uniformity and quality

The cutting systems considered were:

- fixed blades or renewable blades of either metal or ceramic. Continuous strip blades.
- water
- mechanical shearing - eg. breaking, tearing, homogenising
- laser

Of these options mechanical shearing would not give the precision considered necessary, although an application to several plant types could be envisaged. Cutting with high pressure water presented a multitude of engineering and biological problems and was dismissed. The use of various types of mechanical cutting devices was a real possibility, although maintenance of sterility, without rapid deterioration of the cutting edges, presented a problem which was not solved satisfactorily.

The laser approach offered a series of positive potentials such as speed, precision, computerised control and with no sterilisation problems, which could not be matched by the alternatives. The major question had to be answered. What effect the laser would have on the explant and subsequent development?

Trials established that a low power laser (up to 32 watts) would be unsuitable as the cutting time was relatively long and thermal damage to surrounding tissue could be severe. The higher the moisture content of the tissue, the greater the problem. The use of a high power beam, up to 500 watts, supplied by an Electrox M500 CO_2 laser (manufactured by Electrox Industrial Lasers, Hitchin, Herts, UK) resulted in very rapid cutting. No significant spread of thermal damage to surrounding tissues was experienced, providing the power employed was adjusted to the nature and width of tissue being cut. The small stem of Betula required 100 to 125 watts. More voluminous tissues, like the scales of Lilium, required much higher energy levels. Explants, of several species, prepared by laser cutting, showed normal to enhanced growth over more conventionally cut material, with no significant lag phase and no adverse effect on multiplication or rooting.

Having established the feasibility and acceptability of cutting with the laser the major problems to be overcome were:

- presentation of plants to be cut to a recognition system
- recognition of points to be cut
- translation of these to the laser control
- presentation of the tissue to the laser cutting site in the correct order and orientation

- serial recording and collection of the cut explants
- disposal of cut-waste
- transport of the collected explants according to type
- recognition of the orientation of the explants
- instruction to the pick and place tool
- picking up and planting in the new container of medium
- rapid sterilisation of all mechanical parts coming in contact with plant tissue during the sequence

For internal and general demonstration purposes the company chose a Betula cultivar 'Golden Cloud' as the test plant. Within the culture system single nodal explants are grown as single shoots. The well spaced nodes can be identified, dissected and used for further shoot development. The shoot tip may be used either for rooting or for further ongrowth to provide nodal cuttings. The culture period is 4 weeks and results in 4 to 5 nodes plus the shoot tip.

Grown shoots were carried to the camera recognition system by simple grippers mounted on a circulating conveyor, holding the shoot at the base. En route to the recognition position the grippers are inverted so that the whole shoot is hanging in clear view. Each shoot is subjected to image analysis via a vision recognition camera viewing from three directions effected by the gripper being rotated 90° twice. A three dimensional recognition system was chosen to prevent nodes being hidden by leaves, as may occur in a two dimensional system. The three images are presented on a screen. Using a mouse or pen tip indicator the desired cutting line for the shoot tip is indicated on two of the three images and each node is also pinpointed on any two of the three images. The cutting lines for the marked nodes are automatically established by an algorithm specific for the culture form. The software presents all proposed cutting lines on the screen images (figure 2), these may be checked, and if required overridden manually.

After camera recognition procedures the shoot is transferred by the gripper carrier to the cutting area. The shoot tip, followed in sequence by the nodal areas, is cut by the laser under computerised command using a flying-spot to direct the beam. The shoot tip and nodal sections (explants) to be retained are collected singly in sterilised cups, brought to and removed from the cutting area by a conveyor, whilst unwanted pieces fall to the waste collection area.

After cutting is completed the gripper moves on, discards the held basal tissue and then passes through a sterilisation stage prior to returning to the point for collection of a new shoot. Several gripper units and cup containers are employed to ensure the continuous supply of transport units to operate without delays.

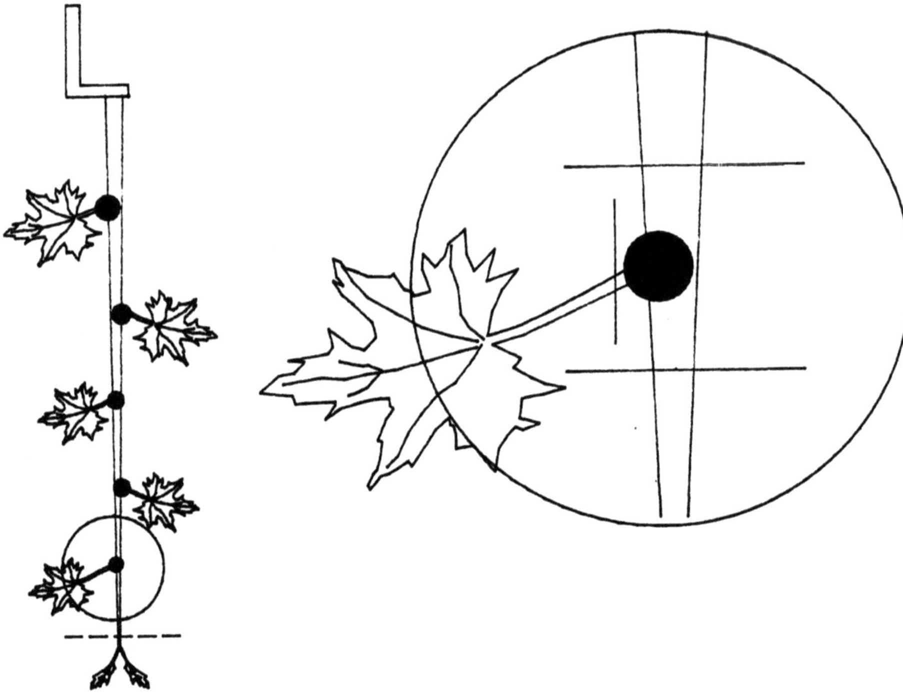

Figure 2. Schematic representation of gripper held shoot
Left: with pointed cutting line ---- for the shoot tip and marked nodes ●
Right: cutting lines generated by algorithm.

Each nodal explant now held in a cup is conveyed to a pick and place area where the orientation is determined using a vision camera. The explants are picked up, using a robotised vacuum pipette and placed in the semi solid medium in a predetermined position. After filling, the box is closed and passed to trays which are despatched to the growth room.

After removal of the explant the cup passes to a sterilising unit prior to returning to the laser cutting area to receive another piece of tissue. The pick and place tool is sterilised during which period a second tool is employed. At present the shoot tips by-pass the nodal pick and place unit for hand planting in vitro or in vivo. Development of the mechanised planting of the shoot tip material into plugs suitable for in vivo rooting has not been completed.

3.4. CONTROL SYSTEMS

All stages of the system are controlled and coordinated by software. Each and every step is automatically recorded, stored and integrated to provide a continuous mechanised production process with performance schedule checks. As each event is collected there is a managerial opportunity to check biological performance against those which have been established in research and pilot plant studies.

The speed of the vision recognition system linked with a computer controlled laser for cutting facilitates mechanical collection, recording and tagging of each explant through the application of unique numbers. This links each explant to their original parent plant source, and their subsequent cutting, after the growing cycle, to further explants. This provides total data which can be exported as ASCII files for total production control and progressive planning of necessary future activities - eg. instructions for media type and volume - to maintain the continuous production process, see figure 3.

4. Concluding Remarks

The development of a fully automated micropropagation system incorporating cutting, transportation, planting and recording of performance in respect to multiplication rates has been demonstrated; patent applications have been made.

The system developed has entered production trials. Further development programmes are directed to modification and extension of the software and some of the mechanical aspects to extend the range of plant types which can be handled. Algorithms already developed for fully automated two dimensional recognition systems will be upgraded to the three dimensional level.

The design characteristics provides for modular extension from current rates of 3.4 million units to 10 million units or more per annum based on a 16 hour operational day. The input of labour is reduced to between 10 and 15% of current levels, but with an increased level of supervisors and maintenance personnel to control the overall system.

The automated laser cutting system linked with automated medium delivery system to the pick and place unit provides the opportunity to produce plants within a more industrialised concept and with greater control of uniformity, quality and production progress to meet orders. It should be appreciated that the use of such industrial procedures will involve much higher capital investments, than have been required to date, by micropropagators. Eventhough, or because of, the offset by reducing labour levels and securing greater management control of the day to day processing, such investments may well prove to be more attractive to larger integrated businesses than to the horticultural entrepreneur.

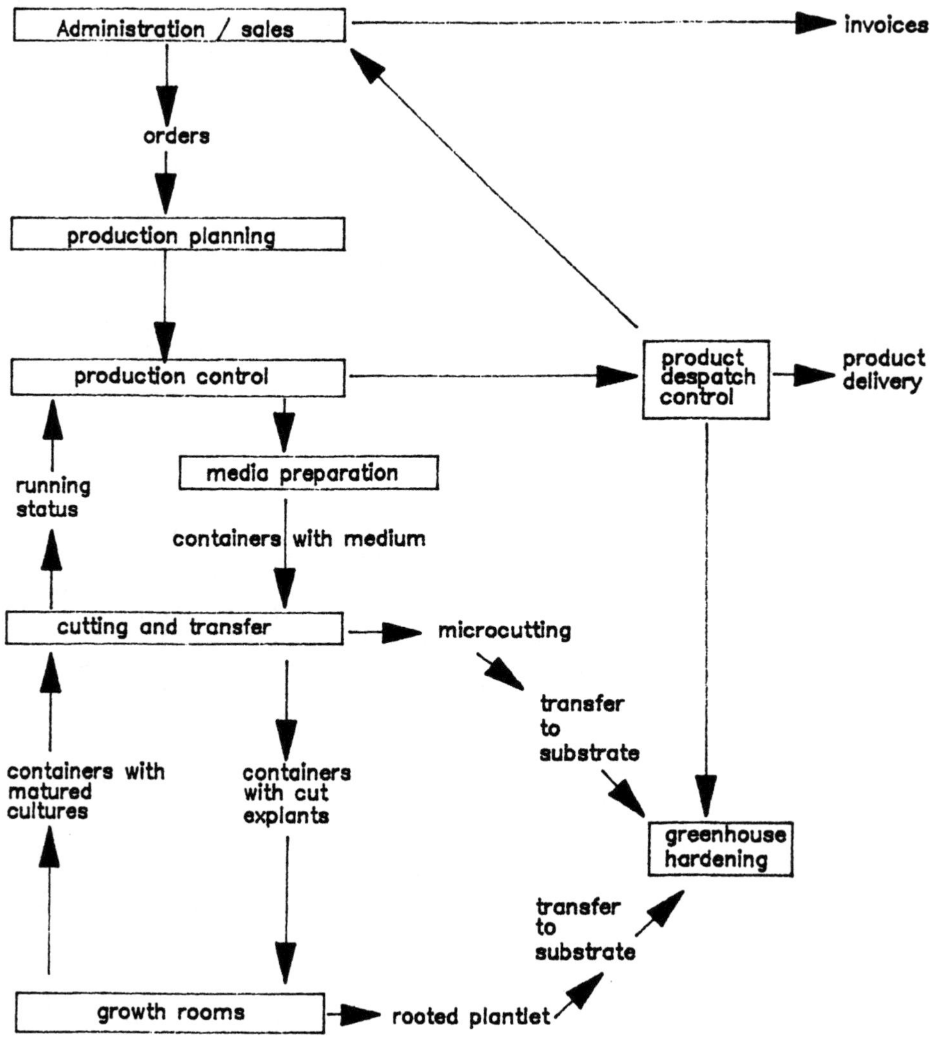

Figure 3. Integrated control system via data collection and analysis from automated micropropagation system

5. Acknowledgement

The development of the PhytoNvoa automated micropropagation concept was achieved by the combined efforts of several companies. FACE B.V. (P.O. Box 2226, 5500 BE, Veldhoven, The Netherlands) generated the major software, Plant Production Systems B.V. (P.O. Box 3032, 5700 JC, Helmond, The Netherlands) developed the mechanised system and the biological inputs were made by PhytoNova B.V.

REFERENCES

Aitken-Christie, J., Jones C.(1987), 'Towards automation: radiata pine shoot hedges in vitro', Plant Cell Tissue Organ Culture 8: 185-196.

Fari, M. (1987), 'Development of technical and biological bases of micropropagation towards automation and industrialisation', Moët-Hennessy conference 1987, Monaco, Electronique et Pilotage des Plants, p55.

Fujita, N. (1989), 'Applications of robotics to mass propagation system', In Vitro Cell Dev. Biol. 25(3): 22A-47.

Jones, O. (1977), East Malling Research Institute, AFRC, U.K., Personnel communication.

Maene, L., Debergh, P. (1985), 'Liquid medium additions to estabished tissue cultures to improve elongation and rooting in vivo', Plant Cell Tissue Organ Culture 5:23-33.

Pierik, R.L.M. (1990), 'Commercial micropropagation in Western Europe and Israël', in: P.C. Debergh and R.H. Zimmerman (eds), Micropropagation, Technology and Application, Kluwer Academic Publishers, Dordrecht, The Netherlands, pp 155-167.

Schonstein, D.N., Johnson, B.J. (1986) 'Method and apparatus for dividing plant material', Australian patent PCT/A-U086/00136.

Takayama, S., Swedlund, B., Miwa, Y. (1991) 'Automated propagation of microbulbs of lilies', in: I.K. Vasil (ed), Cell culture and somatic cell genetics of plants vol. 8, Scale up and automation in plant propagation. Academic Press, New York.

TRANSPLANT PRODUCTION ROBOTS IN JAPAN

Kenji Kurata
University of Tokyo
Yayoi 1-1-1, Bunkyo-ku
Tokyo 113
Japan

ABSTRACT. A great amount of effort has been made to develop automated transplant production systems in Japan. A total of eight transplant production systems are selected and introduced: five micropropagation systems, two grafting systems and one transplanting system. Their basic concepts and technical characteristics are presented. Most are prototypes and are presently under development.

1. Introduction

In recent years, much effort has been devoted in many countries to developing automated, robotized and more efficient transplant production methods. The importance of mass production of healthy transplants has been widely recognized due to the emergence of global environmental problems, the predicted population explosion, and people's desire for more amenities. Production of transplants in more efficient and more labor saving ways will play an important role in solving these problems and in meeting the people's desires. Automation of transplant production procedures is expected to reduce the required labor and thus promote the transplant production industry. In Japan, prototype robots have been developed, some of which were demonstrated in the International Garden and Greenery Exposition (EXPO '90) held in Osaka, Japan in 1990. This report presents some of the Japanese activities in this field including those demonstrated in EXPO '90.

One of the fields in which robotization may be effective in saving labor is micropropagation. The micropropagation methods widely employed at the present time require much manual labor in picking up plantlets out of the culture vessels, cutting them into segments, and transferring the segments into new culture vessels. Robotization of these procedures has the additional advantage of reducing the risk of contamination in comparison with manual handling. For these reasons, many companies and researchers have become involved in developing micropropagation robots.

Another labor intensive procedure, which is common in Japan, is grafting of fruit-vegetable seedlings onto rootstock stems. Transferring transplants onto new vessels is also a labor intensive process. Examples of activities in robotization of these procedures are also presented.

2. Micropropagation robot of KIRIN BREWERY CO.,LTD.

KIRIN BREWERY CO.,LTD. has developed a micropropagation robot, TOMOCA (Figure 1),

K. Kurata and T. Kozai (eds.), Transplant Production Systems, 313–329.

for bushy plants such as *Ficus beñjamina* L. based on concepts from Twyford International Inc., U.S.A. (Iwasaki, 1991). Figure 2 shows the top and side views of TOMOCA (2 m wide, 1 m long, and 2 m high). Capped vessels (60 mm x 60 mm x 75 H mm) containing plantlets are supplied via a conveyor to a clean box. The clean box has a filter unit at the top to maintain aseptic conditions (NASA 100) within it. Capped destination vessels (same size as the source vessels) which contain sterilized agar are also supplied via a conveyor to the clean box. In the clean box, the vessels are uncapped and the plantlets are dissected and transplanted into destination vessels, which are then capped and transported out of the clean box to Point D in Figure 2. Pressurized air for actuator control is supplied to the clean box through filter sterilization. In this way, the risk of contamination during the transplanting process is minimized.

Figure 3 shows the cutter head. The blades are arranged so that the plantlets are dissected into 36 cubical segments. The cutter head is pushed down into the uncapped vessel, presses the plantlets down, and cuts them into 36 segments. The cutter head is then lifted up, holding dissected plantlets in each section. Figure 4 shows the method of planting of the 36 segments into 4 destination vessels (9 segments per vessel). The segments in 9 sections are simultaneously pushed down by rods into agar in a destination vessel. Planting depth can be easily adjusted by changing the position of a limit switch. After planting the segments into the 4 vessels, the cutter is sterilized. One cycle takes about 1 minute, which is equivalent to transplanting 2160 segments per hour.

TOMOCA was constructed with possibly the simplest structure, and thus is low-priced and can be applied to a practical transplant production industry.

3. Micropropagation system of Ministry of Agriculture, Forestry and Fisheries and KOMATSU LTD.

Ministry of Agriculture, Forestry and Fisheries and KOMATSU LTD. have developed an automated micropropagation system (Hereafter referred to as KOMATSU system; Nagaoka and Sei, 1990). In designing the system, intensive investigation was conducted on how to reduce the risk of contamination and how to minimize the loss due to contamination. Figure 5 shows a schematic diagram of the system components. The culture vessel ("culture box") is a transparent plastic box of length 60 cm and width 60 cm. This size was determined by considering both the environmental control efficiency and the scale of loss in case of contamination. Water-absorbing material such as ceramic wool is used for plantlet support, and liquid nutrient is fed into or drained from the box. Five culture boxes are mounted in a culture rack (Figure 8). A temperature controlled culture room contains a number of culture racks.

For multiplication, a culture box is taken out of the culture room and transported to the sterilization unit (upper part in Figure 5). The outer surface of the box is sterilized by spraying it with a mist of peracetic solution. The box then moves into the cutting and transplanting unit, where the plantlets are dissected and transplanted into destination boxes under aseptic conditions (Figure 9). The box lid is removed and a computer-controlled cutter, moving horizontally, dissects the plantlets. The cutter dissects the upper part of the plantlets first and gradually lowers itself for dissecting the lower parts. A number of dissected segments are held by the cutter itself and transplanted into destination boxes. A lid is put on the new box full of transplanted segments. The outer surface of the new box is also sterilized. Then, the sterilized box is attached to the culture rack and transported into the culture room.

Each culture rack has a nutrient solution supply unit, a waste solution drainage unit, and a

Figure 1. Micropropagation robot TOMOCA developed by KIRIN BREWERY CO.,LTD. (courtesy of KIRIN BREWERY CO.,LTD.)

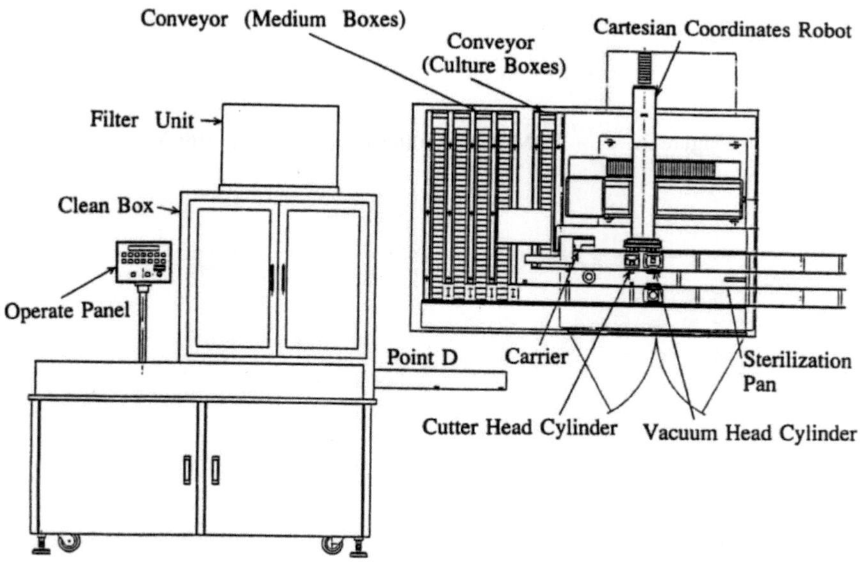

Figure 2. Top and side views of TOMOCA (Iwasaki, 1991).

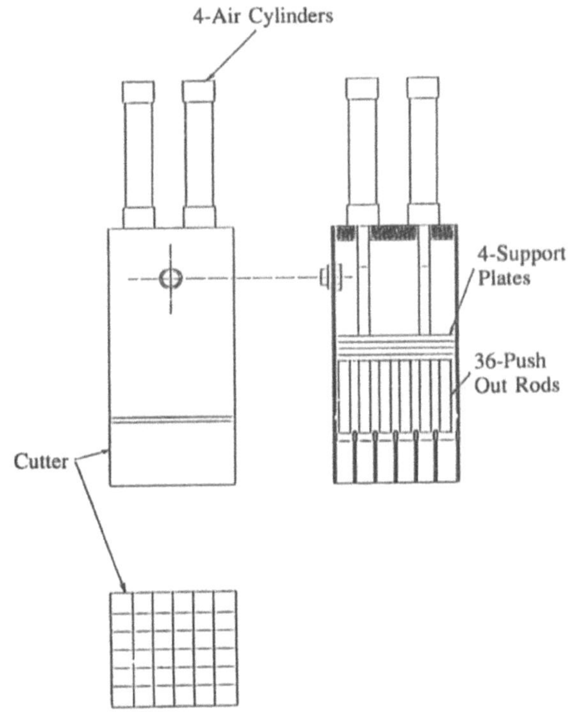

Figure 3. TOMOCA's cutter head (Iwasaki, 1991).

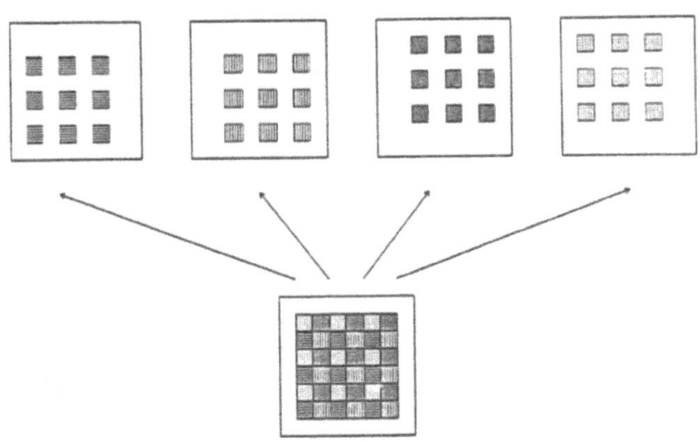

Figure 4. Transplanting method of TOMOCA (Iwasaki,1991).

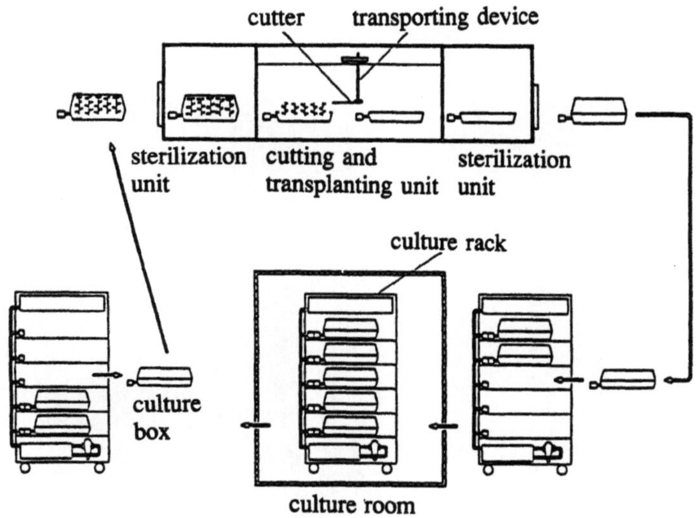

Figure 5. Schematic diagram of the KOMATSU system (Nagaoka and Sei, 1990).

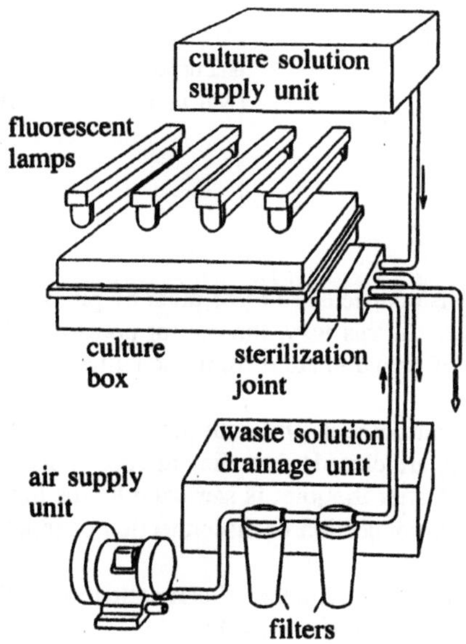

Figure 6. Culture rack of KOMATSU system (Nagaoka and Sei, 1990).

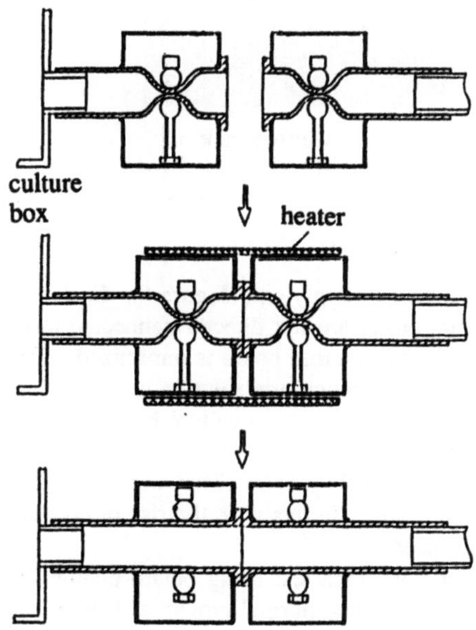

Figure 7. Sterilization joint of KOMATSU system (Nagaoka and Sei, 1990).

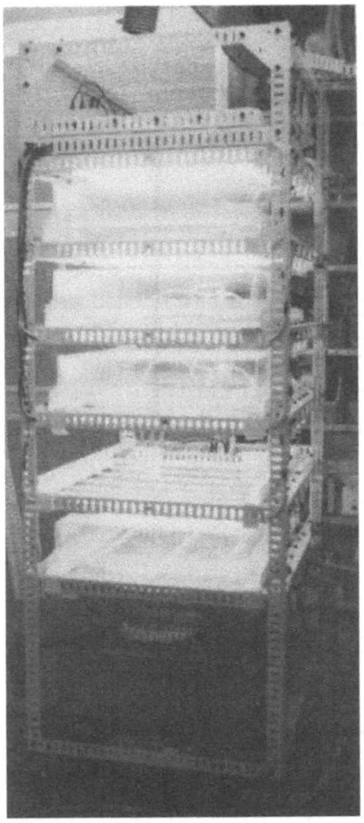

Figure 8. Culture rack of KOMATSU system (Nagaoka and Sei, 1990).

Figure 9. Cutting and Transplanting unit of KOMATSU system (Nagaoka and Sei, 1990).

sterilized air supply unit (Figure 6). Each culture box is connected to these units by pipes. The culture boxes are not directly connected to each other. In this way, the risk of contamination spreading into other boxes is minimized. The spread of contamination from one culture rack to another through culture solution or air is not possible.

Figure 7 shows how a culture box is connected to a culture rack unit with pipes. The ends of the pipes, both from the culture box and the rack units, are closed when they are not connected (upper part of Figure 7). They are opened (lower part) after the joint is sterilized by the heater (middle part). In this way, the risk of contamination by connection or disconnection of pipes is minimized.

A performance test using potato plantlets resulted in no occurrence of contamination. The plantlets showed normal growth and thus, indicated that the system was feasible.

4. MIWA's robot for lily micropropagation

Miwa, Waseda University, and his colleagues developed a robot for lily bulb micropropagation which fully automates the process of removing roots from a bulb, separating bulbscales, and transplanting bulbscales into a destination vessel (Miwa, 1991). The robot consists of a bulb-supply unit, root-removing unit, bulbscale-separating unit, bulbscale-transfer unit, image-processing unit, transplanting unit, and microcomputer.

4.1. BULB-SUPPLY UNIT

When a large number of bulbs are supplied to the robot, the bulbs are generally entangled and difficult to separate. In order to feed individual bulbs to the root-removing unit, bulbs are dropped gradually through rotating sheets onto a V-shaped vibrating tray, which allows the bulbs to be separated completely.

4.2. ROOT-REMOVING UNIT

The bulb usually has 10 to 30 roots with a length 10 to 20 mm. It is difficult to remove these roots by means of computer vision. Although the computer vision can be used to detect the 3-dimensional shape of the bulb having roots, mechanical removal tasks based on data obtained by this method may be difficult. A completely mechanical method was thus investigated.

In the root-removing unit, a bulb is fed onto two parallel rollers, which are in contact with each other and rotating in opposite directions. The bulb also rotates parallel to the rollers' axes and the roots are engulfed by the rollers. A vertical plate then moves in the axial direction and removes the roots. Tests showed a combination of an 11 mm diameter roller and 4 mm one was effective for the removal task for various bulbs.

4.3. BULBSCALE-SEPARATING UNIT

A simple but effective structure was devised for separating individual bulbscales. A bulb is fed into a cylinder (about 100 mm in diameter), which is equipped with a rotating disk at its bottom. The disk has small projections on its surface and can rotate up to a speed of 5000 rpm. The bulb hits the inside surface of the cylinder several times and is separated into bulbscales by the mechanical shock. Culture tests of separated bulblets showed that the mechanical shock did not affect their growth.

4.4. BULBSCALE-TRANSFER UNIT

Bulbscales are transferred to the transplanting unit through two rotating cylinders. The cylinders have an inner diameter of 25 mm and are tilted 6 degrees from the horizontal plane. In this process, bulbscales are separated from each other and are fed to the transplanting unit one by one.

4.5. TRANSPLANTING UNIT

The image processing unit is used to calculate positions of the bulbscale centroid and of the farthest point on the bulbscale profile from the centroid, thus analyzing a 2-dimensional profile of the bulbscale. Based on these data, the robot sucks up each bulbscale and passes it to the planting arm. The number of planting arms determines the maximum number of bulbscales which can be planted into the destination vessel simultaneously. The depth of planting is controlled by detecting the resistance in pushing the bulbscales into the agar.

5. MIWA's robot for chrysanthemum micropropagation

The same group developed a prototype of an automated system based on a robot for chrysanthemums micropropagation (Yamashita et al., 1991). In this system, image processing is applied to dissected explants to select and transplant appropriate segments.

The robot cuts off the bottom part of an explant's stem and transfers it to a dissection unit. The explant is fed onto a V-shaped tray, and a cutter with 9 blades presses it down and cuts it into segments. The distance between blades is adjustable from 9 to 15 mm. These dissected segments are transferred to an image-processing unit where a 2-dimensional image of each segment is taken by a CCD camera and fed into a computer. It is judged based on the 'thinning' of the image whether a segment has a node or not: if the thinned image has a Y-shaped part, it has a node. From angles defined by three lines composing the Y-shape, the planting direction, and gripping position of a planting robot hand are determined. The number of leaves on each segment is estimated from the image data and the number of Y-shapes by using fuzzy logic. The number is used to adjust the distance of the cutter blades again by applying fuzzy logic.

Tests showed that transplanting one segment takes about one minute and all the segments transplanted by this system showed normal growth.

6. Micropropagation robot of TOSHIBA CORPORATION

TOSHIBA CORPORATION developed a multiplication robotic system for plants with upright shoots (Watake, 1991), which was demonstrated in EXPO '90. The system automated the whole process of loading plantlet trays, recognizing each plantlet, cutting internodes, transferring each segment onto a destination tray, and unloading the destination tray full of segments into a culture room. Thus, the system basically mimics the manual operation of the multiplication procedure. The system has two robots: a sensing robot and cutting robot.

6.1. RECOGNITION OF 3-DIMENSIONAL PLANTLET SHAPE

Many of the current micropropagation robots utilize a CCD camera for recognition of plantlet profiles. However, because the visual image obtained is 2-dimensional, there are difficulties in identifying 3-dimensional profiles. The Toshiba system uses a laser beam instead of a camera. The sensing robot has a laser emitter, a scanning mirror, and a position sensitive detector. Based on the triangulation method for measuring distances, the position of the object is detected. Horizontal scanning of the laser beam is repeated at different heights from the root to the top, so that the three-dimensional profile of the plantlet can be measured. To speed up the process, the plantlet position is first roughly detected by scanning with wider intervals. Fine interval scanning follows based on the data obtained by rough scanning.

A node is defined as a part of the stem where the diameter exceeds a predetermined value or where the stem branches. The cutting position is determined so that each cutout segment contains one node. If the distance between two nodes is less than a predetermined value, that segment is not cut out. The sensing robot transfers 3-dimensional cutting position data and gripping position data to the cutting robot.

6.2. CUTTING ROBOT

The cutting robot is a 6-axis articulated robot which has a hand equipped with a tweezer-like gripper and a scissor-like cutter, both at the tip. To minimize the size of the hand tip, the motor unit is separated from the hand, and force is transmitted to the gripper and the cutter via a wire rope. The gripping force is measured by strain gauges on the gripper finger and is fed back to the motor in order to enable gentle gripping of plantlet segments. The gripping force is set to 30 gf according to the results of preliminary experiments.

6.3. SPEED OF THE ROBOTS

In the demonstration in EXPO '90, the average time required for recognition, cutting, and planting of a node was set to 15 seconds, although a slight increase in speed was possible. However, to make the system feasible, one cycle time must be reduced to less than 5 seconds, and efforts are being made to this end.

7. BRAIN's grafting robot

Grafting is a common technique applied in Japanese fruit vegetable culture. Detailed statistics on grafted fruit vegetable culture are presented in Ito's article in this volume. In protected cultivation of watermelon and cucumber, 99% and 85% of the cultivation area are for grafted crops, respectively. Other vegetables which are commonly grafted include musk melon, cantaloupe, tomato, and eggplant. The primary objective of grafting is to prevent soil-borne diseases, but recently grafting has been conducted with additional objectives such as higher yield, higher quality product, extending the potential cultivation period and saving energy by giving seedlings chilling tolerance.

However, grafting is a very labor intensive process. Growers must manually graft a large number of seedlings within a limited amount of time, and they often must hire part-time helpers. As the average age of growers increases, the growers' demand for grafted seedlings increases. Nursery industries which produce grafted seedlings, however, are suffering from a labor shortage.

Bio-oriented Technology Research Advanced Institution (BRAIN) has been involved in developing automated grafting systems. BRAIN has developed first and second prototypes and is now in the process of completing the third one. This section presents the second prototype (Suzuki, 1991; Kobayashi, 1991).

The adopted grafting method is illustrated in Figure 10. A growing point and one cotyledon are cut off from the rootstock plant. The root part is removed from the scion plant. The rootstock and the scion are joined and held in place by a clip.

Figure 11 shows the whole system. The two wheels in the foreground are plant supply units.

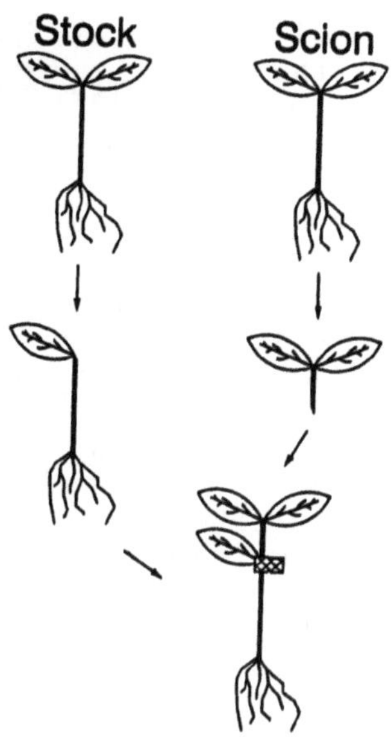

Figure 10. Grafting method of BRAIN system (after Suzuki, 1991).

Each wheel has 20 U-shaped openings on the perimeter from which plants are hung. The left wheel is for scion plants, while the right one is for rootstock plants. The wheels can be easily removed from or attached to the system. Both wheels rotate and feed plants to the gripping and transporting units (two small discs behind the wheels in Figure 11).

Figure 11. BRAIN grafting system (Prototype 2) (courtesy of BRAIN).

Figure 12 shows the gripping and transporting units. Both units have the same structure but turn in opposite directions. Each disc has 3 plant grippers equally spaced along the perimeter. The discs rotate and stop at 120 degree intervals. Three functions are performed as the plants are rotated around

Figure 12. Gripping and transporting unit (left) and cutting unit (right) of BRAIN system (courtesy of BRAIN).

the discs. At one stop, a gripper receives a plant. At the next stop, the plant is cut. Finally, at the third stop, the plant is joined to another plant and released from the gripper. The rootstock plant cutting unit (on the right part in Figure 12) has an arm with a razor blade at the tip. The arm turns and cuts off one cotyledon and a growing point from the plant. The scion plant cutting unit has the same structure and cuts off the root. In the space between the two discs of the gripping and transporting unit, the rootstock and the scion are joined and fixed with a clip (Figure 13).

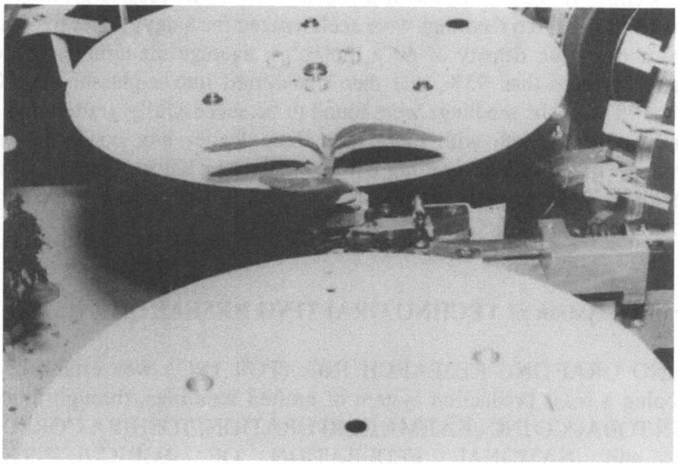

Figure 13. Fixing a grafted seedling with a clip (courtesy of BRAIN).

Figure 14. Grafted seedling (courtesy of BRAIN).

The system is regulated by a programmable controller. The system has three modes of operation: manual mode (each part operates independently), individual grafting mode (the system stops after producing one grafted seedling), and successive grafting mode (the system stops after producing 20 grafted seedlings). An example of a grafted seedling is shown in Figure 14.

Performance test results of grafting cucumber (cultivar: Sharp 1) scions onto pumpkin (cultivars: Tosashinichigo and Kitora) rootstocks are as follows. Successive graftings of Sharp 1 onto

Tosashinichigo stems (3.0 mm diameter) resulted in 59 out of 60 (98%) seedlings jointed and fixed by clips. In these tests, the time required for 20 successive graftings was set to 59 seconds. The successfully fixed seedlings were acclimatized for 8 days under the conditions of 12 hour day length, photon flux density of 60 μ mol/m^2 s, average air temperature of 24 °C, and relative humidity of more than 95%, and then transferred into a glasshouse. On the 34th day after grafting, 95 % of the seedlings were found to be successfully grafted and 87 % of the seedlings showed normal growth with more than three leaves and could be transplanted for further cultivation. In the case of grafting onto the narrower Kitora stems (2.2 mm diameter), 97 % of the seedlings were properly joined and fixed, 88 % were found to be successfully grafted after acclimatization, and 83 % showed normal growth.

8. Grafting system of TECHNO GRAFTING RESEARCH INC.

TECHNO GRAFTING RESEARCH INC. (TGR INC.) was established in March 1990, for developing a mass production system of grafted seedlings, through investments from BRAIN, JAPAN TOBACCO INC., KAJIMA CORPORATION, TOSHIBA CORPORATION, KOMATSU LTD., and NATIONAL FEDERATION OF AGRICULTURAL CO-OPERATIVE ASSOCIATION. TGR INC. has been involved in investigations on grafting methods which best utilize the advantages of mechanical grafting. TGR also investigates methods for managing grafted seedlings.

Figure 15 schematically illustrates the adopted grafting method. Rootstock plants are not removed from the tray at the time of grafting. They are fixed slightly below the cutting points by supporting plates, and then they are cut. Similarly, scion plants are fixed slightly above the cutting points by supporting plates. Then the scion plants are cut and transported with the plates onto the rootstocks, and joined to the rootstocks. As illustrated in Figure 15, a number of seedlings are grafted simultaneously in this way, enabling mass production of grafted seedlings.

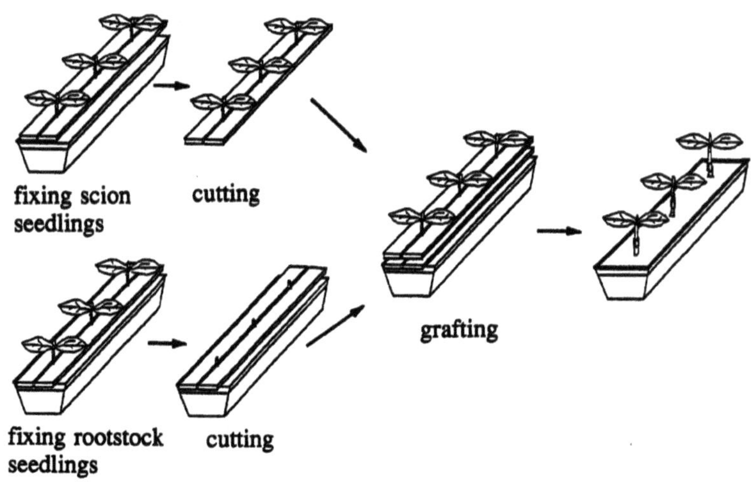

Figure 15. Grafting method of TGR INC. (courtesy of TECHNO GRAFTING RESEARCH INC.)

Two technical problems need to be solved for realization of this grafting method: how to align the scions with the rootstocks, and how to support the joints of the scions and the rootstocks. To investigate these points, experiments are being carried out using the device shown in Figure 16, which grafts a single seedling.

Additional topics to be investigated include allowable error in positioning of the rootstock and scion stems, required strength for supporting the joint, and allowable error for the joint supporting strength considering variations in diameters of the rootstock and scion stems.

For completion of a mass production system of grafted seedlings which includes the automatic grafting device, the following points need to be considered: how to germinate seeds at the desired positions at the desired rate, how to uniformly raise the seedlings, and how to acclimatize the grafted seedlings.

Figure 16. Experiments using a single-grafting device of TGR INC. (courtesy of TECHNO GRAFTING RESEARCH INC.)

9. Transplanting robot of JAPAN TOBACCO INC.

JAPAN TOBACCO INC. (JT) has developed a transplanting system for carnation seedlings (Fujiwara, 1991). A robot picks up plug seedlings one by one, judges whether the seedling is 'good' one or not, and transplants 'good' seedlings onto rockwool cubes (Figure 17). 'Bad' seedlings are discarded. Unlike industrial products, there are no straightforward criteria for discriminating 'good' seedlings from 'bad' ones. However, experts can judge 'good' and 'bad' seedlings

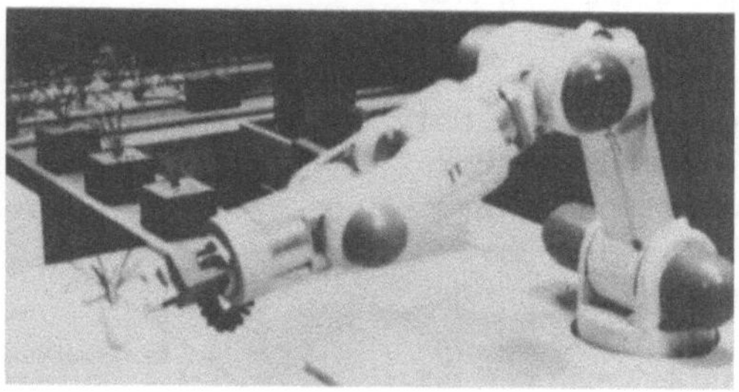

Figure 17. Transplanting robot of JT (courtesy of JAPAN TOBACCO INC.).

visually, based on knowledge obtained through experience. One of the characteristics of the JT system is that it incorporates the experts' criteria for discrimination by applying fuzzy logic. The system was demonstrated in EXPO '90 and is equipped with two CCD color cameras for image processing.

Discrimination criteria were collected through interviews with many experienced growers. Their expertise was expressed in their natural language (Japanese). Examples of criteria for 'good' seedlings are as follows: 'Jujitsu shite gacchiri shiteiru' (solid and firm), and 'aoao toshiteiru' (freshly green). These somewhat vague expressions were translated into terms suitable for computer vision analysis. As a result, the following 7 terms were derived for describing 'good' seedlings: projected leaf area seen from above (x_1), number of leaves (x_2), average projected leaf width seen from above (x_3), average projected leaf length seen from above (x_4), projected leaf area seen from the side (x_5), seedling height (x_6), and 'leafiness' (x_7). The first four terms are analyzed from the top view image and the rest from the side view image. Information on the stem and the internode length were not considered because they were less important than the other criteria according to the experts.

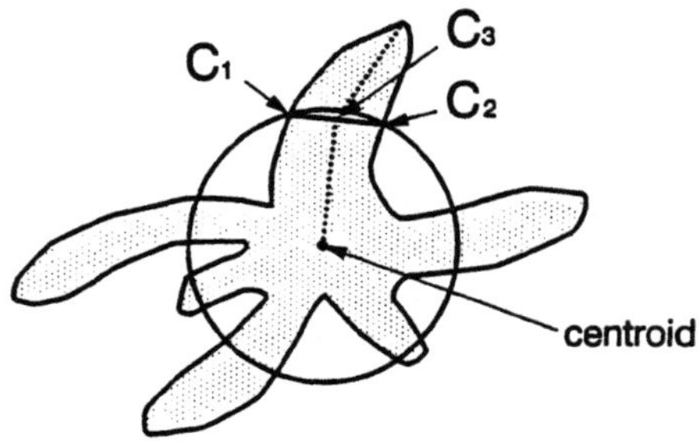

Figure 18. Scematic illustration of top view image of plantlet (after Fujiwara, 1991).

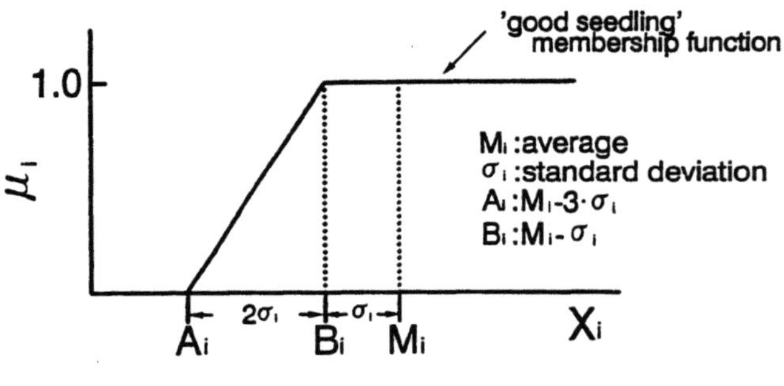

Figure 19. 'Good seedling' membership function (after Fujiwara, 1991).

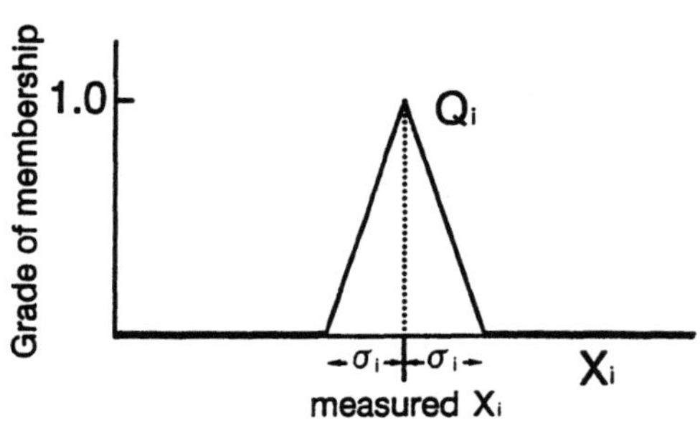

Figure 20. Definition of Q_i (after Fujiwara, 1991).

Color information could not be included in the present system.

The top view image is analyzed by using the following procedure (see Figure 18). First, noise from the background is eliminated by applying a filter. A binary image is generated by using a threshold value which is automatically determined. The pixels corresponding to the leaf are summed to determine the projected leaf area. The position of the centroid is calculated and the image contour is extracted. The distance, L, between the centroid and the contour is measured for each point along the contour. The maximum (L_{max}) and the minimum (L_{min}) values of L are used to calculate L_{mid}, where L_{mid} is defined as $(L_{max}+L_{min})/2$. Following along the contour, if the L values are larger than L_{mid} for a predetermined number of contiguous points on the contour, the radial area from the centroid to those contour points is considered to be a leaf. The leaf width is calculated as the distance between two points, C_1 and C_2, where the leaf contour and the circle with radius L_{mid} centered at the image centroid intersect. The leaf length is defined as the distance between the centroid and the midpoint (C_3) of C_1 and C_2 plus the distance between C_3 and the leaf tip (dotted line in Figure 18). The side view image analysis follows a similar procedure. 'Leafiness' is calculated as follows. A fan-shaped area is defined which approximately covers the seedling's side image: the pivot of the fan is positioned at the base of the seedling, and the radius of the circular arc is equal to the seedling's height. 'Leafiness' is defined as the ratio of the number of seedling pixels within this area to the total number of pixels within this area.

Two step fuzzy calculations are conducted based on the measured x_i-values, to judge whether the seedling is 'good' or not. The two steps are fuzzy matching and fuzzy integral. Before performing fuzzy matching in real-time, a number of seedlings were analyzed in advance to obtain the average, M_i, and the standard deviation, σ_i, of the x_i-value. Additionally, a membership function, μ_i, for a 'good seedling' was defined for every x_i as illustrated in Figure 19.

The real-time fuzzy matching was conducted based on the 'converse truth qualification' (e.g. Honda and Ohsato, 1989). The function $h(x_i)$ expresses the 'likelihood' that a fuzzy set Q_i belongs to a fuzzy set P_i and is defined as:

$$h(x_i)=\{sup(P_i \cap Q_i)+inf(P_i \cup {}^{-}Q_i)\}/2,$$

where ${}^{-}Q_i$ is the complement of Q_i. Q_i is defined by σ_i and the measured x_i-value, as illustrated in Figure 20, and P_i corresponds to μ_i in Figure 19. Thus, $h(x_i)$ expresses the degree of 'goodness' based on the measured x_i-value.

Final evaluation of the seedling is accomplished by calculating the 'fuzzy integral' of each $h(x_i)$-value (Honda and Ohsato, 1989). Before the real-time fuzzy integral step, 'degree of importance' of each term in the seedling evaluation were determined through interviews with experts, resulting in the following values: 0.3 (important) for projected leaf area seen from above and number of leaves; 0.2 (moderately important) for average projected leaf length seen from above and seedling height; and 0.1 (less important) for average projected leaf width seen from above, projected leaf area seen from the side, and 'leafiness'. The fuzzy integral, e, introduces the concept of λ-fuzzy measure and is calculated as:

$$e = \int h(x) \circ g\lambda = \bigvee_{i=1}^{7} [h(x_i) \wedge g\lambda(H_i)]$$

where the values of the subscript, i, of x_i are rearranged so that $h(x_k) \geq h(x_{k+1})$, and $g\lambda(H_i)$ is the λ-fuzzy measure calculated by the following equation:

$$g\lambda(H_i) = g^i + g\lambda(H_{i-1}) + \lambda \cdot g^i \cdot g\lambda(H_{i-1})$$
$$= [\prod_{k=1}^{i} (1 + \lambda \cdot g^k) - 1]/\lambda,$$

where g^i corresponds to the above-mentioned 'degree of importance' and λ must satisfy $g\lambda(H_7)=1$. Solving the equation: $g\lambda(H_7)=1$ gives $\lambda=-0.5$. The threshold value of the fuzzy integral, e, for 'good seedlings' was found through tests to be 0.6.

Performance tests of the system were conducted using 500 seedlings and resulted in 97% agreement with human judgements.

10. Conclusions and considerations

A total of eight automated transplant systems in Japan were selected and introduced. Some were developed for exhibition and relatively little consideration was made for cost effectiveness at the time of development. Others were developed or are presently under development with the basic concept of reducing cost to make them economically feasible. Further analyses are needed to reduce the costs of transplant production. The factors to be considered include costs of robots, market demands for transplants, labor costs, labor availability, and factors besides automation such as environmental controls for improved plant growth. In particular, prediction of future trends in these factors will be important, because they are rapidly changing.

In some systems, efforts to introduce emerging technologies such as application of fuzzy logic were made. Considering the differences between requirements for industrial robots and agricultural robots, not only introducing industrial robots to the areas where they are applicable, but also introducing or developing new concepts for handling living materials will be important.

Acknowledgements

The author would like to thank KIRIN BREWERY CO.,LTD., JAPAN TOBACCO INC., BIO-ORIENTED TECHNOLOGY RESEARCH ADVANCEMENT INSTITUTION, and TECHNO GRAFTING RESEARCH INC. for providing photographs for this article and Nancy Okamura for assistance in preparation of this article.

References

Fujiwara, H. (1991) 'Intelligent robot for discriminating seedlings with fuzzy logic', Proceedings of 1st National Symposium on Recent Development of Transplant Production Systems, Organizing Committee of International Symposium on Transplant Production Systems, Tokyo, pp. 35-47. (in Japanese)

Honda, N. and Ohsato, A. (1989) Introduction to fuzzy engineering, Kaibundo, Tokyo. (in Japanese)

Iwasaki, A. (1991) 'Development of a robot for plant tissue culture', Proceedings of 2nd National Symposium on Recent Development of Transplant Production Systems, Organizing Committee of International Symposium on Transplant Production Systems, Tokyo, pp. 61-66. (in Japanese)

Kobayashi, K. (1991) 'Development of a grafting robot for fruits-vegetables' Plant Cell Technology, 3(6), pp. 477-482. (in Japanese)

Miwa, Y. (1991) ' Micropropagation robot for lily bulb', Proceedings of 2nd National Symposium on Recent Development of Transplant Production Systems, Organizing Committee of International Symposium on Transplant Production Systems, Tokyo, pp. 74-79. (in Japanese)

Nagaoka, M. and Sei, S. (1990) 'Transplant mass production system by micropropagation', Agriculture and Horticulture, 65(1), pp. 87-92. (in Japanese)

Suzuki, M. (1991) 'State and trends of research activities for automation in BRAIN - with emphasis on developing a grafting robot for fruits vegetables -, Proceedings of 1st National Symposium on Recent Development of Transplant Production Systems, Organizing Committee of International Symposium on Transplant Production Systems, Tokyo, pp. 27-34. (in Japanese)

Watake, H. (1991) 'Micropropagation robot', Proceedings of 2nd National Symposium on Recent Development of Transplant Production Systems, Organizing Committee of International Symposium on Transplant Production Systems, Tokyo, pp. 67-73. (in Japanese)

Yamashita, T., Murase, T. and Miwa, Y. (1991) 'Automation of plantlets dissection in tissue culture', Proceedings of JSPE Autumn Meeting, pp. 353-354. (in Japanese)

SUBJECT INDEX

MIX
Papier aus verantwortungsvollen Quellen
Paper from responsible sources
FSC® C105338

If you have any concerns about our products,
you can contact us on
ProductSafety@springernature.com

In case Publisher is established outside the EU,
the EU authorized representative is:
Springer Nature Customer Service Center GmbH
Europaplatz 3, 69115 Heidelberg, Germany

Printed by Libri Plureos GmbH
in Hamburg, Germany